D1003468

Perspectives on Technology

Perspectives on Technology

NATHAN ROSENBERG
Professor of Economics, Stanford University

CAMBRIDGE UNIVERSITY PRESS

CAMBRIDGE
LONDON ● NEW YORK ● MELBOURNE

Published by the Syndics of the Cambridge University Press
The Pitt Building, Trumpington Street, Cambridge CB2 1RP
Bentley House, 200 Euston Road, London NW1 2DB
32 East 57th Street, New York, NY 10022, USA
296 Beaconsfield Parade, Middle Park, Melbourne 3206, Australia

First published 1976

Printed in the United States of America
Typeset by Edge Hill Typographic Service Inc., Maywood,
New Jersey; printed and bound by Vail-Ballou Press Inc.,
Binghamton, N.Y.

Library of Congress Cataloging in Publication Data

Rosenberg, Nathan, 1927-
Perspectives on technology.

Includes bibliographical references.
1.   Technological innovations–United States.
2.   Diffusion of innovations.   3.   Economic development.
I.   Title.

HC110.T4R64   338'.06   75-14623
ISBN 0 521 20957 9 hard covers
ISBN 0 521 29011 2 paperback

For Rina - with whom all things are possible

Contents

Introduction   1

Part 1   Some origins of American technology

1   Technological change in the machine tool industry,   9
    1840-1910
2   America's rise to woodworking leadership   32
3   Anglo-American wage differences in the 1820's   50

Part 2   The generation of new technologies

4   Problems in the economist's conceptualization of   61
    technological innovation
5   Neglected dimensions in the analysis of economic   85
    change
6   The direction of technological change: inducement   108
    mechanisms and focusing devices
7   Karl Marx on the economic role of science   126

Part 3   Diffusion and adaptation of technology

8   Capital goods, technology, and economic growth   141
9   Economic development and the transfer of   151
    technology: some historical perspectives
10   Selection and adaptation in the transfer of   173
    technology: steam and iron in America 1800-1870
11   Factors affecting the diffusion of technology   189

**Part 4   Natural resources, environment and the
                growth of knowledge**

12   Technology and the environment: an economic   213
    exploration

viii Contents

13 Technological innovation and natural resources: the   229
    niggardliness of nature reconsidered
14 Innovative responses to materials shortages   249
15 Science, invention, and economic growth   260

Epilogue   280
Notes   290
Index   343

Acknowledgments to reprinted articles

"Technological change in the machine tool industry, 1840-1910," *Journal of Economic History,* vol. XXIII:4, December 1963. Reprinted by permission of the Economic History Association.

"America's rise to woodworking leadership," *America's Wooden Age,* Brooke Hindle, ed. Proceedings of a conference sponsored by Sleepy Hollow Restorations.

"Anglo-American wage differences in the 1920s," *Journal of Economic History,* vol. XXVII:2, June 1967. Reprinted by permission of the Economic History Association.

"Problems in the economist's conceptualization of technological innovation," *History of Political Economy,* Neil DeMarchi, ed.

"Neglected dimensions in the analysis of economic change," *Bulletin of the Oxford Institute of Economics and Statistics,* vol. 26:1. Copyright 1964 by Basil Blackwell.

"The direction of technological change: inducement mechanism and focusing devices," *Economic Development and Cultural Change.* Copyright 1969 by the University of Chicago Press.

"Karl Marx on the economic role of science," *Journal of Political Economy,* July-August 1974. Copyright 1974 by the University of Chicago Press.

"Capital goods, technology and economic growth," *Oxford Economic Papers,* vol. 15, 1963, pp. 217-27. Reprinted by permission of the Oxford University Press, Oxford.

"Economic development and the transfer of technology," *Technology and Culture* 11:4, pp. 550-75. Copyright 1970 by the University of Chicago Press.

"Selection and adaptation in the transfer of technology: steam and iron in America 1800-1870." Paper presented at a conference sponsored by the International Cooperation in the History of Technology Committee, Pont-à-Mousson, France, July 1970.

"Factors affecting the diffusion of technology," *Explorations in*

Economic History, Fall 1972. Copyright 1972 by Academic Press, New York.

"Technology and the environment," *Technology and Culture,* vol. 12:4, pp. 543-61. Copyright 1971 by the University of Chicago Press.

"Technological change and natural resources: the niggardliness of nature reconsidered." Paper prepared for the Georgia Technological University Innovation Project, a National Science Foundation project.

"Innovative responses to materials shortages," *American Economic Review,* vol. LXIII:2, May. Copyright 1973 by the American Economic Association.

"Science, innovation, and economic growth," *The Economic Journal,* March 1974, vol. 84:333. Copyright by the Cambridge University Press.

Introduction

The chapters that comprise this book were written over a ten-year period running from the early 1960s to the early 1970s. They had their origin in my interest in long-term economic growth processes, especially the behavior of industrializing societies. The form and directions these chapters have taken reflect two basic influences: (1) the growing awareness of the centrality of technological phenomena in generating economic growth, an awareness that had been sharply reinforced for economists by the publication of two articles by Moses Abramovitz and Robert Solow in the mid-1950s; (2) a growing sense that, in spite of the basic orientation and genuine insights into technological phenomena provided by the neoclassical economics in which I had been trained, a deeper and richer understanding of these phenomena would involve a willingness to step outside of the limited intellectual boundaries of this mode of reasoning. In the course of this pursuit, I suppose it is fair to say that I became an economic historian, a very amateurish historian of technology, and an avid (but highly selective) reader of the literature of neighboring disciplines. With apologies to Clemenceau it might be said that if technological change is not too important a subject to be left to the economist, it certainly is too diverse a subject to be left to the economist who refuses to step across narrow disciplinary boundaries. I hope that these chapters will offer some modest demonstration of the potential rewards to such intellectual trespassing.

Technological change encompasses, of course, a highly complex and wide-ranging collection of human activities. Moreover, not only do technologies change over time, but there are, in fact, numerous technologies that coexist in a society at any moment. This heterogeneity renders distinctly suspect all attempts to speak about technology and its consequences in highly aggregated ways. It is not possible to come to grips with the complexities of technology, its interrelations with other components of the social system, and its

1

social and economic consequences, without a willingness to move from highly aggregated to highly disaggregated modes of thinking. One must move from the general to the specific, from "Technology" to "technologies." One must even be prepared to "dirty one's hands" in acquiring a familiarity with the relevant details of the technology itself. Only in this way is it possible to develop an appreciation for the characteristics of particular technologies and the consequences that flow (or fail to flow) from them. It is not possible to analyze the effects of technological change independent of the particular context within which it appears, for the availability of the same technology will exercise very different kinds of consequences in societies that differ with respect to their institutions, their values, their resource endowments, and their histories.

The following chapters are presented in this spirit. They accept the diversity and complexity of technology as axiomatic, and they attempt to provide angles of vision for the examination of technologies that are apt to be overlooked from more highly aggregated and ahistorical perspectives.

Part 1 is entirely historical. Chapters 1 and 2 deal with the historical emergence of two key sectors of the American economy in the early stages of industrialization - machine tools and woodworking machines. They attempt to identify the forces accounting for the growth of these industries and to trace in some detail the factors that shaped them and that influenced the course of their further development, as well as the ways in which they interacted with and influenced the performance of other sectors of the economy. Together with Chapter 3, they also raise some larger conjectures in attempting to account for those features of early American technologies that appeared to be most distinctive against the backdrop of European, and especially British, industrialization.

Part 2 explores the conceptual apparatus within which economists have analyzed the creation of new technologies. Chapter 4 addresses itself directly to the basic concepts currently employed by economists, particularly the distinctions drawn among the separate stages of the innovation process. It is argued that the concepts distort our perception of technological events in ways that have serious implications for our ability to relate technological change to economic growth. (The last section of this chapter, which was prepared for delivery at a conference devoted to the subject of innovation in the social sciences, explores briefly the question of what the study of technological innovation can teach us about innovation in the social sciences.) Chapter 5, which is much wider in scope, considers a whole range of feedback phenomena, specifically the manner in which participation in current economic processes shapes and modifies the behavior of the human agent in ways that significantly affect future economic growth possibilities. Among the most important of these feedbacks is the capacity to generate technological change. It is shown that this class of phenomena received much more attention in the hands of classical than of later neoclassical economists. Chapter 6 examines the question, much discussed in the recent theoretical literature, of the factor-saving bias of technological improvements. It considers technological change as a problem-solving activity and searches for mechanisms that have pushed such activities in some specific directions rather than in others. Chapter 7 examines Marx's views on the causes of the remarkably high degree of technological dynamism that historically was associated with the rise of European capitalism. The analysis turns in a crucial way upon the interrelationship between science and technology and the extent to which the growth of scientific knowledge has been responsive to changing economic needs.

Part 3 takes up the question of the determinants of the speed and extent of the diffusion of new technologies, once their technical feasibility has been established. The economic importance of this question is obvious: Inventions exercise an economic impact only as a result of their superiority over old techniques and only once they have been introduced into the productive process. The speed of diffusion is therefore a critical economic matter. Chapter 8 considers the problem in the specific context of underdeveloped economies. It provides a partial explanation for the failure of underdeveloped economies to create capital-saving techniques that would be more appropriate to economies in which labor is abundant and capital is scarce. The analysis focuses primarily upon the role played by capital goods industries in the spread of industrial technologies, that is, both in developing new techniques and in modifying techniques that have been developed elsewhere. Chapters 9 and 10 are concerned with the international transfer of technology. Chapter 9 considers the mechanisms of transfer in the nineteenth century. It attempts to identify critical factors in effectuating such transfers, such as the need for the movement of skilled personnel, the composition of demand, and the extent of final product standardization. It then goes on to consider the possible implications of this experience for the transfer of technology in the present-day context. Chapter 10 considers the specific experiences of the transfer from Britain to the United States of the steam engine and the new iron-making technology. Whereas the steam engine was transferred very rapidly for certain uses such as transportation, America experienced a prolonged lag of several decades in the adoption of the major innovations in iron making, especially those connected with the introduction of mineral fuels into the blast furnace. These experiences highlight the related features in the transfer process of a high degree of selectivity and of the common need to undertake substantial modifications in a technology before it can be successfully employed in a new resource environment. Chapter 11 starts from the observation that most of the writing in the history of technology has been undertaken by people trained in technical and engineering disciplines and not in economics. As a result, we often have a detailed knowledge of purely technical developments while at the same time we know very little about the train of events that lent economic significance to these developments - specifically the speed of the diffusion process. The paper attempts to organize a wide range of observations at the technological level in a way that will illuminate the economic process of diffusion.

Part 4 looks at technological change as an activity that goes on in specific natural resource contexts and that therefore involves dealing with unique patterns of environmental features. At the same time the success of technological activities is contingent upon a growing body of knowledge, including scientific knowledge, of a kind that is applicable in these specific contexts. Each chapter focuses upon an important dimension of the interactions among these variables. Chapter 12 examines the impact of technological change upon the environment. Primary concern is with the nature of the tradeoffs that all societies confront in the economic exploitation of their environment. Concerns are grouped under three categories: the urban context, the ecological context, and the poor-country context. Chapter 13 surveys the changing treatment of natural resources as a constraint upon economic growth, beginning with the publication of Malthus's *Essay on Population* in 1798. Malthusian and neo-Malthusian models are criticized for their consistent deficiency in failing adequately to incorporate technological phenomena into their frameworks. Chapter 14 is an extension of the criticisms launched in

Chapter 13. It addresses itself to the historical record of the American experience, which clearly shows a declining economic importance of natural resources. It demonstrates the vital role of technological change in bringing about this decline through a wide range of activities that have had the effect of increasing the productivity of specific resources or of developing new substitute inputs as older, traditional resources become increasingly scarce. Chapter 15 attempts to forge a link between the foregoing concerns with technological change and the growth of scientific knowledge. It is argued that the observed patterns of inventive activity reflect not only changes in the composition of demand as it has shifted over time but also the differential rates of growth of the separate subdisciplines of science. These differences in the state of scientific subdisciplines constitute an important determinant of society's technical competence. Attempts to account for the changing composition of inventive activity must therefore be closely linked to the unique combination of opportunities and constraints afforded by the state of science at any time.

I am painfully aware of the preliminary - and in some cases even exploratory - nature of much that follows. But the state of our ignorance on the subject persuades me that an authoritative treatment will not soon be forthcoming. In the meantime (with apologies once again to Clemenceau) technological change is much too important a subject to be left to some treatise writer of the twenty-first century.

Part 1

Some origins of American technology

1

Technological change in the Machine Tool Industry, 1840-1910

I

Technological change has come to absorb an increasing share of the attention of the economist in recent years. Several attempts have been made to assess the quantitative importance of technological change, as opposed to increases in factor supplies, in accounting for the secular rise in per capita incomes in the United States. It appears, in all these studies, that technological changes (shifts in the production function) have been far more important than has the mere growth in the supplies of capital and labor inputs, as conventionally measured (movement along an existing production function).[1] In a sense, this should be cause for deep concern, since the comparative neglect of the process of technological change (with the major exceptions until very recent years, of the works of Marx, Schumpeter, and Usher) suggests a serious malallocation of our intellectual resources. If the studies of such people as Abramovitz and Solow are even approximately correct with respect to orders of magnitude, then the contribution of technological change to rising per capita incomes absolutely dwarfs the contribution from a rising but qualitatively unchanging stock of capital. It would appear that we have indeed been playing Hamlet without the Prince.

Even the recent quantitative studies referred to in the previous paragraph provide the beginnings of only a very partial corrective to this neglect. For what they attempt to establish are, essentially, the quantitative *consequences* of technological innovation. From the point of view of our understanding of the process of economic change, the critical, unanswered question is: what are the major causes of technological change? Why do some firms, industries, regions, countries, show an apparently greater readiness and ability to undertake technological innovations than do other firms, industries, regions, countries? The question may be given a further historical dimension if we seek to explain the apparent variations in

9

innovative activity for the same firm (industry, region, country) at different points of time.[2]

The present paper constitutes a modest attempt to examine and to explain the rapidity of technological change in a sector of the American economy which played a strategic role in the industrialization process.

In the past decade or so, the attempt to formulate a theory of the "take off" into economic growth has centered upon the process whereby an economy, in a relatively short time period, sharply accelerates the annual rate of net additions to its capital stock.[3] This has been an important and fruitful line of inquiry, if only because it has generated historical research which now reveals the considerable diversity of different economies during the growth process. Historical data accumulated by Kuznets for twelve countries show no individual case of abrupt increases in net investment of the magnitude implied by the take-off hypothesis, and cast overwhelming statistical doubt upon any attempt to link precisely the acceleration in the rate of growth of per capita income with rising capital formation proportions.[4]

It seems apparent, then, that changes in rates of investment are only a part of the process by which economic growth is initiated and maintained, and that attention may be usefully directed to other aspects of the transition to growth. Recent work by Gallman and Kuznets makes it clear that growth has been associated with major changes in sector shares of commodity output and with an associated sharp *compositional* shift in the pattern of investment activity.[5] It is with the latter point that we are primarily concerned.

The development process in the United States has been characterized by a significant increase in the importance of manufactured producers' durables and a decline in the relative importance of construction goods. This suggests that an important aspect of industrialization may be illuminated by examining the changing historical role of the capital goods industries, and more particularly that growing portion of them which is devoted to the production of producers' durable goods. The role of these industries in introducing and in diffusing technological change is obviously multidimensional, but two aspects at least may be singled out: (1) All innovations - whether they include the introduction of a new product or provide a cheaper way of producing an existing product - require that the capital goods sector shall in turn produce a new product (capital good) according to certain specifications. We may usefully look upon the capital goods sector as one which is, in effect, engaged in custom work. That is, firms in this industry have typically become highly specialized, in the sense that most firms produce a relatively narrow range of output (at least in industrialized economies) in response to technical specifications laid down by a wide range of customers in the consumer goods or other capital goods industries. (2) In addition to this "external" adaptation there is an important "internal" one. Quite simply, members of the producers' durables industry have an internal motivation to improve their own techniques in the production of the durable goods themselves. Their success in accomplishing this improvement in turn affects the price of their machinery output and therefore is an important determinant, first, of investment activity throughout the economy and second, of the rate at which technological innovations, once made, will be diffused - that is, of the speed with which the economy will install and apply new techniques of production once they have been discovered. Moreover, third, cost reduction by the capital goods industries is thus capital saving for the economy, and cost reduction here also raises the marginal efficiency of capital of other industries. For these reasons it is suggested that a significant - and so far largely unexplored - dimension of the transition to economic growth lies in the ability of the capital goods sector to assimilate and develop proficiency in the new machine technology and thus both to generate, and to adapt itself to, the continually altering technological requirements of an industrializing economy. It will be argued that the machinery-producing industries possess certain unique characteristics which played a major role in accounting for the rapid production and diffusion of technological innovations, which were such a well-known and outstanding feature of the period under consideration.

Machine tools are the most important members of the larger classification of power-driven metalworking machinery. The basic distinction is that machine tools shape metal through the use of a cutting tool and the progressive cutting away of chips, whereas other metalworking machinery shapes metal without the use of a cutting tool - by pressing (forming, stamping, punching), forging, bending, shearing, etc. There is considerable complementarity and substitution between the two classes of tools, and occasional reference will necessarily be made in what follows to developments outside the category of machine tools as defined here.

II

In 1820 or so, there was no separately identifiable machine-tool or machinery-producing sector in the American economy. Although machines of varying degrees of complexity were, of course, being used, the production of the machines had not yet become a specialized function of individual firms.[6] Machines were, by and large, produced by their ultimate users on an *ad hoc* basis. For many years, the most intractable problems associated with the introduction of techniques of "machinofacture" lay in the inability to produce machines which would perform according to the special and exacting requirements and specifications of the machine user.[7] A major episode, then, in the process of industrialization lay in the emergence of a specialized collection of firms devoted to solving the unique technical problems and mastering the specialized skills and knowledge requisite to machine production.

It is useful to examine the growth of the machinery-producing sector from the point of view of the learning process involved. For, as we shall see, most machinery production poses a broadly similar set of problems and involves a broadly similar set of skills and technical knowledge in their solution. Moreover, the pace of industrialization was, in large measure, determined by the speed with which technical knowledge was diffused from its point of origin to other sectors of the economy where such knowledge had useful applications.

The growth of independent machinery-producing firms occurred in a continuing sequence of stages roughly between the years 1840-1880.[8] These stages reflect both the growth in the size of the market for such machines and the accretion of technical skills and knowledge (and growth in the number of individuals possessing them) which eventually created a pattern of product specialization by machine-producing firms which was closely geared to accommodating the requirements of machine users.

In the earliest stages, machinery-producing establishments made their first appearance as adjuncts to factories specializing in the production of a final product. Thus the first machine-producing shops appeared in the textile firms of New England attached directly to such firms as the Amoskeag Manufacturing Company in Manchester, New Hampshire, and the Lowell Mills in Lowell, Massachusetts. As such shops achieved success as producers of textile machinery, they gradually undertook, not only to sell textile machinery to other firms, but to produce a diverse range of other types of machinery - steam engines, turbines, mill machinery and (most important) machine tools - as well. In the early stages then, skill acquired in the production of one type of machine was transmitted to the production of other types of machines by this very simple expedient whereby a successful producer of one type of machinery expanded and diversified his operations. Thus, with the introduction of the railroads in the 1830's, the Lowell Machine Shop (which became an independent establishment in 1845) became one of the foremost producers of locomotives.[9] Similarly, locomotives were produced by the Amoskeag Manufacturing Company; and the locomotive works in Paterson, New Jersey, grew out of the early cotton textiles industry in that city. The most successful of all American locomotive builders, the Baldwin Locomotive Works in Philadelphia, grew out of a firm previously devoted, *inter alia,* to textile-printing machinery.[10]

Whereas the production of heavier, general-purpose machine tools - lathes, planers, boring machines - was initially undertaken by the early textile machine shops in response to the internal requirements of their own industry and of the railroad industry, the lighter, more specialized high-speed machine tools - turret lathes, milling machines, precision grinders - grew initially out of the production requirements of arms makers. Somewhat later, the same role was played by the manufacturers of sewing machines and, toward the end of the period under consideration, by the demands of bicycle and automobile manufacturers.[11]

Thus the machine tool industry itself was generated as the result of the specific production requirements of a sequence of industries which adopted techniques of machine production throughout the period. In each case, the introduction of a new process or a new product required an adaptation and adjustment in the capital goods industries to new technical requirements and specifications which did not initially exist. There took place, as it were, a period of technical gestation at the intermediate stages of production, during which time the appropriate accommodations were made to the specific technical needs of the new process or product. As the demand for particular kinds of machines became sufficiently great, reflecting the fact that the same machines came to be employed in a progressively increasing number of industries, the production of that machine itself came to constitute a specialized operation on the part of individual establishments.[12]

The machine tool industry, then, originated out of a response to the machinery requirements of a succession of particular industries; while still attached to their industries of origin, these establishments undertook to produce machines for diverse other industries, because the technical skills acquired in the industry of origin had direct application to production problems in other industries; and finally, with the continued growth in demand for an increasing array of specialized machines, machine tool production emerged as a separate industry consisting of a large number of firms most of which confined their operations to a narrow range of products - frequently to a single type of machine tool, with minor modifications with respect to size, auxiliary attachments, or components.

In late years . . . manufacturers starting in this branch of industry [metalworking machinery] have very generally limited their operations to the production of a single type of machine, or at the most to one class embracing tools of similar types. For example, there are large establishments in which nothing is manufactured but engine lathes, other works are devoted exclusively to planers, while in others milling machines are the specialty.

This tendency has prevailed in Cincinnati perhaps more than in any other city, and has been one of the characteristic features of the rapid expansion of the machine-tool industry in that city during the past ten years. During the census year there were in Cincinnati 30 establishments devoted to the manufacture of metal-working machinery, almost exclusively of the classes generally designated as machine tools, and their aggregate product amounted to $3,375,436. In 7 shops engine lathes only were made, 2 were devoted exclusively to planers, 2 made milling machines only, drilling machines formed the sole product of 5 establishments, and only shapers were made in 3 shops.[13]

In 1914, there were 409 machine tool establishments in the United States producing an output of $31,446,660.[14] In the same year the *American Machinist,* in a survey which was admittedly incomplete, published a map showing the locations of 570 firms engaged in the production of "machine tools, small tools, machinist's tools and machine tool appurtenances. . . ." These firms were all in the northeast quadrant of the country, with Ohio leading with 117 and then Massachusetts with 98, Connecticut with 66, Pennsylvania with 60, New York with 57, and Illinois with 42.[15]

III

A proper understanding of the "portentously rapid" rate of technological innovation which accompanied American industrialization during the period under consideration requires that we focus attention upon a particular aspect of the changing nature of manufacturing activity. For this purpose it is necessary to discard the familiar Marshallian approach, involving as it does the definition of an industry as a collection of firms producing a homogeneous product - or at least products involving some sufficiently high cross-elasticity of demand. For many analytical purposes it is necessary to group firms together on the basis of some features of the commodity as a final product; but we cannot properly appraise important aspects of technological developments in the nineteenth century until we give up the Marshallian concept of an industry as the focal point of our attention and analysis. These developments may be understood more effectively in terms of certain functional processes which cut entirely across industrial lines in the Marshallian sense.[16]

It is a common practice to look upon industrialization as involving not only growing specialization but also growing complexity and differentiation.[17] While this is certainly true in the sense that there takes place a proliferation of new skills, facilities, commodities, and services, it also overlooks some very important facts. The most important for present purposes is that industrialization was characterized by the introduction of a relatively small number of broadly similar productive processes to a large number of industries. This follows from the familiar fact that industrialization in the nineteenth century involved the growing adoption of a metal-using technology employing decentralized sources of power.[18]

If we look at the vertical dimension of productive activity in the sense of the sequence of stages involved in the production of a final product, it appears that in preindustrial economies, skills and techniques tended to be much more specific and tied down to individual vertical sequences than was the case in industrial economies. The central role, in industrial economies, of the application of decentralized power sources in the working of metals has meant the employment of similar skills, techniques, and facilities at some of the "higher" stages of production for a wide range of final products. Thus, in contrast to sequences of parallel and unrelated activities, we find a phenomenon which we will call "technological convergence." This convergence exists throughout the machinery and metal-using sectors of an industrial economy. Throughout these sectors there are common processes, initially in the refining and smelting of metal ores, subsequently in foundry work whereby the refined metals are cast into preliminary shapes, and then in the various machining processes through which the component metal parts are converted into final form preparatory to their assembly as a finished product. It is with the machinery stages, of course, that we are primarily concerned here.

The use of machinery in the cutting of metal into precise shapes involves, to begin with, a relatively small number of operations (and therefore machine types): turning, boring, drilling, milling, planing, grinding, polishing, etc. Moreover, all machines performing such operations confront a similar collection of technical problems, dealing with such matters as power transmission (gearing, belting, shafting), control devices, feed mechanisms, friction reduction, and a broad array of problems connected with the properties of metals (such as ability to withstand stresses and heat resistance). It is because these processes and problems became common to the production of a wide range of disparate commodities that industries which were apparently unrelated from the point of view of the nature and uses of the final product became very closely related (technologically convergent) on a technological basis - for example, firearms, sewing machines, and bicycles.

This technological convergence had very important consequences for both (1) the development of new techniques and (2) their diffusion, once developed.[19] The intensive degree of specialization which developed in the second half of the nineteenth century owed its existence to a combination of technological convergence plus what Stigler has called vertical disintegration - that is, a tendency for individual sequences in the production of a final product to be undertaken as separate operations by separate firms. Stigler suggests that vertical disintegration, and therefore increasing process specialization by firm, is likely to be characteristic of growing industries.

If one considers the full life of industries, the dominance of vertical disintegration is surely to be expected. Young industries are often strangers to the established economic system. They require new kinds of qualities of materials and hence make their own; they must overcome technical problems in the use of their products and cannot wait for potential users to overcome them;

they must persuade customers to abandon other commodities and find no specialized merchants to undertake this task. These young industries must design their specialized equipment and often manufacture it, and they must undertake to recruit (historically, often to import) skilled labor. When the industry has attained a certain size and prospects, many of these tasks are sufficiently important to be turned over to specialists.[20]

It seems clear that the extraordinary degree of specialization achieved in the machine-producing sector of the American economy is attributable, not only to the growth of individual industries experiencing vertical disintegration in Stigler's sense, but also to the simultaneous growth of several industries which were technologically convergent in our sense. The extent of machinery specialization which was achieved would not have been possible if there were only vertical disintegration *without convergence.* For the degree of specialization achieved owed its existence in large part to the fact that certain technical processes were common to many industries. Individual firms producing nothing but milling machines would not have emerged in an economy where only firearms manufacturers employed milling machines, nor would specialized grinding machine producers have emerged in an economy where only bicycle manufacturers employed grinding machines. With technological convergence, however, milling and grinding became important operations in a large number of metal-using industries, thus permitting a degree of specialization at "higher" stages of production which would not otherwise have been possible. Since, as Adam Smith, Allyn Young, and George Stigler have taught us, "the division of labor is limited by the extent of the market," the unique degree of specialization developed in the American machinery-producing sector owed as much to technological convergence as it did to the expansion in the demand for individual final products.

The importance of this specialization must be conceived, not only in a static sense, but in a dynamic sense as well. For there is an important learning process involved in machinery production, and a high degree of specialization is conducive not only to an effective learning process but to an effective *application* of that which is learned. This highly developed facility in the designing and production of specialized machinery is, perhaps, the most important single characteristic of a well-developed capital goods industry and constitutes an external economy of enormous importance to other sectors of the economy.

Metal-using industries, therefore, were continually being con-fronted with similar kinds of problems which required solution and which, once solved, took their place in short order in the production of other metal-using products employing similar processes. Using Usher's useful terminology, metal-using industries were continually engaged in a "setting of the stage" for particular problems which, once they were solved, produced free technological inputs to other metal-using industries.

In all of this the machine tool industry, as a result of technological convergence, played a unique role both in the initial solution of technological problems and in the rapid transmission and application of newly-learned techniques to other uses. We suggest that the machine tool industry may be regarded as a center for the acquisition and diffusion of new skills and techniques in a machinofacture type of economy. Its chief importance, therefore, lay in its strategic role in the learning process associated with industrialization. This role, as I have asserted and will now elaborate further, is a dual one: (1) new skills and techniques were developed or perfected here in response to the demands of specific customers; and (2) once they were acquired, the machine tool industry was the main transmission center for the transfer of new skills and techniques to the entire machine-using sector of the economy.

IV

In this section we will examine the nature and the consequences of technological convergence. An exhaustive cataloguing of specific instruments is neither possible nor, fortunately, is it necessary for our purposes. The basic pattern which I wish to emphasize with respect to both the origin and diffusion of machine tool innovations will be explored by reference to its historical role in four industries: firearms, sewing machines, bicycles, and automobiles. If time and space permitted, a more comprehensive account would include also a wide spectrum of machine-tool-using industries ranging from watches and clocks, scientific instruments, hardware, and typewriters to agricultural implements, locomotives, and naval ordnance.

What is important here is an historical sequence in which the need to solve specific technical problems in the introduction of a new product or process in a single industry led to exploratory activity at a vertically "higher" stage of production; the solution to the problem, once achieved, was conceived to have immediate applications in producing other products to which it was closely related on a technical basis; and this solution was transmitted to such other industries via the machine tool industry. The machine tool industry may be looked upon as constituting a pool or reservoir of skills and technical knowledge which are employed throughout the entire machine-using sectors of the economy. Because it dealt with processes and problems common to an increasing number of industries, it played, during this period, the role of a transmission center in the diffusion of the new technology. The pool of skill and technical knowledge was added to as a result of problems which arose in particular industries. Once the particular problem was solved and added to the pool, the solution became available, with perhaps minor modifications and redesigning, for employment in technologically related industries. Thus, as a result of technological convergence, external economies of enormous importance were rapidly generated.

Throughout the whole of the first half of the nineteenth century and culminating perhaps with the completion of Samuel Colt's armory in Hartford in 1855, the making of firearms occupied a position of decisive importance in the development of specialized, precision machinery. The notion that the system of interchangeable parts sprang full-blown from Whitney's genius in musket manufacture has now been accorded a decent burial.[21] What is clear is that the new machinery and technology were the joint product of efforts to overcome the same group of problems, not only by Whitney, but by men employed at such places as Robbins and Lawrence, Ames Manufacturing Company, Colt's armory, and the government armories at Springfield and Harper's Ferry as well.

The introduction of Thomas Blanchard's stocking lathe for the shaping of gunstocks, in 1818, represents an interesting transitional innovation, inasmuch as it was originally developed for the shaping of wooden materials but involved a principle which eventually found wide applications in other materials for reproducing irregular patterns. Blanchard's lathe, which replaced the tedious and time-consuming hand techniques of shaping the gunstock by whittling, boring, and chiseling, was introduced at the national armories at Springfield and Harper's Ferry during the 1820's.[22] The principle embodied in the machine was quickly applied to such sundry items as hat blocks, handles, spokes of wheels, sculptured busts, oars, and shoe lasts.[23]

The firearms industry was instrumental in the development of the whole array of tools and accessories upon which the large-scale production of precision metal parts is dependent: jigs (originally employed for drilling and hand-filing), fixtures, taps and gauges, and the systematic development of die-forging techniques.[24]

The milling machine, perhaps with the turret lathe one of the two most versatile of all modern machine tools, owed its origin in the United States to the attempt of arms makers to provide an effective machine substitute for highly expensive hand filing and chiseling operations. Although here, as in so many other cases, its exact origins are shrouded in obscurity, it is clear that both Eli Whitney and Simeon North employed crude milling machines in their musket-producing enterprises in the second decade of the nineteenth century, as did John H. Hall at the Harper's Ferry Armory.[25] Its subsequent development was largely the work of the national armories, especially the highly important work of Thomas Warner at the Springfield Armory,[26] and such gun-producing firms as Robbins and Lawrence, of Windsor, Vermont. The design of the plain milling machine was stabilized in the form which came to be known as the Lincoln miller, in the 1850's, and rapidly assumed a prominent place in all the metal trades. Fitch states that, between 1855 and 1880, "... nearly 100,000 of these machines or practical copies of them, have been built for gun, sewing-machine and similar work."[27]

The final major contribution of the arms makers was the role played in the development of the turret lathe which, together with the milling machine, was indispensable to the production of all commodities based upon interchangeable parts. The turret lathe, holding a cluster of tools placed on a vertical axis, made it possible to perform a sequence of operations on the work piece without the need for resetting or removing the piece from the lathe. It therefore revolutionized all manufacturing processes requiring large volumes of small precision components such as screws - which were, in short order after the development of the turret lathe, produced on turret lathe machines.[28]

The origin of the turret lathe (initially employing a horizontal axis for the turret) has been attributed to Stephen Fitch, of Middlefield, Connecticut, in 1845, while he was engaged on a government contract for the production of percussion locks for an army horse pistol.[29] The turret lathe principle was employed and improved at the Colt armory (where Root introduced a double-turret machine in 1852) and by Frederick W. Howe while superintendent of the

Robbins and Lawrence Company at Windsor, Vermont; and turret lathes were built and sold commercially by that company in 1854. A turret screw machine, designed in 1858 by H. D. Stone, was sold commercially by the Jones and Lamson Company.[30] From this point on, the machine was adapted and modified for innumerable uses in the production of components for such products as sewing machines, watches, typewriters, locomotives, bicycles and, eventually, automobiles. Its most important subsequent improvement was introduced by Christopher Spencer, a former Colt employee and the inventor of the Spencer repeating rifle. As a result of a machine which he invented for turning sewing-machine spools, Spencer went on to explore methods for making metal screws automatically and, in so doing, invented the automatic turret lathe.[31] The importance of this innovation is difficult to exaggerate, since the self-adjusting feature of the cam cylinder with adjustable strips, through which automaticity was achieved, was eventually to make possible all modern automatic lathe operations. Together with the subsequent perfection of multiple spindle techniques, it was instrumental in a major acceleration in the pace of machine tool operations.

From the 1850's through the 1870's, the technical requirements of the sewing-machine industry played a major role as a source of machine tool innovations. Although sewing-machine production was virtually nonexistent in 1850, it constituted a flourishing industry in 1860,[32] and grew with remarkably swift strides, nationally and internationally, in the following decade.[33] Out of the innumerable modifications of the sewing machine grew the vast boot-and-shoe and men's and women's ready-to-wear clothing industries; and the machine, by 1890, was used extensively in the production of such items as awnings, tents and sails, pocketbooks, rubber and elastic goods, saddlery and harnesses, etc., and in bookbinding. The rapid diffusion of the sewing machine after 1860 was due to the fact that it provided a highly effective mechanical device for performing an operation common to many industries. It therefore constitutes a major historical example of what we have called technological convergence.[34]

The machining requirements and processes of sewing-machine manufacturing were broadly similar to those of firearms production, and sewing-machine manufacturers were quick to adopt these processes.[35] However, just as in the case of firearms, the solution of technical problems in sewing-machine production resulted in major additions to the stock of machine-cutting instruments which, in turn, were applied to the production of other metal-using products. The most important innovations in the sewing-machine industry were products of the remarkable Brown and Sharpe Manufacturing Company of Providence, Rhode Island.

The Brown and Sharpe Company was founded in 1833 by David Brown and his son, Joseph R. Brown. Until 1850, the firm was engaged in the production and repairing of clocks, watches, and mathematical instruments. In 1850, the firm introduced a fully automatic linear dividing engine and shortly thereafter a vernier caliper and then, in 1855, a precision gear-cutting machine.[36] In 1858, the firm commenced production of the Willcox and Gibbs sewing machine, which was an immediate success and resulted in a very considerable plant expansion. The unique machine tool contributions of this firm were generated primarily by the necessity to provide the appropriate machinery for their sewing-machine operation. But the results of these efforts were machine tools of a general usefulness far surpassing the industry of origin.[37]

The first of these machines, a turret screw machine devised for sewing-machine parts, was impressed into other uses upon the outbreak of the Civil War and thereafter became a major tool in machine shop practice generally.[38] The primary purchasers of Brown and Sharpe's screw machine were producers of hardware and tools, sewing machines, shoe machinery, locomotives, rifles and ammunition - and machine tools.[39]

Another machine tool extensively developed by Brown and Sharpe arose out of a major production problem in the building of sewing machines and one which was to assume even greater proportions in the future production of automobiles - that is, the precision grinding, to a fine finish, of hardened steel parts. This problem was encountered by Brown and Sharpe in providing components of the Willcox and Gibbs sewing machine - needle bars, foot bars, and shafts.[40] Brown finally produced a cylindrical grinding machine which was employed within his own firm and was sold to other firms (including foreign firms) beginning in 1865.[41] After several years of extensive modification and redesigning, the firm introduced a universal grinding machine of far greater versatility which was exhibited at the Centennial Exhibition in Philadelphia in 1876. This machine is the direct ancestor of the modern heavy production grinding machines.[42]

The development of the universal milling machine by Brown and Sharpe is, perhaps, the most outstanding example of a machine which was initially developed as a solution to a narrow and specific range of problems and which eventually had enormous unintended ramifications as the technique was applied to similar productive processes over a wide range of metal-using industries.

The universal milling machine had its immediate stimulus in the production of Springfield muskets by the Providence Tool Company at the outbreak of the Civil War. One of the gun parts (the nipple) required a hole to be drilled in it, and for this purpose the Providence Tool Company employed twist drills, the twist drills in turn being made by a crude process of hand filing the spiral grooves in tool-steel rods or wire. Frederick W. Howe, at the time superintendent of the company, brought the matter to the attention of Joseph Brown, whose appreciation for the problem was heightened by the fact that Brown and Sharpe employed similar drills in the production of the Willcox and Gibbs sewing machine. Brown's solution was the universal milling machine, the first of which was sold to the Providence Tool Company in March 1862. It was an amazingly useful machine, which would not only cut the grooves of spiral drills but could be employed in all kinds of spiral milling operations and in gear cutting, as well as in the cutting of all sorts of irregular shapes in metal.[43] Within the first ten years, Brown and Sharpe sold universal milling machines to manufacturers of hardware, tools, cutlery, locks, arms, sewing machines, textile machinery, printing machines, professional and scientific instruments, and locomotives, to machine shops and foundries, and of course to machine tool manufacturers. Later on in the century, with each successive product innovation, universal milling machines were sold to a succession of firms producing cash registers, calculating machines, typewriters, agricultural implements, bicycles, and automobiles. Even this impressive list of users is far from exhaustive. Toward the end of the nineteenth century, heavy-duty milling machines increasingly undertook machining operations previously performed by planing and shaping machines.

The Brown and Sharpe sales records show, furthermore, that the largest single group of buyers of their universal milling machine was other machine tool producers. Thus, the creation of a new machine tool to solve technical problems in the production of a final product resulted in a significant source of increased productive efficiency in the machine tool industry itself.

By 1880, the proliferation of new machine tools in American industry had begun to reach torrential proportions.[44] Although there were relatively few dramatically new machines comparable to the milling machine or turret lathe, the period from 1880 to 1910 was characterized by an immense increase in the development of machine tools for highly specialized purposes, by a continuous adaptation of established techniques such as automatic operation to new uses, and by a systematic improvement in the properties of materials employed in machine tool processes. The introduction of high-speed steel in machine-cutting tools and the use of superior artificial abrasives such as silicon carbide in grinding processes are the outstanding examples of the last development. In all of this the emergence of the new forms of transportation, most notably the bicycle and automobile, played a vital role.

Although high-wheeled English bicycles were exhibited at the Philadelphia Exposition in 1876 and at that time engaged the interest of Colonel Albert A. Pope, who was to play a pioneering role in their eventual introduction in the United States, they did not achieve large-scale popularity until they assumed their modern "safety" form in the early 1890's.[45] The industry's spectacular growth during the 1890's and its subsequent abrupt decline are indicated by the fact that there were 27 establishments producing bicycles and tricycles in 1890, 312 in 1900, and 101 in 1905. The value of the industry's output was $2,568,326 in 1890, $31,915,908 in 1900, and $5,153,240 in 1905.[46]

Many of the unique problems associated with the production of a satisfactory bicycle - given its source of locomotion - revolved around the need for lightness,[47] hardened precision parts, and efficient power transmission and friction reduction. In solving these problems, bicycle manufacturers and machine tool makers not only introduced novel techniques but redesigned, perfected, and popularized techniques which antedated the bicycle and thus made them available for numerous new uses. The most important direct beneficiaries of the innovations in bicycle production were the automobile makers. But, in some measure, these innovations were transferred and made an important impact in all forms of manufacturing where friction reduction and power transmission constituted serious problems and wherever the newly designed machine tools had useful applications.

The problems posed by large-scale bicycle production were instrumental in improving and popularizing two highly important machining techniques and applying them to new uses: the forming tool and the oil-tube drill. Although the forming tool was employed previous to 1890, its use was confined to metals of soft composition, such as were used in making caps for salt and pepper boxes. Its much more important application to hardened metals, which made of it a standard machine shop practice, resulted from the transfer of the technique to the production of bicycle-wheel hubs. Similarly, the oil-tube drill, which had an oil channel leading to or near the point, and which made possible the lubrication and cooling of cutting edges as well as the removal of chips, had been employed before 1890 in drilling gun barrels. Its rapid diffusion after 1890 resulted from its extended application, together with the forming tool, in the drilling of holes in bicycle-wheel hubs.[48]

The requirements of bicycle production played a crucial role in the development of effective techniques for making ball bearings which, in turn, had an incalculable impact through reducing the effects of wear and friction on all machine processes. The highly exacting requirements of the ball bearing, as well as of the hardened cup and cone on which the bicycle balls roll, necessitated grinding operations of great precision.[49] In some cases the grinding machines which had been designed by Brown and Sharpe for grinding sewing-machine needle bars were adapted for this new use.[50] The eventual solution to the grinding problems involved, however, as they pertained both to the bicycle and automobile, relied heavily upon the improvements in grinding from the work of men like Charles H. Norton[51] and Edward G. Acheson, who revolutionized grinding operations through the introduction of artificial abrasives on the grinding wheel itself.

The bicycle industry was responsible also for numerous other innovations and modifications whose ultimate use extended far beyond the bicycle industry. The flat-link chain, an integral part of the safety bicycle, was applied to numerous other uses as a convenient device for the transmission of power. The chainless bicycle had focused attention on the need for hardened bevel gears, and the resulting improvements in gear-cutting machinery, such as those of Leland and Faulconer Company, were of considerable importance in the automobile.[52] The production possibilities of the turret lathe and of the automatic screw machine in mass production operations were extended by the "demonstration effect" of their application to new uses in the bicycle industry.[53] The need for metals with specific properties, such as the light tubular steel employed in the frame, and the high-tensile-strength steel wires employed in the wheel spokes, led to metallurgical explorations which were of great benefit in other metal-using industries.[54]

The automobile was in the earliest stages of its phenomenal growth in the first decade of the present century. The value of automobile output in 1900 was less than $5,000,000 and - although there were 57 establishments engaged in automobile manufacturing - their work was still essentially experimental.[55] In 1909, there were 265 establishments manufacturing complete automobiles and 478 manufacturing automobile bodies and parts; the corresponding figures for 1914 were 300 and 971, respectively. The value of automobile output rose from $26,645,064 in 1904 to $193,823,108 in 1909 and to $503,230,137 in 1914. For the same years, the number of automobiles made was 21,692, 126,570 and 568,781.[56]

The massive requirements for heavy, high-speed, and increasingly automatic tools generated by this growth in automobile output quickly made of the automobile industry the largest single buyer of machine tools, and in so doing they exerted a profound effect on the industry and on the designing of machine tools.[57] But while it is easy to look upon the automobile industry as something *sui generis,* such an attitude would reflect an immature appreciation of the technological basis underlying automobile production. For by 1900, as we have seen, there existed an extensive accumulation of technological and engineering experience in the production of machine tools and a highly developed sophistication in designing and adapting basic types of machine tools for special production purposes.

The problems of large-scale automobile production involved the extension to a new product of skills and machines not fundamentally different from those which had already been developed for such products as bicycles and sewing machines. Underlying the discontinuity of product innovation, then, were significant continuities with respect to productive processes. The transition to automobile production for the American economy after 1900 was therefore *relatively* easy, because the basic skills and knowledge required to produce the automobile did not themselves have to be "produced" but merely transferred from existing uses to new ones. This transfer was readily performed by the machine tool industry.

The transfer process is seen most clearly in the further evolution and adaptation of the grinding machine, which, as we have seen, had been developed to an advanced state before the advent of the automobile. The automobile, however, far surpassed the relatively modest needs of the sewing machine and bicycle in its need for precision-finished, hardened steel parts. Until confronted with the compelling needs of the automobile, the grinder had been used either for relatively light operations or for finishing components which had acquired their basic shapes upon a lathe. In response to the needs of automobile production, the grinding machine was converted into a tool capable of heavy production operations in the course of which it frequently replaced entirely the lathe and other machine tools. Its role here was indispensable in that it provided, for its time, the only way of undertaking the precision machining of the stronger and lighter alloy steels which played such a prominent part in automobile components.

Thus, within a few years after 1900, specialized grinding machines were devised for vital parts of the automobile engine and transmission, including special ones for crankshafts, for camshafts, for piston rings, and for cylinders. Perhaps most far reaching of all, because of their importance elsewhere, were the contributions to gear cutting. The automobile generated a demand for strong, durable gears which was quite unprecedented. Here the technological interrelations between the bicycle and the automobile are particularly clear, since the most important innovator in the grinding of gear teeth was the Leland and Faulconer Company. "Faulconer was, in 1899, the first to design a machine for production grinding of hardened bevel gears for bicycles."[58] This was the same firm which was later to become the Cadillac Automobile Company. The earliest automobile firms drew very heavily upon the business and technical leadership, plant facilities, and skilled labor of the bicycle industry, the decline in which coincided exactly (in the first decade of the century) with the rapid growth of automobiles.[59]

The requirements of automobile production induced innovations or substantial improvements across the whole range of machine tools, in drilling and tapping, in milling, in lathe work generally, etc. Moreover, it brought about a significant substitution of one machine process for another, most particularly as a result of the development of power presses and dies. Intricate automobile components which would once have been produced by the lathe, drill press, milling machine, or casting or forging, were increasingly stamped directly out of sheet metal - a technique which had been given considerable impetus in the production of bicycles.[60]

The relations between the machine tool builders and the automobile industry also provided compelling evidence of the manner in which technological convergence produces learning experiences which generate diffuse and unanticipated benefits - for the automobile, itself a machine of considerable complexity, encounters many problems in its operation similar to those of the machines which produce it. As a result, not only were existing machine tool techniques adapted to the production of this new product, but important features of the automobile itself were actually transferred and embodied into machine tools.[61] Thus, the transmission for the drive and feed mechanisms of machine tools was considerably improved when machine tool builders adopted the alloy steel sliding gears and integral keyshafts developed by automobile designers. Moreover, the introduction of antifriction bearings into key points of the machine tool resulted from the demonstration of their usefulness in automobiles. Finally, the whole approach to the lubrication of machine tools was radically revised as a result of the automobile. An important problem in the maintenance of machine tools had been the frequent breakdowns when inexperienced or negligent operators failed to attend properly to the numerous separate lubrication points of their machines. The solution, of course, was the eventual adoption of the centralized, self-acting lubrication system of the automobile, which went into operation automatically as soon as the machine was activated.[62]

V

We have attempted to show how, with the growing volume of manufacturing output, increasing vertical disintegration from the point of view of a single industry was accompanied by technological convergence of larger groups of industries. The result of this convergence was a growth in a relatively small number of process specializations the consequences of which, for the production and diffusion of new technical knowledge, we have examined by focusing attention on a sequence of industries which played a most important role in this development. A few further comments appear to be called for.

An explanation of many of the technological changes in the manufacturing sector of the economy may be fruitfully approached at the purely technological level. This is not to deny, of course, that the ultimate incentives are economic in nature; rather, the point is that complex technologies create internal compulsions and pressures which, in turn, initiate exploratory activity in particular directions. The notion of imbalances in the relation *between machines* is virtually *de rigueur* in any treatment of the English cotton textile industry in the eighteenth century (Kay's flying shuttle leading to the need for speeding up spinning operations, etc.). We suggest that, *within a single complex machine or operation,* even more important imbalances frequently exist among its component parts. A concept of technological disequilibrium may be helpful here. At any time, the component parts of a machine vary in their ability to exceed their present level of performance, which is determined by the capacity of some limiting component. Any important improvement in the operation of a component, whether it be the currently limiting one or not, is likely to create new obstacles, in the form of limitations imposed by another component, to the achievement of a higher level of performance. Thus single improvements tend to *create* their own future problems, which compel further modification and revision.

The interdependence between the forming tool and the oil-tube drill in the machining of bicycle hubs, referred to earlier, is an important case in point. The introduction of the forming tool for the outside of bicycle hubs created a disequilibrium between the operations carried on for the outside and the inside of the hub. Since the forming tool worked more rapidly on the outside of the hub than the old-fashioned drills worked on the inside, the fullest gains from the use of the forming tool required a speeding up of drilling operations. This imbalance was corrected by the oil-tube drill which, in speeding up drilling operations, brought about a closer synchronization between the two operations. Numerous other instances of the role of disequilibrating forces could easily be cited,[63] but the induced improvements in machine tool design following the introduction of high-speed steel around the turn of the century are easily the most important within our period.

When Frederick W. Taylor and his associates introduced high-speed steel (a steel alloy which drastically improved the ability of a cutting tool to maintain its hardness at high temperatures) it became possible at once to remove metal by cutting operations at dramatically higher speeds. But it was impossible to do so on machines designed for the older carbon steel cutting tools, because they could not withstand the stress or provide sufficiently higher speeds in the other components of the machine tool. As a result, the availability of high-speed steel for the cutting tool quickly generated a complete redesign in machine tool component - the structural, transmission, and control elements:

During the first decade of the 20th century we see high-speed steel revolutionizing the lathe - as it does all production machine tools. Beds and slides rapidly become heavier, feed works stronger, and the driving cones are designed for much wider belts than of old. The legs of big lathes grow shorter and shorter, and finally disappear as the beds grow down to the floor. On these big machines massive tool blocks take the place of tool posts and multiple tooling comes into vogue.[64]

The final effect, then, of this redesigning which was initiated by the use of high-speed steel in cutting tools was to transform machine tools into much heavier, faster, and more rigid instruments which, in turn, enlarged considerably the scope of their practical operations and facilitated their introduction into new uses.

Many aspects of technological change, in order to be adequately understood, must be examined in terms of particular historical sequences, for in technological change as in other aspects of human ingenuity, one thing often leads to another - not in a strictly deterministic sense, but in the more modest sense that doing some things successfully creates a capacity for doing other things. We have indeed already explored this theme at some length: given what we have called technological convergence, experience in the production of firearms made it a relatively simple matter to produce sewing machines,[65] just as the skills acquired in producing sewing machines and bicycles greatly facilitated the production of the automobile. This is even more apparent, in microcosm, in the chronological history of certain individual firms, some of which, over a period of several decades, ran the entire gamut of the sequence of products - or of the machinery for producing the products - which we have considered here. This was true of the Pratt and Whitney Company. Beginning with the Civil War and over the next fifty years, Pratt and Whitney introduced in succession machinery for the production of firearms, sewing machines, bicycles, and automobiles, as well as numerous other kinds of high-precision specialized macinery.[66] A machinery plant in Hartford which was originally owned by the Robbins and Lawrence Company, and later acquired by the Sharps Rifle Manufacturing Company, ran through a succession of owners (including the showman Phineas T. Barnum) and produced successively machine tools, guns, sewing machines, bicycles, motorcycles, and automobiles.[67] The Leland, Faulconer and Norton Company (later the Cadillac Automobile Company) of Detroit, which was founded in 1890 as a producer of machine tools and special machinery, introduced machinery for producing bicycle gears during the brief heyday of the bicycle, switched to building gasoline engines for motor boats when the bicycle industry began to decline, and by 1902 had undertaken the production of automobile engines.[68]

An examination of these continuities in technology provides a basis for understanding historical events which otherwise appear to be random or capricious. The interesting thing about the group of industries discussed here is that they were all dependent, in their development, upon technological changes dealing with a limited number of processes and that the solution to problems posed by these processes eventually became the specialized function of a well-organized industry. A question of more contemporary interest is whether similar technological convergences are occurring in twentieth-century conditions; whether, for example, the chemicals and electronics industries are playing the same roles of information production and transmittal that machine tools played at an earlier stage in our history.[69] The answer to the question may be very important, even from the point of view of pure theory. For a theory which assumes that most technological change enters the economy "through a particular door," so to speak, might turn out to be much simpler, and therefore more elegant, than one which assumes that technological changes may be initiated, with equal probability, anywhere in the economy.

2

America's rise to woodworking leadership

I

One of the greatest difficulties confronting the writer of economic history is to convey to his audience a full sense of the kinds of problems which plagued and confounded his ancestors. This is particularly so in a country such as the United States, which prides itself upon its technological versatility and which tackles the most spectacular technological problems with an exuberant - not to say brash - self-confidence. In a society which now routinely practices such arcane crafts as "molecular architecture" and "genetic engineering" - activities which put to shame the medieval alchemist who would have been satisfied merely to turn dross into gold - it requires a great mental leap to understand the limitations confronting the colonial craftsman or the early nineteenth-century machinist. In a society which takes for granted a remarkably wide range of substitutability among material inputs in the production process - indeed, whose members would be hard-pressed even to identify the materials composing their table tops or sweaters - it is difficult in the extreme to appreciate the constraints confronting early Americans as they went about their ordinary productive activities. For our technological versatility is a recent acquisition - essentially a product of the past century or so. As we go farther back in historical time we enter a period one of the most distinctive characteristics of which was an extreme dependence upon the raw facts of the natural environment.

This dependence provides a central underlying theme of my paper, which concentrates upon the emergence of woodworking machinery in America between 1800 and the 1850's. Preindustrial societies - let us say America in 1800 - were heavily dependent upon the particular combinations of raw materials presented by their environments because their capacity to manipulate or transform these materials was extremely modest, and therefore the number of technological options open to them was typically very small.[1] An American builder in 1800 knew nothing of aluminum or prestressed concrete, nor was the textile manufacturer even dreaming of synthetic fibers. Their choices were first of all, therefore, dictated by their small stock of technological knowledge and the severely restricted portions of the natural environment which could be made economically productive within the confines of that limited technology.

The constraints upon their productive activities, however, were not solely technological. They were economic as well. It was technologically possible in 1800 to produce a wide range of objects - for example, machinery - out of either wood or iron. But, in America at least, although iron was a technologically feasible alternative, the retarded state of the iron industry, and therefore the high cost of iron, rendered that material prohibitively costly in many uses. The extent of American dependence upon wood in the first half of the nineteenth century was a reflection of the economic fact that wood was a far cheaper raw material. The profusion of forest resources, therefore, must be seen as an environmental fact which significantly accelerated the industrial growth of the American economy once Americans had developed appropriate techniques for their exploitation.

The extent of early American dependence upon wood is difficult to exaggerate. It was the major source of fuel, it was the primary building material, it was a critical source of chemical inputs (potash and pearlash),[2] and it was an industrial raw material par excellence. Lewis Mumford's assertion that "wood was the universal material of the eotechnic economy" was still true at the beginning of the nineteenth century. The following statement would also apply without serious modification to America in 1800:

As for the common tools and utensils of the time, they were more often of wood than of any other material. The carpenter's tools were of wood, but for the last cutting edge: the rake, the oxyoke, the cart, the wagon, were of wood: so was the washtub in the bathhouse: so was the bucket and so was the broom: so in certain parts of Europe was the poor man's shoe. Wood served the farmer and the textile worker: the loom and the spinning-wheel, the oil presses and the wine presses were of wood, and even a hundred years after the printing press was invented, it was still made of wood. The very pipes that carried water in the cities were often tree-trunks: so were the cylinders or pumps. One rocked a wooden cradle; one slept on a wooden bed; and when one dined one "boarded." One brewed beer in a wooden vat and put the liquor in a wooden barrel.

Stoppers of cork, introduced after the invention of the glass bottle, begin to be mentioned in the fifteenth century. The ships of course were made of wood and pegged together with wood; but to say that is only to say that the principal machines of industry were likewise made of wood: the lathe, the most important machine-tool of the period, was made entirely of wood - not merely the base but the moveable parts. Every part of the windmill and the water-mill except for the grinding and cutting elements was made of wood, even the gearing: the pumps were chiefly of wood, and even the steam engine, down to the nineteenth century, had a large number of wooden parts: the boiler itself might be of barrel construction, the metal being confined to the part exposed to the fire.[3]

The importance of wood in the American economy may be gauged in various ways. If American manufacturing industries are ranked by value added by manufacture in 1860, the lumber industry component of wood use alone ranks a close second behind cotton goods. Lumbering was the largest single manufacturing industry, on this basis, in the South and West.[4] Table 1 presents estimates of lumber consumption in the United States and compares them with estimates for the U.K. Two propositions emerge clearly from the table, and deserve emphasis: (1) Per capita lumber consumption begins to rise

Table 1. *Lumber consumption for the U.S. and U.K. (specified years)*

	U.S.		U.K.	
Year	Consumption in Board Feet, thousands[a]	Per Capita Consumption	Consumption in Board Feet, thousands[b]	Per Capita Consumption
1799	300,000	58	102,703	10
1809	400,000	57	121,916	10
1819	550,000	59	244,745	17
1829	850,000	67	319,306	20
1839	1,604,000	98	430,267[c]	23
1849	5,392,000	239	1,024,565[d]	50
1859	8,029,000	259	1,796,596	79
1869	12,755,543	328	2,419,390	95

[a]Henry B. Steer, *Lumber Production in the United States, 1799-1946* (Washington, D.C., U.S.G.P.O., 1948), p. 10.

[b]U.K. figures: To 1839 "Accounts and Papers: Reports of the Commissioners, Estimates for the House of Commons," *Great Britain Customs and Excise Department Statistical Office: Annual Statement of Trade,* H.M.S.O.

[c]Figures to this year are only labeled as "Timber." This excludes Wooden Hoops, Deals, Battens, and Hardwoods imported for furniture. For an approximation of total consumption, add about 10 percent to all figures to 1839.

[d]U.K. figures after 1843 are from *Annual Abstracts of Statistics: Great Britain: 1840-1946* (London: Kraus-Thomson Ltd.), Kraus Reprint Nos. 1 to 84.

in the 1820's and rises very sharply in the 1830's and after; (2) Per capita lumber consumption in the United States was several times as great as in the U.K.

The second proposition should not be surprising. The U.K. in the early nineteenth century was much farther along the road to industrialization than the United States. It had begun to encounter serious shortages of wood as far back as the days of Elizabethan England. Indeed, in terms of raw material inputs, the industrial revolution in the U.K. in the late eighteenth century and after may be characterized as a successful bypassing of constraints upon productive activity imposed by a severely limited wood supply. The early technological breakthroughs in fuel and power sources and metallurgy substituted coal and iron for the intensive reliance upon wood both as a fuel and as an industrial raw material.

In the United States the situation was vastly different. Whereas the early technological breakthroughs in the U.K. constituted an attempt at developing substitutes for scarce forest products, the United States, by contrast, possessed a rich abundance of such resources. In fact, much of what was unique in the American industrialization experience was attributable to the fact that the United States began her industrialization in a much more favorable resource position than western Europe generally. The innovative process in the United States in the early nineteenth century (and earlier) was a direct reflection of this circumstance. Whereas much of Britain's early industrialization effort needs to be understood as a deliberate attempt to overcome the constraints imposed by a dependence upon organic materials, Americans possessed no similar inducement. In fact, a key to much of early American industrialization - certainly until at least the middle of the nineteenth century - needs to be understood in terms of a technology specifically geared to the intensive exploitation of natural resources which existed in considerable abundance relative to capital and labor. This background information is critical to the explanation of the fact that, in spite of America's late industrial start relative to Britain, she quickly established a worldwide leadership in the design, production, and exploitation of woodworking machinery. By the middle of the nineteenth century a good deal of this initial advantage had already been dissipated. Subsequently the rising price of wood and the falling price of iron begin to put the American economy back on a somewhat more conventional "European" track, and we observe a shift from cordwood to coal in fuel, and from wood to iron as a raw material.[5]

The purely technical problems involved in the development of woodworking machinery should not be underestimated. There were many special problems involved which had no counterpart in metalworking. Richards has stated: "It is safe to assert that with their high speed and endless modification, wood machines demand a higher grade of ingenuity and skill in their construction than machines for cutting and shaping metal."[6] In contrast to metalworking machinery, woodworking machines were operated at very high rates of speed, and the nature of the materials upon which they worked was subject to a much wider degree of variability. "It is easy to calculate the strain and provide for the proper performance of cutting tools moving at sixteen feet a minute, but when these cutting edges are moved *five to ten thousand* feet in the same time, a new set of conditions are involved, conditions that cannot be predicated upon the ordinary laws of construction."[7]

The high speeds of woodworking machinery created unique problems with respect to the operation of shafts and cutters. Special attention had to be paid to such matters as lubrication, balancing, centrifugal strain, and bearings. Whereas metalcutting machinery was designed to deal with metals which did not vary significantly in their physical characteristics and therefore machining properties from one place to another, woodworking machines had to deal with enormous variability. They worked materials ranging from soft fir-timbers, on the one hand, to woods of a hardness approaching cast iron - such as ebony - on the other. As a result, "some of the leading manufacturers make as many as eighty different machines and modifications, which is the more surprising when we consider that they are all cutting machines, that is, working with sharp edges, and not to be contrasted with those for metal work, which include punching, forging and grinding machines."[8]

II

Victor Clark has pointed out that "the first patent issued in America for a mechanical invention was given, in 1646, by the colony of Massachusetts to Joseph Jenks, for improved sawmills and scythes."[9] This was a peculiarly appropriate prolog. Both inventions heralded - one in raw material processing and the other in agriculture - the development of a technology geared to the maximum utilization of natural resource inputs and the substitution, wherever possible, of abundant resources for scarce labor. Sawmills in colonial America, in fact, long antedated their introduction in England. Sawmills may possibly have been constructed by the Dutch on Manhattan Island as early as the 1620's, but certainly by 1633. The first sawmill in England is reputed to have been built in 1663, but it was quickly destroyed, and it was a century or so before they began to be seriously used in England. But even as early as 1663 there were already hundreds of sawmills in New England. Their construction, often in conjunction with a gristmill, generally followed quickly upon the settlement of a new community.[10] The colonial emphasis upon the export of forest products is evident from the very beginning. The first ships returning from Jamestown carried "clapboards and wainscott," and the pilgrim settlers in Plymouth, after the harvest of 1621 collected lumber, which they sent back to England.[11]

The abundance of wood in the American environment meant that the economic payoff to inventions which facilitated the exploitation of that abundant resource would also be likely to be very high. In particular, anything which reduced the price of inputs which were complementary to wood were especially welcome - for example, nails. In fact, cost-reducing nailmaking machinery was one of the earliest classes of American inventions to excite the interest of Europeans. Apparently 23 patents for nailmaking machinery had been granted by the patent office even before 1800. Perhaps the most notable figure in this development was Jacob Perkins of Newburyport, whose water-powered machinery for cutting and heading nails was patented on January 16, 1795. It "was said to be capable of turning out 200,000 nails in a day."[12] The resulting reduction in the cost of cut nails in the first half of the nineteenth century was a significant contribution to the intensive utilization of wood for building purposes. The price of wrought nails, mostly imported, was about 25 cents a pound when the nailmaking machinery was first devised. The price declined to 8 cents a pound in 1828, to 5 cents by 1833, and to 3 cents a pound by 1842.[13] As a result, the cost of products made of wood was substantially reduced. This reduction in the relative cost of products fashioned out of wood in turn increased even further the large and rapidly growing market for woodworking machinery.[14]

Although no reliable basis for an accurate estimate is available, it is probable that most of the wood products fashioned by woodworking machinery consisted of building materials.[15] Indeed, the combination of cheap, machine-made nails, abundant wood supplies, and an expanding and improving armory of woodworking machines was directly responsible for a major American building innovation in the 1830's: the balloon frame house. The distinctive structural feature of the balloon frame house is that it systematically eliminated all the heavy members of the traditional New England frame house (or barn). It did away, furthermore, with the awkward and time-consuming mortising and tenoning method of joining. As Siegfried Giedion has put it, "The principle of the balloon frame involves the substitution of thin plates and studs - running the entire height of the building and held together only by nails - for the ancient and expensive method of construction with mortised and tenoned joints."[16]   The critical importance of this technique is that its lightness and simplicity - the house was essentially nailed together with light 2"x 4" studs - sharply reduced the total labor requirements of construction and made it possible to substitute relatively unskilled labor for the skilled carpenter. What the invention lacked in elegance was more than compensated for by its highly utilitarian qualities and, above all, by its cheapness. Its characteristics were admirably suited to the American environment, and its use spread rapidly across the country.[17]

It is, unfortunately, impossible to follow the growth in the number of sawmills in the United States with any pretensions of accuracy until the end of the period. It is not until the Census of 1840 that the data were tabulated with any sort of care.[18] The returns for 1840, 1850, and 1860 are presented in Appendix A. It will be seen that the Census of 1840 reported 31,650 sawmills and a value of product of $12,943,507; for 1850 the number of sawmills was 17,475 and the value of product was $58,611,976; the Census of 1860 reported 20,658 sawmills and a value of product of $96,699,856. The apparent decline in the number of sawmill establishments between 1840 and 1850 may have been due to alterations in coverage. Defebaugh suggests that "it is probable that the 1840 report included independent shingle mills, cooperage shops, planing mills, etc., in the total."[19]

Although Americans, as we have seen, applied power to the sawing of timber at a very early date, improvements in the saw itself came much later and more slowly. The simple up and down sash saw or frame saw, consisting "of but a single blade surmounted in a frame of wood which surrounded the log,"[20] was the standard equipment from early colonial times until the middle of the nineteenth century. The sash saw began to give way to the gang saw and the muley saw in the 1840's. Although the muley saw was much faster than the old-fashioned sash saw, all these saws suffered from the functional limitations inherent in their intermittent, reciprocating action. It was therefore inevitable that they would eventually give way to the continuous cutting action of saws operated on different principles. Samuel Miller of Southampton, England, had patented the first circular saw in 1777, but it was not introduced into the United States until 1814.[21] Although its greater speed was an important advantage, it generated considerable heat at high velocities. "At high speeds the saw expanded and wobbled, and the sawyer found it difficult to follow true lines. Moreover, the saw could not cut logs larger than half of its diameter, and the size of the saw was limited. In order to cut large logs two or three saws were employed, one above the other."[22] The use of the circular saw therefore was for many years confined to specialized uses such as veneer cutting.[23]

All the saws mentioned so far possessed a common feature: they converted a distressingly large proportion of the timber into sawdust rather than lumber.

The saws were thick and seven feet long, with large teeth, and would bear heavy feed. The boards sawed in the single mills looked rough, as the saws cut from one-half to three-quarter inches at the stroke, and made coarse sawdust. The gang saws had finer teeth, cut more slowly, and made finer sawdust, leaving the boards smooth even from knotty logs. Gang boards were sometimes used without planing. The quantity of sawdust shoved into the outlet from these mills in a year was enormous. The mill ponds below, the willow bars, eddies, etc., received these deposits; and the accumulation of years is still to be seen along the outlet, in bends and other places.[24]

The blade of the sash saws which still reigned supreme in the forests of Maine, New York and Pennsylvania as late as 1840 was usually five-sixteenths of an inch thick, and produced a kerf of over three-eighths of an inch. The later saws, whatever their other features, were no less wasteful of wood.

Outside of the best gang mills, which form but a small share of the whole, it is safe to assume that one-fifth of all the timber sawed is converted into sawdust. Considering that the lumber of commerce in America consists mainly of one-inch boards, it might even be set down at one-fourth, after the stock is squared. Circular and muley saw mills make a kerf of about five-sixteenths of an inch wide, which, with the irregularity of the lines, may be counted as three-eighths of an inch, in the manufacture of a stock into one-inch boards, and gives us five-eighths lumber and three-eighths sawdust.[25]

American circular saws used blades of thicker gauge than the English, and had their teeth spaced wide apart.[26] These were characteristics admirably suited for high speed of operation, but at the same time produced a very large kerf.

As lumber became increasingly expensive, toward the mid-nineteenth century, more and more inventive effort was directed toward reducing the waste.[27] The technical solution had been available, in principle, for some time, in the band saw. William Newberry of London had patented a band mill as early as 1808, and it seems to have been subsequently reinvented independently in the United States.[28] The band saw was a very simple invention in principle. It consisted of an endless band of steel which passed over two wheels, thus giving the saw continuous rather than intermittent action. Not the least of its advantages was that it could cut logs of greater size than most circular saws. Its use, however, was held up until after the Civil War because of intractable technical problems. These included the difficulty in constructing a blade which would not snap or alter its shape when subjected to an unusual combination of torsional strains, as well as high temperatures. In this respect, improvements in metallurgy eventually proved to be of vital importance, and the band saw was widely adopted after the solution of problems of steel chemistry in the 1870's and 1880's.

The savings which were finally achieved by the adoption of the band saw were highly significant to an economy in which the price of lumber was rising.

The circular saw whose bits averaged five-sixteenths of an inch would turn 312 feet into sawdust for every thousand feet of inch boards. If the saw could be reduced to one-twelfth of an inch, which was the thickness of the early band saws, only 83 feet would be lost in sawdust.[29]

III

Planing machines were second only to saws in a ranking of woodworking machines by their relative importance, and they also typically followed the sawing operation in the processing of wood. In the first half of the nineteenth century a variety of planing machines were introduced, operating on different principles (carriage-planing machines, parallel-planing machines, and surface-planing machines[30]). The planing machine which easily had the greatest impact, however, was the Woodworth planing machine, first patented by William Woodworth of Poughkeepsie, New York, on December 27, 1828. The invention represented a highly effective combination of feeding rolls and rotary cutting cylinders. Most references to the Woodworth planer are preoccupied with its tortuous history of litigation continuing into the 1850's, as the patent rights were extended and reissued. The patent did not finally expire until 1856. The many attempts to invent around this "notorious monopoly," as it was frequently called, led to numerous suits for patent infringement. This was hardly surprising, since planing was a critical woodworking operation and Woodworth's patent protection was defined so broadly that it was, indeed, most difficult to circumvent.

The planing machine had evolved, by 1850, into numerous specialized forms, each one well adapted to a particular product. The Woodworth machine was widely adopted for the manufacture of flooring boards, and variants of the Woodworth machine were employed in boxmaking. Where an accurate, smooth surface was required, Daniels' traverse planing machine was far more satisfactory. "It consists of an upright frame, in which a vertical shaft revolves, having horizontal arms, at the ends of which are fixed the cutters. The work is carried along on a travelling bed under the cutters, which are driven at a very high speed."[31] The Daniels planing machine dominated the heaviest classes of work on the railroads and in the manufacture of wagons, and also where the nature of the timber presented special problems, as in cases where it was twisted or warped.[32]

It is difficult in a short scope to follow woodworking invention through the subsequent stages of processing after sawing and planing, because from this point machinery design branched out in a multiplicity of directions to accommodate the highly specialized needs of a wide range of users. Mortising and tenoning machines were probably the most important subsequent categories, since they were the usual method of joining.[33] Mortising machines were much more complex than tenoning machinery, and were divided into reciprocating and rotary mortising machines. In America much greater reliance was placed upon the reciprocating machine than was the case elsewhere.[34]

Beyond this, however, there existed a wide array of specialized machinery for boring, slotting, dovetailing, edging, grooving, etc. Further, within each of these categories of machines were modifications and alterations in design to meet the unique needs of particular classes of final products - shingles, laths, clapboards, staves, wagons, agricultural implements, boxes, stairs, sashes, doors, blinds, furniture, veneers, etc. These were often complex machines involving a high degree of technological imagination and sophistication.

The parliamentary committee which visited the United States in 1854 was astonished at "the wonderful energy that characterizes the wood manufacture of the United States."[35] They were particularly impressed by precisely this extensive specialization of machinery. Whitworth stated in his report:

Many works in various towns are occupied exclusively in making doors, window frames, or staircases by means of self-acting machinery, such as planing, tenoning, mortising, and jointing machines. They are able to supply builders with the various parts of the woodwork required in buildings at a much cheaper rate than they can produce them in their own workshops without the aid of such machinery. In one of these manufactories twenty men were making panelled doors at the rate of 100 per day.[36]

There was one special machine not so far mentioned which deserves to be singled out. This is Thomas Blanchard's lathe for turning irregular forms. Blanchard's lathe was originally devised for the shaping of gunstocks. The gunstock had been the most serious bottleneck in the transition from the handicraft technology to the production of guns by machinery.[37] The highly irregular shape of the gunstock had long been an extremely tedious operation, since it involved separate hand activities of whittling, boring, and chiseling. Before Blanchard invented his first gunstocking machine in Millbury, Massachusetts in 1818, a skilled man could produce only one or two gun stocks per day. Blanchard's lathe made it possible to reproduce any irregular shape by machinery by copying a model.

A pattern and block to be turned are fitted on a common shaft, that is so hung in a frame that it is adapted to vibrate toward or away from a second shaft that carries a guide wheel opposite and pressing against the pattern, and a revolving cutter wheel of the same diameter opposite the block to be turned. During the revolution of the pattern the block is brought near to or away from the cutting wheel, reproducing exactly the form of the pattern.[38]

Blanchard's lathe was an immediate success and it was introduced into the national armory at Springfield during the 1820's under Blanchard's personal supervision.[39] It too quickly evolved from a single machine into a series of highly specialized machines. In fact,

By 1827 Blanchard's stocking and turning machinery had been developed into 16 machines, in use at both national armories, and for the following purposes: sawing off stock, facing stock and sawing lengthwise, turning stock, boring for barrel, turning barrel, milling bed for barrel-breech and pin, cutting bed for tank of breech-plate, boring holes for breech-plate screws, gauging for barrel, cutting for plate, forming bed for interior of lock, boring side and tang-pin holes, and turning fluted oval on breech.[40]

Although Blanchard's lathe had been devised specifically for the production of gunstocks, it embodied a principle of far wider applicability in the development of automatic machinery. Indeed, it could be used to produce any irregular shape, and it was shortly introduced into making shoe lasts, hat blocks, spokes of wheels, oars, etc., in addition to gunstocks. But, in the longer run, it also provided vital design elements in the construction of specialized machinery for the manufacture of products of an interchangeable nature.

IV

This is not the appropriate place to speculate upon the reasons for the high aggregate level of inventive activity which Americans displayed in the first half of the nineteenth century.[41] But it does seem important to say something about the direction which such activity took, since that is a central concern of this paper.

Contemporaries as well as later writers have emphasized the labor-saving aspect of American technology, from the reports of the British parliamentary committees[42] - and others long before them - to Habakkuk's seminal work, *American and British Technology in the 19th Century* (Cambridge, 1962) which bears the subtitle, "The Search for Labour-saving Inventions." What has been insufficiently stressed is that the technology both developed and adopted in the United States was a *resource-intensive* technology. The preoccupation with capital and labor in a two-factor view of the world has served to obscure the fundamental point that Americans pushed the technological frontier in directions where it was possible to substitute abundant natural resources *for either labor or capital.* Moreover, many of the American inventions which were so obviously labor-saving were also, necessarily and less obviously, resource-intensive in their operation.

In a highly resource-abundant environment such as the United States, it made excellent economic sense to trade off large doses of abundant raw material inputs for the scarcer factors of capital and labor. This was done in many ways. The "wastefulness" of wood which seemed to characterize American woodworking technology and to which Europeans so often called attention, was a general feature of America's technological adaptation.[43] Just as we developed and employed woodworking machinery which utilized wood more wastefully than did, say, European handicraft methods, so did we, at an early stage, build entire road surfaces out of wood, and construct large domestic fireplaces which were highly inefficient in fuel utilization but economized on the labor-intensive processes of chopping or sawing wood in order to accommodate smaller fireplaces or stoves (The stoves, of course, eventually came[44]). But, similarly, at an early stage we employed "inefficient" wooden pitchback water wheels, which utilized only a small proportion of the water power available to them, since these wheels were very cheap to construct and therefore economized on scarcer capital. The same trade-off characterized our utilization of high-pressure steam engines aboard steamboats on western rivers. Such engines, as Dr. Hunter has pointed out, were highly wasteful of fuel but considerably cheaper to construct than low pressure engines. The latter were much more popular in the east where cordwood prices were considerably higher.[45] And finally, a major thrust of American agricultural invention in the nineteenth century was toward the development of a technology which maximized the acreage which could be cultivated by a single farmer. Here again is a substitution of an abundant input - agricultural land - for other scarcer factors of production. Doubtless a visitor from Japan or the Indian subcontinent would have been appalled at such "wastefulness," as were some observers of American woodworking technology.[46] The essential point, however, is that, when resource endowments differ significantly, one economy's "criminal wastefulness" may be another economy's optimal resource allocation.

All of this does, in fact, shed at least a glimmer of light upon the apparently high level of American inventive activity. The American economy entered into the process of industrialization at a point when her resource endowment - certainly in the cases of forest resources and agricultural land - was substantially more favorable than was the case in western Europe. This difference in endowment made it economically rational to search for inventions along portions of the invention possibility frontier which were not being carefully explored by Europeans on the grounds that they were inefficient. Thus, the Americans were the first people whose resources made it worthwhile to explore systematically the realm of highly resource-intensive inventions. And, as her woodworking machinery abundantly demonstrated, there was a rich harvest of inventions available to an economy which could afford to trade off large quantities of natural resources for other factors of production.[47] Perhaps we also find here the roots of American profligacy with her natural environment about which there is now so much intense concern. But that is a big as well as a separate subject which must await another occasion.

Appendix

Table 1. *Census of 1840*

States and territories	Number of sawmills	Value of product ($)
Alabama	524	169,008
Arkansas	88	176,617
Connecticut	673	147,841
Delaware	123	5,562
District of Columbia	1	–
Florida	65	20,346
Georgia	677	114,050
Illinois	785	203,666
Indiana	1,248	420,791
Iowa	75	50,280
Kentucky	718	130,329
Louisiana	139	66,106
Maine	1,381	1,808,683
Maryland	430	226,977
Massachusetts	1,252	344,845
Michigan	491	392,325
Mississippi	309	192,794
Missouri	393	70,355
New Hampshire	959	433,217
New Jersey	597	271,591
New York	6,356	3,891,302
North Carolina	1,056	506,766
Ohio	2,883	262,821
Pennsylvania	5,389	1,150,220
Rhode Island	123	44,455
South Carolina	746	537,684
Tennessee	977	217,606
Vermont	1,081	346,939
Virginia	1,987	538,092
Wisconsin	124	202,239
Total, United States	31,650	12,943,507

Table 2. *Census of 1850*

States and territories	No. of establish- ments	Capital ($)	No. of wage- earners	Wages ($)	Cost of raw material ($)	Value of products ($)
Alabama	173	952,473	937	164,268	529,976	1,103,481
Arkansas	66	113,575	275	42,828	31,719	122,918
California	10	147,200	114	175,080	38,050	959,485
Connecticut	239	308,150	371	97,392	277,831	534,794
Delaware	83	158,180	224	50,640	118,322	236,863
District of Columbia	1	5,000	14	2,400	22,500	29,000
Florida	49	271,400	499	99,072	121,216	391,034
Georgia	333	1,008,668	1,221	238,356	377,766	923,403
Illinois	468	843,535	1,306	298,524	591,508	1,324,484
Indiana	928	1,502,811	2,265	513,216	858,634	2,195,351
Iowa	144	204,475	344	81,348	225,135	470,760
Kentucky	466	1,029,980	1,490	306,324	721,889	1,592,434
Louisiana	138	892,785	948	226,452	273,694	1,129,677
Maine	732	3,009,240	4,439	1,301,376	3,609,247	5,872,573
Maryland	123	237,850	357	77,892	298,715	585,168
Massachusetts	448	1,369,275	1,337	396,576	834,847	1,552,265
Michigan	558	1,880,875	2,730	740,076	987,525	2,464,329
Minnesota	4	92,000	62	18,300	23,800	57,800
Mississippi	259	711,130	1,079	221,628	332,141	913,197
Missouri	334	633,109	1,220	289,092	623,518	1,479,124
New Hampshire	545	859,305	969	265,068	622,564	1,099,492
New Jersey	324	928,500	665	177,180	646,209	1,123,052
New Mexico	1	5,000	4	864	10,000	20,000
New York	4,625	8,032,983	10,840	2,863,188	6,813,130	13,126,759
North Carolina	299	1,057,685	1,135	198,984	480,907	985,075
Ohio	1,639	2,600,361	3,756	924,084	1,693,688	3,864,452
Oregon	37	536,200	242	331,980	190,000	1,355,500
Pennsylvania	2,894	6,913,267	7,052	1,787,520	3,869,558	7,729,058
Rhode Island	51	138,700	134	60,252	142,768	241,556
South Carolina	353	1,106,033	1,431	203,220	525,844	1,108,880
Tennessee	451	707,280	1,229	192,612	283,607	725,387
Texas	89	300,075	426	96,912	156,148	466,012
Utah	5	12,400	18	. . . .	. . . .	14,620
Vermont	326	438,025	606	153,288	303,306	618,065
Virginia	2	17,000	10	2,460	419,536	977,412
Wisconsin	278	1,006,892	1,569	419,340	538,237	1,218,516
Total, United States	17,475	40,031,417	51,218	13,017,792	27,593,535	58,611,976

Table 3. *Census of 1860*

States and territories	No. of establishments	Capital ($)	No. of wage-earners	Wages ($)	Cost of materials used ($)	Value of product ($)
Alabama	339	1,756,947	1,686	428,268	692,027	1,875,628
Arkansas	178	583,690	969	268,716	303,137	1,158,902
California	295	1,948,327	1,924	1,474,626	1,215,244	4,003,431
Connecticut	208	386,800	311	89,878	377,580	589,456
Delaware	71	247,760	176	48,132	154,500	276,161
District of Columbia	1	20,000	4	1,680	17,000	21,125
Florida	87	1,282,000	1,222	316,292	541,531	1,476,645
Georgia	411	1,639,717	1,872	438,828	1,211,807	2,414,896
Illinois	463	1,446,088	1,798	497,280	1,153,237	2,681,295
Indiana	1,331	2,544,538	3,631	1,001,034	1,734,483	4,451,114
Iowa	561	1,656,535	1,762	478,080	1,071,285	2,185,206
Kansas	124	395,940	497	204,920	538,882	1,563,487
Kentucky	482	1,405,835	1,665	439,080	990,021	2,495,820
Louisiana	161	1,213,726	1,039	286,956	548,647	1,575,995
Maine	926	4,401,482	4,969	1,453,739	4,504,368	7,167,762
Maryland	187	472,800	377	101,208	239,808	609,044
Massachusetts	611	1,419,473	1,408	421,548	1,570,362	2,353,153
Michigan	986	7,735,780	6,980	1,895,162	3,425,613	7,303,404
Minnesota	163	1,349,620	1,175	371,988	603,095	1,257,603
Mississippi	229	1,049,910	1,441	436,116	653,157	1,832,227
Missouri	548	1,809,725	1,753	477,372	1,398,564	3,085,026
Nebraska	46	127,800	155	43,648	113,750	335,340
New Hampshire	567	1,185,126	1,195	341,160	702,111	1,293,706
New Jersey	268	1,163,100	591	188,752	942,706	1,623,160
New Mexico	9	45,100	42	14,520	12,950	45,150
New York	3,035	7,931,708	8,798	2,369,720	5,531,704	10,597,595
North Carolina	349	941,880	1,354	296,952	510,379	1,176,013
Ohio	1,911	3,708,153	4,327	1,209,386	2,521,481	5,279,883
Oregon	126	430,400	378	210,312	189,925	690,008
Pennsylvania	3,078	10,978,464	9,419	2,485,103	5,211,990	10,994,060
Rhode Island	26	66,000	79	21,828	46,027	76,114
South Carolina	361	1,145,116	1,263	219,361	498,290	1,125,640
Tennessee	546	1,492,013	1,867	435,536	880,595	2,228,503
Texas	194	1,278,080	1,211	365,376	530,545	1,754,206
Utah	28	151,656	67	46,460	61,973	145,505
Vermont	415	862,060	939	244,551	477,798	928,541
Virginia	784	1,292,886	2,139	464,182	911,714	2,218,962
Washington	33	1,168,000	653	383,130	424,671	1,194,360
Wisconsin	520	5,785,355	4,703	1,227,385	2,067,816	4,616,430
Total, United States	20,658	74,519,590	75,852	21,698,365	44,580,773	96,099,856

Table 4. *U.S. lumber wholesale price index (1910-14 = 100)*

Year	Index	Year	Index	Year	Index
1798	24	1822	25	1846	42
1799	23	1823	26	1847	41
1800	24	1824	26	1848	41
1801	27	1825	27	1849	40
1802	27	1826	28	1850	43
1803	24	1827	29	1851	43
1804	26	1828	29	1852	46
1805	27	1829	28	1853	47
1806	27	1830	27	1854	48
1807	27	1831	29	1855	51
1808	26	1832	29	1856	52
1809	26	1833	30	1857	53
1810	26	1834	31	1858	48
1811	25	1835	31	1859	46
1812	24	1836	32	1860	46
1813	25	1837	45	1861	45
1814	27	1838	45	1862	48
1815	37	1839	45	1863	58
1816	35	1840	42	1864	74
1817	31	1841	43	1865	79
1818	28	1842	40	1866	87
1819	28	1843	37	1867	83
1820	27	1844	39	1868	80
1821	26	1845	43	1869	75

Source: Historical Statistics of the United States: Colonial Times to 1957 (Washington, D.C., U.S.G.P.O., 1960), p. 317.

3

Anglo-American wage differences in the 1820's

In 1829 Zachariah Allen, a lifelong resident of Providence, Rhode Island, published his book, *The Science of Mechanics*. Neither the title nor a casual glance at the contents of the book suggests that it contains material of major interest to economists or economic historians. Allen's book was intended as a manual for American mechanics and manufacturers. It summarized that portion of the industrial arts of its day which the author considered most useful and relevant for the edification of his American readers. The book contains elementary tables of conversion, arithmetic and geometric rules and formulas, a good deal of simple physics, and extensive descriptions of the workings of machinery - especially water wheels, steam engines, and millwork generally. In particular, Allen attempted to summarize much of what he had learned during a recent tour of the major manufacturing districts of England and France, so as to bring Americans up to date on the "latest improvements in mechanical invention in those countries."[1] Unobtrusively placed in the back of this book, however, is a chapter, "Comparative View of the Relative Advantages Possessed by England, France and the United States of America as Manufacturing Nations," which records his more strictly economic impressions of his tour, taken in 1825. His observations in this short chapter, often trenchant and occasionally pungent, invite comparison with some of the better known European travelers to the United States. Anyone who can write, as Allen does (p. 355), that "An industrious New-England mechanic commonly appears to take pleasure in his business; but the French mechanic is rather inclined to make a business of his pleasures," is obviously entitled to a sympathetic hearing.

Our major concern here, however, will be with a table, reproduced below as Table 1, which contains economic observations of great interest for a period which was not only of critical importance in the industrialization process, but for which also our information is

50

	England		France		United States
	s.d.	d.c.	francs	c.	d.c.
A common day labourer earns *per day*	3.0 stg. =	.74	about 2.	37 to 40	1.00
A Carpenter	4.0 stg. =	.97	about 3-4.	55 to 75	1.45
A Mason	4.6 stg. =	1.10	about 3½-4½	60 to 80	1.62
A Farm-Labourer (*per* month and found)	27.0 stg. =	6.50	about	400 to 600	about 8.00 to 10.00
A Servant maid (*per* week and found)	2.9 stg. =	.67			about 1.00 to 1.50
Best Machine Makers, Forgers, &c. *per* day	8.0 stg. =	1.94			about 1.50 to 1.75
Ordinary Machine Makers, Forgers, &c. *per* day	4.6 stg. =	1.10	about 5.	92	about 1.25 to 1.42
Common Mule Spinners in Cotton Mills	4.2 stg. =	1.02	about	80 to 90	about 1.08 to 1.40
Common Mule Spinners in Woollen Mills	3.1 stg. =	.94	about	40 to 50	about 1.08
Weavers on hand looms	3.0 stg. =	.74	about	37 to 50	.90
Boys 10 or 12 years of age, *per* week	5.6 stg. =	1.30	about	85 to 100	about 1.50
Women in Cotton Mills *per* week, average	8.0 stg. =	1.96	about	148 to 200	about 2.00 to 3.00
Women in Woollen Mills *per* week, average	8.0 stg. =	1.96	about	150	about 2.50
In Holland a day labourer earns about 35 cents					
In Holland Carpenters and Masons, about 60 cents					
In Holland Ship Carpenters, about 80 cents					N.York Pittsburgh
Average price of Wheat *per* bushel in 1827	7.4 stg. =	1.79	about	117	.96 .49
Average price of good coals for steam engines *per* ton	9.0*a* =	2.20	about *b*	600 to 700	7.00 1.06

Note: This Table was formed with great care from the result of personal inquiries made in the most important Manufacturing Districts of England and France, and the prices are taken at an average, as nearly as practicable. Since the year 1825, at which period these notes were taken, there have been considerable fluctuations in the price of labour in England, resulting probably in a general depreciation of wages. The value of the Spanish dollar is estimated at about 4 shillings 1½ penny Stg. when the exchange between England and the United States is 10 per cent in favour of the former country, making the shilling sterling about 24½ cents. The Spanish dollar is not a current circulating coin in England, and has no standard value in that country; but is bought and sold as bullion. By a Statute Law of the U. States the Spanish dollar was made a standard coin for the currency of the country, and was arbitrarily rated at the value of 4s. 6d. stg. for the purpose of assessing the duties upon all articles imported from England, and paying a certain impost upon the first cost. By thus underrating the value of the sterling currency, the American duties on English Manufactures are in effect reduced about 10 *per* cent.

a In Manchester.   *b* Near Louviers and Paris.

Source: Reproduced, with small stylistic alterations, from Zachariah Allen, *The Science of Mechanics*, Providence, R.I.: Hutchens and Cory, 1829, p. 347.

fragmentary in the extreme.[2] Moreover, these observations were made by a well-informed and keen-eyed observer who was anxious to reduce his information on several countries to a comparable basis.

Allen's data have not been entirely lost from view. Victor Clark cited Allen several times in his *History of Manufactures in the United States* and drew particularly on Allen's table (Table 1). It will be argued that Clark drew inferences concerning Anglo-American wage differences from Allen's table which, because of the wide influence of Clark's book, have received wide currency but were in fact quite unwarranted. Furthermore, it will be suggested that Allen's table is of considerable interest for reasons which no one, to my knowledge, has yet noted.

In his discussion of wages in America and England before 1860 Clark relies upon Allen in a very important particular. He uses Allen's data as the basis for asserting that wage differentials between the United States and England were greater in unskilled occupations than in skilled occupations. Citing Allen's table, he stated:

A comparison of wages in England and the United States, made about 1825, which has been accepted as authoritative, estimated the average pay of an unskilled laboring man in the former country at 74 cents a day as compared with $1 in the United States. Difference in wages between the two countries was greater in unskilled than in skilled occupations.[3]

The authority with which Clark's statements are accepted in some circles suggests that the basis for this important proposition be examined. Most recently, for example, Habakkuk, whose arguments concerning Anglo-American differences in technology turn upon wage differentials (and supply elasticities), cited this passage in Clark. Habakkuk asserts that wage differentials between skilled and unskilled labor in England were greater than such differentials in the United States, and an important component of his argument depends on the existence of such differences in relative wages in the two countries.[4]

Solely on the basis of Allen's table, what can be said concerning wage differences? Suppose, first, we look at wage differentials within each country. The following rearrangement of Allen's table takes the wages of common labor in each country as a base and expresses all other wages in each country in terms of that country's common labor, whose index equals 100. Farm laborers and servant maids are, unfortunately, omitted here because of my inability to impute a money value to the room and board which they received from their employers. Wages which were given on a weekly basis were converted to a daily equivalent by assuming that the workers were employed six full days a week and dividing through by six. Table 2 results.

Table 2. *Index of wage differences within countries (wage of common labor = 100)*

Occupation	England	France	United States	Holland
Common laborer	100	100	100	100
Carpenter	131	149 to 187	145 ⎱	
Mason	149	162 to 200	162 ⎰	171
Ships carpenter	–	–	–	228
Best machine makers, forgers, etc.	262	–	150 to 175	–
Ordinary machine makers, forgers, etc.	149	230	125 to 142	–
Common mule spinners in cotton mills	138	216 to 225	108 to 140	–
Common mule spinners in woolen mills	127	108 to 125	108	–
Weavers on hand looms	100	100 to 125	90	–
Boys - 10 to 12 years old	29	38 to 42	25	–
Women in cotton mills	44	67 to 83	33 to 50	–
Women in woolen mills	44	63	42	–

Expressed in this form, the figures show much that is of interest about the wage structure in each country. The premium for skilled building workers - carpenters and masons - is somewhat greater in the United States than in England, and in France the differential is given as a range just above the American values. The combined figure for carpenters and masons in Holland, 171, is just slightly below the average which we get for these skills in France if we average the midpoints of the ranges given by Allen.

What is truly startling in these figures, and has so far gone unnoticed, is the small size of the premium above the wages of common labor which was received by machine makers in the United States. In the case of ordinary machine makers the range for the United States is 125 to 142 by comparison with an English 149. Both countries, it should be noticed, are far below the comparable French figure of 230. But it is in the category of best machine makers that a remarkable difference occurs between England and the United States. Although this group in the United States received only 50 to 75 per cent more than common labor, in England they received a wage over two and one-half times as high as common labor! If correct, this is a fact of extraordinary importance, since this was the class of workers upon whom the construction of complex machinery depended. It is tempting to interpret this as a striking confirmation of a widely held view that, for reasons connected with differences in cultural and physical environment, and social and educational systems, the supply of highly skilled mechanics was greater in the United States than in Europe generally.[5] This may well have been the case, but we shall need to learn a good deal more about the nature of the demand for such labor before we can confidently attribute the wage differentials primarily to supply forces. Moreover, Allen's figures strongly suggest that much more attention ought to be focused on particular categories of labor rather than dealing with such highly aggregated concepts as "skilled" and "unskilled" labor.

Allen's figures seem to confirm the views expressed frequently before the Select Committee on Artizans and Machinery the year before his trip to England, that skilled machine makers were particularly hard to find, whereas all other workers, including machine tenders, were available in plentiful supply.[6] If it is true that in the industrialization process "The most intractable shortage was of the skill to construct and service machines,"[7] these figures suggest that, comparatively speaking, the United States in this period at least had much less difficulty than the English in overcoming this intractability. Furthermore, the relatively lower cost of machinery construction in the United States which would seem to follow from these figures, must have constituted a powerful force behind the greater readiness of American than of British manufacturers to adopt relatively capital-intensive techniques of production.

Do these figures show a greater differential between skilled and unskilled labor in England than in the United States? Suppose we employ the following classification.

Skilled workers	Unskilled workers
Carpenter	Common laborer
Mason	Common mule spinners in cotton mills
Best machine makers, forgers, etc.	Common mule spinners in woolen mills
Ordinary machine makers, forgers, etc.	Weavers on hand looms
	Boys - 10 to 12 years old
	Women in cotton mills
	Women in woolen mills

We have, then, only four observations for skilled workers and seven for unskilled. However, if we take a simple, unweighted arithmetic average of each classification, using the midpoint in each case in which the observation is given as a range, we find that, for England,

$$\frac{\text{Average wage index of English skilled workers}}{\text{Average wage index of English unskilled workers}} = 208.$$

The corresponding ratio for America is 199. What we have here, of course, is crude in the extreme, and the limitations of both the data and the procedure need not be belabored. Nevertheless, it is clear that Allen's data do not show a significant over-all difference in the wage differentials between skilled and unskilled labor in the two countries. The difference, indeed, is so small that any number of slight, reasonable adjustments to the data might easily reverse the relationship. Allen's American figures, for example, are probably drawn from New England, the part of the country where skilled labor was relatively most abundant. His American figures, therefore, probably understate the American differential.[8] A similar alteration in the relationship could easily be achieved by dropping a single observation, or by a variety of defensible weighting procedures. Allen's figures, then, hardly provide a satisfactory - or, indeed, even a tenuous - basis for asserting that wage differentials between skilled and unskilled workers in England were greater than in the United States.

Suppose we now compare wage differentials between England and the United States, employing Allen's exchange rates, and letting the English wage represent an index relative equal to 100. We classify occupations in the same way as before. However, if we assume that the money wage of farm laborers and servant maids was the same proportion of total wages in both countries, we can use their money wages as an index for comparative purposes, hence Table 3.

If we take a simple, arithmetic average of all thirteen observations, American wages are 30 per cent higher than English wages. Within the skilled worker classification we encounter again the remarkable case of the best machine makers, the only instance in which the American wage is lower than the English in absolute terms.

If we compare the differential for skilled workers between the two countries with the differential for unskilled workers, again using simple, arithmetic averages, the American excess in the former case is 25.5 per cent and in the latter case 32 per cent. Here too, given the number of observations and the quality of the data, the difference is

Table 3. *Index of wage differentials between England and the United States (English wage = 100)*

Workers	England	United States
Skilled		
Carpenter	100	150
Mason	100	147
Best machine makers, forgers, etc.	100	77 to 90
Ordinary machine makers	100	114 to 129
Unskilled		
Common laborer	100	135
Farm laborer	100	123 to 154
Servant maid	100	149 to 224
Common mule spinners in cotton mills	100	106 to 137
Common mule spinners in woolen mills	100	115
Weavers on hand looms	100	122
Women in cotton mills	100	102 to 153
Women in woolen mills	100	128
Boys - 10 to 12 years old	100	115

quite insignificant, and can hardly be used to support with any confidence whatever Clark's proposition that wage differentials between the two countries were greater in unskilled than in skilled occupations.

A t test has been applied to test the significance of differences in the wage differential between skilled and unskilled labor. At a 10 per cent level of significance t would have to be greater than 2.2 in order to reject the Null hypothesis that the two sets of wage figures came from the same universe. The magnitude of t was in fact 0.82.

The American differential for unskilled labor is strongly influenced by the inclusion of servant maids in the comparison. The servant maid category is, by a wide margin, the one in which the American differential over the English is greatest. In fact, if the servant maids are omitted, it turns out that the American differential is the same in both the skilled and unskilled classification - at just a fraction over 25 per cent above the English levels. Allen's figures, then, yield no difference whatever between the two classifications if we confine our attention to nondomestic labor.

Suppose we adopted a different "rough-and-ready" sort of approach to Allen's data. Let us omit not only the servant maids but also all of the textile workers and young boys, on the grounds that the numbers so employed were small by comparison with the two remaining categories - common laborers and farm laborers. If these two groups alone are taken as representative of American unskilled labor, we get an average differential of 37 per cent above similar English workers.[9] Our final result will now depend upon our weighting or selection procedure with respect to the skilled labor category, since this 37 per cent differential is greater than the American differential for machine makers but less than the differential for the more numerous carpenters and masons. We do not have estimates of the numbers of carpenters, masons, or machine makers, and do not know how to weight them. It may be of interest to note, however, that if carpenters and masons were assigned weights equal to twice those of best machine makers and ordinary machine makers, the American skill differential would be 33 per cent; if the weights were three to one, the American skill differential would rise to 37 per cent, exactly the same as the unskilled differential. When the weights of carpenters and masons exceed three to one - as perhaps they might - then of course the American differential for skilled labor actually exceeds the differential for unskilled labor. Approached in this way, then, Allen's data certainly do not provide unambiguous evidence for Clark's proposition that wage differentials were greater in unskilled than in skilled occupations.

Finally, does Allen's table shed any light - even that of a flickering candle - on Habakkuk's interpretation of Anglo-American differences in the first half of the nineteenth century? Habakkuk argued that although "labour in general was more abundant in England than the U.S.A. in the early nineteenth century, skilled labour in relation to unskilled was scarcer in England." As a result, the differential between the wages of skilled and unskilled labor was greater in England than in the United States. Therefore, Habakkuk argues, "...while the scarcity of labour in general relative to capital provided the Americans with a stronger incentive than the English had to replace labour by capital, the English had a stronger incentive than the Americans to replace skilled by unskilled labour."[10] Habakkuk goes on to cite some English cases in which the introduction of machinery was attributed to the desire to replace skilled by unskilled labor. But an important difficulty encountered in this approach was, "The improved machines...almost invariably required more skill to make them than did the simpler machines; and where it was impossible to substitute unskilled labour in their production the greater inelasticity of supply of skilled labour raised the price of machines (relatively to general labour) including the price of those machines which in their operation substituted unskilled for skilled labour."[11]   And Habakkuk's conclusion is, "On the whole . . . it seems doubtful whether the need in England to replace skilled by unskilled labour was a stimulus to mechanization at all comparable with the need to replace labour by capital in the U.S.A."[12]

Allen's table does not provide evidence to support Habakkuk's argument insofar as it turns on the proposition that the ratio of skilled to unskilled wages in England was greater than that in the United States. But it does strongly suggest that Habakkuk may be right in his general emphasis upon the difficulties which the mechanization process encountered in England. The index for best machine makers' wages in England by comparison with common labor (262) is so much higher than the comparable figure for the United States (150 to 175) as to suggest that this was a much more serious constraint in England than in the United States. The impact of such high wages upon English machinery prices must have been a pervasive factor in limiting the substitution of capital for labor in British industry during this period. Perhaps the large difference between the two countries in the wages of skilled machine makers is an important part of the answer to the question which Habakkuk posed early in his book: "If it paid American entrepreneurs to replace expensive American labour by machines made by expensive American labour, why did it not pay English entrepreneurs to replace the cheaper English labour by machines made with that cheaper labour?"[13]   If Allen can be believed, skilled American machine makers were not only relatively but absolutely cheaper than their English counterparts.

Can he be believed? A possibility which cannot, of course, be precluded, is that Allen's observations on wages of best English machine makers were not accurate - that his figure was based on a small number of very unusual observations or that, for whatever reason, he included in this category in England only men of highly unusual talents.[14]   We do not know, of course, how good a machine maker had to be to be classified as "best" by Allen. In view of the importance of these wage differentials to the path of each country's development, high priority ought therefore to be attached to future research which will provide more reliable information on the nature of these differentials - not only by country but by region, by more precisely defined skilled category, and over time.

Part 2

The generation of new technologies

4

Problems in the economist's conceptualization of technological innovation*

The purpose of this paper is to examine how economics has conceptualized the process of technological innovation. The issues involved are, clearly, very substantial, since there is widespread (although by no means universal) agreement that this process is the primary cause of long-term economic development. Yet, with a few exceptions, it is only in the past twenty years or so that economists have attempted to relate the subject in a systematic way to their analysis of long-term development. Some significant progress has been made. However, I will argue that our cognitive framework causes us to misconceive the process and, as a result, to ignore or to understate drastically the importance of many forms of technological change. I will argue, in particular, that our reasoning on these matters has been seriously flawed by a strong prejudice against recognizing the full economic importance of many forms of valuable knowledge which are intrinsic to activities of a technological nature.

Where this prejudice came from is a matter of interesting speculation. Indeed, its subtle intrusion into our thinking might form a fascinating chapter in the history of economic thought. I suspect it has been related to the status anxieties of the economist and his determination to associate his activities with the high prestige of the scientist rather than the much more modest prestige of the engineer. For my immediate purposes, however, its origins are not nearly so urgent or important as its existence and its consequences in shaping our thinking on matters technological.

The prejudice with which I am concerned may be simply stated. Economists, when working with the process of technological change, typically reveal a hierarchical conceptualization of different forms of knowledge quite similar to that of the natural scientist. That is to say, they attach the greatest importance to "pure" forms of

*Helpful suggestions by Professors Stanley Engerman and W. Lee Hansen are gratefully acknowledged. The author also wishes to thank NSF for financial support.

knowledge, scientific knowledge which purports to be of the highest and widest degree of generality. Conversely, they hold "mere" technological or engineering knowledge in low esteem for being too specific and particularistic in nature, and show little interest in the manner in which such knowledge is generated or diffused.

This perspective strikes me as highly appropriate to the activities of, let us say, the committee which determines the annual award of Nobel Prizes in the natural sciences. But it is entirely inappropriate - indeed, it is positively mischievous - in a discipline which is concerned with the principles of optimal resource use and the understanding of the manner in which the productivity of resources has grown over historical time. The factors which influence the productivity of resources in economic activity are numerous and prosaic. It is time to recognize that the intellectual division of labor has given the economist a subject matter in which relatively grubby and pedestrian forms of knowledge play a disconcertingly large role.

The production function

The objects of my complaint emerge very clearly when we examine the basic analytical apparatus of production theory. The central concept is, of course, that of a production function, which specifies in a rigorous way the relationship between inputs and outputs. The basic distinction is that between factor substitution - a movement along an existing portion of the production surface in response to an alteration in factor prices - and technological change which, when it occurs, shifts the production function itself - that is, increases the amount of output obtainable from given quantities of factor inputs. Movements along a single isoquant involve shifts from one known technique to another known technique as the optimal combination of factor inputs is altered in response to changing conditions of supply and price.

As is so often the case, the usefulness of a particular analytical framework depends upon the collection of problems which one wants to analyze. The notion of a production function has been highly fruitful in addressing certain kinds of problems - for example, the study of relative income shares, technological unemployment, and some aspects of factors influencing economic growth.[1] But when we turn our attention to other kinds of problems the concept is distinctly less helpful. And, among these other problems is the problem of technological change itself. The framework provides no illumination when we ask questions dealing with how the existing range of technological alternatives came into existence, and what specific forces generated them.

Indeed, the historical questions confront the analytical questions as soon as we inquire more closely about the precise meaning of a single portion of a production surface: the isoquant. Individual isoquants are usually presented for analytical convenience as smooth and continuous curves, representing a wide range of alternatives of varying factor intensities. They constitute a spectrum of what Schumpeter called "eligible choices."[2] The entrepreneur is visualized as being confronted with these alternatives in the blueprint stage, and making his least cost decision by putting together these *technological* data with *economic* information derived from the market place.

And yet, although the *analytical* meaning of such isoquants is simple and clear, troublesome questions arise as soon as one introduces pertinent economic facts as they are likely to present themselves in any historical context. In particular: in what precise sense is it likely to be the case that a wide range of technological alternatives will ever be "known"? Since - to introduce a primary assumption of my argument - the production of knowledge is itself usually a costly activity, why should technological alternatives representing factor combinations far from those justified by present prices be known? Why should a society, where the price of capital is low relative to that of labor, have available detailed information about labor-intensive techniques of production?

We may, of course, derive such knowledge from observation of other societies, or even, in some cases, from our own earlier history. We may, for example, learn about labor-intensive techniques for constructing large dams by consulting the Chinese. Such possibilities clearly exist, and yet experience suggests that the notion of a wide range of alternatives readily available, as implied by the drawing of smooth, continuous isoquants, is largely a fiction. This is confirmed by the very modest success of the search for "Intermediate Technologies" for underdeveloped countries in recent years. Such technologies may prove to be *attainable,* but they are certainly not *readily* available. Moreover, profit-oriented business firms do not ordinarily commit resources to the generation of knowledge which has no prospective relevance whatever to their productive activities. In fact, firms which do commit substantial amounts of resources in this way are not likely to survive very long. Why, as a result, should not the known portion of an isoquant typically be a relatively small segment, Fig. 1 (a), and not a smooth, continuous curve, Fig. 1 (b)?

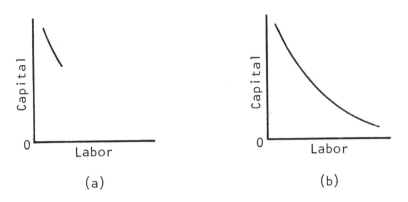

(a)   (b)

And, where only a small range of alternative techniques is known, what becomes of the fundamental distinction between factor substitution and technological change? If, in response to a change in factor prices, a firm has to commit resources to establishing new optimal input mixes, should not the activity leading to the new knowledge be described as technological change and not factor substitution? Clearly that new knowledge is acquired through the same costly process of research which we ordinarily associate with technological change. As Schultz has stated: "I shall define research as a specialized activity that requires special skills and facilities that are employed to discover and develop special forms of *new information*, a part of which acquires the properties of economic information. By this definition, such research is an *economic activity* because it requires scarce resources and because it produces something of value."[3] Once a substantial research expenditure is required for what is called "factor substitution," what is left of the economic basis for the distinction between technological change and factor substitution? Moreover, when a variety of alternative substitution possibilities is in fact known, these possibilities are themselves the outcome of some past research activities. *Today's* factor substitution possibilities, in other words, are the product of *yesterday's* technological explorations.

These problems are not ordinarily posed within the framework of microeconomic analysis, where the central concern is with optimal resource allocation. When interest shifts to productivity analysis, however, and the exact behavior of technological data becomes obviously relevant, such questions inevitably intrude themselves. The late W. E. G. Salter, in his important book *Productivity and Technical Change* (Cambridge University Press, 1960), was one of the few people to address such questions seriously and explicitly. But, after a careful and subtle discussion of the possible meanings of the production function, Salter adopted an interpretation according to which the production function includes "all possible designs" which can be developed with the present stock of knowledge, whether those designs have already been developed or not.[4] A production function then encompasses not just techniques which are immediately available in the sense of blueprints on the shelf but also the "much wider range of techniques which could be designed with the current stock of knowledge."[5]

Now, I want to argue that, when we take this last step with Salter, we are really allowing factor substitution to swallow up much of technological change. To say, with Salter, that the movement along the isoquant to points which are not yet known or attained may be regarded as factor substitution and not technological change, so long as we can get there without an increment to the stock of knowledge, is to say something very strange - at least as an *economic* proposition. For these are points which, by Salter's own admission, are not attainable within the present state of technical or engineering knowledge. Salter means to say, of course, that such points are attainable within the present state of *scientific* knowledge. But if additional knowledge must be acquired through research or other costly learning processes before production can begin at some new factor combination, what is the *economic* importance of whether the requisite knowledge is of a scientific or an engineering nature? To attach great significance, as we currently do, to this distinction, is implicitly to accept the criteria of the natural scientist who esteems new scientific knowledge but who is often contemptuous or at least disdainful of mere engineering or technical knowledge. Given the central concerns of the scientist, his attitudes are perfectly legitimate for him. But it is not at all obvious why the economist, with very different concerns, should share that perspective. It is not, after all, scientific knowledge which is ultimately valued in the economic sphere but, rather, knowledge in a form which is directly applicable to productive activity. Some scientific knowledge, to be sure, will some day be converted into knowledge of economic value. But, equally surely, some will not.

Technological change undoubtedly enters the economy through

many doors. My basic complaint with the present conceptualization - which places great stress upon the distinction between technological change and factor substitution - is that it rather arbitrarily
closes some of these doors. In so doing it overlooks what much of
the day-to-day thrust of technological change is all about. The
analytical justification for the notion of distinct shifts in well-defined
production functions seems to be that some breakthroughs in
scientific knowledge bring with them whole new *ranges* of more
efficient factor combinations for producing a commodity. There is,
in effect, a "spillover" phenomenon, affecting several points on a
hypothetical isoquant and not just one. Clearly this sometimes
happens, although we need to know much more about the empirical
evidence in support of such a view. But the analytical framework at
the same time obliterates those numerous sources of pressure against
production possibility frontiers which cause them to be either
extended or shifted in limited regions. Although the sort of
movements which I have in mind are, individually, very modest,
there is much evidence to suggest that *cumulatively* they are of major
significance.[6]

The Schumpeterian heritage

The bias about which I am complaining, which leads to the neglect of
the analysis of small improvements, was of course reinforced
historically by the great impact of the Schumpeterian system in
shaping our thinking about such matters. Schumpeter himself was
quite explicit that his analysis was intended to apply only to major
innovations of a kind which involved significant shifts to an entirely
new production function.[7] However, his influence has been so great
that, in fact, his model has become the accepted one for the analysis
of *all* innovative activity, not just the major ones. In this way we
have become saddled with an analytical framework which simply
does not explicitly recognize or lead us into a consideration of
sources of productivity improvements other than those emanating
from major innovations.

The consequences of this have been particularly serious in
impeding our understanding of the origin and nature of technological
change. For here too we have inherited Schumpeterian concepts
which draw a sharp distinction between invention, innovation, and
imitation. In his emphasis upon the distinctive nature and social

importance of leadership,[8] Schumpeter placed great stress upon that charismatic figure, the entrepreneur, who possessed the character, courage and, above all, vision, required to depart sharply from accepted routines and practices. The qualities of leadership required to innovate, Schumpeter argued, far surpass the requirements of subsequent imitations. For "as soon as the various kinds of social resistance to something that is fundamentally new and untried have been overcome, it is much easier not only to do the same thing again but also to do *similar* things in different directions, so that a first success will always produce a cluster."[9] Schumpeter was concerned to emphasize the discontinuous nature of innovative activity since the clustering of innovations was at the heart of his business cycle theory. As a result, he drew the sharpest possible contrast between invention and innovation: "the making of the invention and the carrying out of the corresponding innovation are, economically and sociologically, two entirely different things."[10] Within the sequence of (1) invention, (2) innovation, and (3) imitation, Schumpeter's theory had the result of focusing attention upon the circumstances surrounding and influencing the act of innovation. Inventive activity stood as an exogenous factor outside of his framework. At irregular intervals, entrepreneurs *selected* certain of these inventions and carried out the introduction of a new production function with them. At this point, when the technological shape of the invention was such that it was already suitable for commercial introduction and success, it entered the economic arena and generated growth and instability. But inventive activity itself is never examined as a continuing activity whose nature, timing, and special problems are relevant to the subsequent Schumpeterian stages of innovation and imitation. It is an activity carried on offstage and out of sight. Inventions come onto the Schumpeterian stage already fully grown, and not as objects or processes the development of which is a matter of explicit interest; nor are subsequent improvements or modifications of the invention typically treated as significant. As a result, invention appears as an isolated phenomenon, one which is significant for the impact which it acquires through the subsequent innovation and imitation processes. But the *characteristics* of the inventive process, and the stages through which inventions proceed on the way to full commercial application and exploitation, never emerge. Nor, as a result, are these characteristics or stages examined in a way which serves to illuminate the innovation decision or the pace of the diffusion process. The consequence of this sequential isolation of invention and rigid segregation of the technological and economic realms is the failure to exploit technological factors in furthering our understanding of innovation and diffusion. Accordingly, economists who accept this conceptualization are cut off from technological factors which can (1) account for the timing of innovations, (2) link specific innovations with the resulting growth in resource productivity, and (3) account for both the rate and direction of diffusion of innovations throughout the economy.

The innovation process

The difficulties created by our present concepts emerge forcefully when we examine the literature dealing with central questions concerning the innovation process. Since the productivity-increasing effects of technological change begin with the innovation process and not with invention, a central question is: what determines the length of the time interval which separates the making of an invention and its innovation?

Surprisingly, in view of his preoccupation with innovative activity, Schumpeter never subjected this question to systematic examination. The inventive process really did not interest him and he therefore has little to offer in accounting for the interval between invention and innovation. The most oft-cited attempt to do so was a study by John Enos. In the course of a valuable study of "Invention and Innovation in the Petroleum Refining Industry,"[11] Enos brought together information concerning 46 major innovations, eleven of them in petroleum refining. His basic data on innovations outside of petroleum refining are reproduced in Table 1.[12] Enos finds an average interval of 11 years between invention and innovation in the petroleum refining industry and an average interval of 13.6 years for the other 35 innovations. The observed variance, however, is very large: from 79 years in the case of the fluorescent lamp, 56 years for the gyro-compass, and 53 years in the case of the cotton picker, to one year for freon refrigerants and 3 years for DDT, the long-playing record, lucite Plexiglas, and shell moulding. Enos also concludes that "Mechanical innovations appear to require the shortest time interval, with chemical and pharmaceutical innovations next. Electronic innovations took the most time. The interval appears shorter when the inventor himself attempts to innovate than when he is content

Table 1. *Time interval between invention and innovation for thirty-five different products and processes*

Invention			Innovation		Interval between Invention and Innovation (years)
Product	Inventor	Date	Firm	Date	
Safety razor	Gillette	1895	Gillette Safety Razor Company	1904	9
Fluorescent lamp	Bacquerel	1859	General Electric, Westinghouse	1938	79
Television	Zworykin	1919	Westinghouse	1941	22
Wireless telegraph	Hertz	1889	Marconi	1897	8
Wireless telephone	Fessenden	1900	National Electric Signaling Company	1908	8
Triode vacuum tube	de Forest	1907	The Radio Telephone and Telegraph Company	1914	7
Radio (oscillator)	de Forest	1912	Westinghouse	1920	8
Spinning jenny	Hargreaves	1765	Hargreaves'	1770	5
Spinning machine (water frame)	Highs	1767	Arkwright's	1773	6
Spinning mule	Crompton	1779	Textile machine manufacturers	1783	4
Steam engine	Newcommen	1705	English firm	1711	6
Steam engine	Watt	1764	Boulton and Watt	1775	11
Ball point pen	I. J. Biro	1938	Argentine firm	1944	6
Cotton picker	A. Campbell	1889	International Harvester	1942	53
Crease-resistant fabrics	Company scientists	1918	Tootal Broadhurst Lee Company, Ltd.	1932	14
DDT	Company chemists	1939	J. R. Geigy Co.	1942	3
Electric precipitation	Sir O. Lodge	1884	Cottrell's	1909	25
Freon refrigerants	T. Midgley, Jr., A. L. Henne	1930	Kinetic Chemicals, Inc. (General Motors and Du Pont)	1931	1
Gyro-compass	Foucault	1852	Anschütz-Kaempfe	1908	56
Hardening of fats	W. Normann	1901	Crosfield's of Warrington	1909	8
Jet engine	Sir F. Whittle	1929	Rolls Royce	1943	14
Turbo-jet engine	H. von Ohain	1934	Junkers	1944	10
Long playing record	P. Goldmark	1945	Columbia Records	1948	3
Magnetic recording	V. Poulsen	1898	American Tele-graphone Co.	1903	5
Plexiglas, lucite	W. Chalmers	1929	Imperial Chemical Industries	1932	3
Nylon	W. H. Carothers	1928	Du Pont	1939	11
Power steering	H. Vickers	1925	Vickers, Inc.	1931	6
Radar	Marconi; A. H. Taylor, L. Young	1922	Société Francaise Radio Électrique	1935	13
Self-winding watch	J. Harwood	1922	Harwood Self-Winding Watch Co.	1928	6

Table 1 (*Cont.*)

Invention			Innovation		Interval between Invention and Innovation (yea
Product	Inventor	Date	Firm	Date	
Shell moulding	J. Croning	1941	Hamburg foundry	1944	3
Streptomycin	S. A. Waksman	1939	Merck and Co.	1944	5
Terylene, dacron	J. R. Whinfield, J. T. Dickson	1941	Imperial Chemical Industries, Du Pont	1953	12
Titanium reduction	W. J. Kroll	1937	U.S. Government Bureau of Mines	1944	7
Xerography	C. Carlson	1937	Haloid Corp.	1950	13
Zipper	W. L. Judson	1891	Automatic Hook and Eye Company	1918	27

Source: Safety razor: G. B. Baldwin, "The Invention of the Modern Safety Razor: A Case Study of Industrial Innovation," *Explorations in Entrepreneurial History,* December 1951, p. 74.

Fluorescent lamp: A. A. Bright and W. R. Maclaurin, "Economic Factors Influencing the Development and Introduction of the Fluorescent Lamp," *Journal of Political Economy,* October 1943, p. 436.

Television: W. R. Maclaurin, "Patents and Technical Progress - A Study of Television," *Journal of Political Economy,* April 1950, pp. 145-153.

Wireless telegraph, wireless telephone, triode vacuum tube, oscillator: W. R. Maclaurin, *Invention and Innovation in the Radio Industry,* New York, Macmillan, 1949, pp. 15-16, 33, 59, 67, 74, 85, 112.

Spinning jenny, water frame, spinning mule, steam engines: P. Mantoux, *The Industrial Revolution in the Eighteenth Century,* 2nd ed., New York, Macmillan, 1927, pp. 220-223, 228-235, 241-243, 323-324, 327-336.

Remainder: J. Jewkes, D. Sawers and R. Stillerman, *The Sources of Invention,* London, Macmillan, 1958, pp. 263-410.

Reproduced from John Enos, "Invention and Innovation in the Petroleum Refining Industry," in *The Rate and Direction of Inventive Activity,* Princeton University Press, 1962, pp 307-8.

merely to reveal the general concept."[13]   More important for our present purposes, however, is his procedure for dating inventions and innovations, since these procedures obviously determine the nature of his findings. Enos identifies the date of an invention as "the earliest conception of the product in substantially its commercial form" and innovation as "the first commercial application or sale."[14]

The question which emerges, of course, is: what economic significance are we to attach to the interval between invention and innovation, when those terms are defined in this fashion? A common practice is to label the interval as a "lag," to introduce factors accounting for this lag, and to ascertain whether this lag is becoming longer or shorter.[15] This last question is particularly interesting in attempting to confirm or disprove the widely held intuition that the rate of technological progress has been accelerating in recent decades. But "the earliest conception of a product in substantially its commercial form" tells us little if anything about the technical success which has been achieved in *incorporating* the conception in material form, and it tells us absolutely nothing about the *economic feasibility* of the conception even if technical feasibility has been fully established. As a matter of fact, however, such invention dates lack any consistent economic rationale. Even ignoring the obvious ambiguities of the definition - just how substantial is "substantially"? - inventions may fulfill the criterion without remotely approaching technical feasibility. Did Galileo "invent" the pendulum clock when he conceptualized its essential principles, even though no such clock was ever actually constructed during his own lifetime? Whereas for some inventions no serious technical obstacles to their implementation may exist once the basic idea has been established,[16] for other inventions such obstacles are formidable and can be overcome only after much further time-consuming search and experimentation.

This may be readily confirmed by consulting individual case histories in John Jewkes et al., *The Sources of Invention,* 2nd ed. (Norton, 1969), from which Enos has drawn most of his information concerning inventions outside of petroleum refining. To date the invention of the fluorescent lamp in 1859, the gyro-compass in 1852, the cotton picker in 1889, the zipper in 1891, radar in 1922, the jet engine in 1929, or xerography in 1937 is to select years in which significant steps forward were indeed made. But in none of these years was the product concerned even remotely near a state of technical feasibility. To date these inventions from an initial basic conceptualization of a product or process is to repeat the unwarranted practice, decried earlier, of downgrading engineering and technological forms of knowledge. The approach states, in effect, that as soon as the basic conceptual or intellectual breakthroughs have been made, all the "real" problems are solved. One might just as well state that the electric light bulb was invented in 1802 by Humphry Davy when he first demonstrated electrically induced incandescence. For in none of these cases does the year in question coincide with the availability of a new product or process. The varying lengths of the "lags" reflect, in part, differences in the complexity of the technical problems which had to be solved before an invention became commercially feasible. In addition, however, the lengths vary due to the inevitably arbitrary selection of a single date out of a continuum of protracted inventive activity. In some cases the single date with which the invention is identified corresponded fairly closely to the date by which the technical problems were reasonably well solved - in which case the lag appears to be short. In other cases, however, the date selected appears to be based more closely upon "earliest conception," a date at which most of the important technical problems still remained to be solved - in which case the lag appears to be long.

The practice which I am criticizing recurs in many variants. In industries employing chemical processes we encounter over and over situations where some new material is known and where techniques for producing it under laboratory conditions are known, and yet no method for producing it in quantities sufficient for commercial production has been established. This was true, for example, of the first polymers which W. H. Carothers had produced with glass equipment in the Du Pont laboratories.[17] Similarly polyethylene, "one of the most useful of modern plastics," is an excellent example of an invention which could be produced under experimental conditions for many years before means of producing it commercially were developed.[18] The same was true for terephthalic acid, an essential material in the production of terylene, one of the major synthetic fibers,[19] and also of titanium.[20] In other cases an innovation has required the availability of materials already in production, but of a degree of purity not previously available. The development of the transistor, for example, awaited the availability of high-purity germanium and, later, silicon.[21] The oxygen method of steel making had been recommended as far back in time as in Bessemer's basic British patent of 1856.[22] The method was not feasible, however, until it was possible to produce pure oxygen on a large scale - a possibility not realized until some three-quarters of a century later. To neglect the activities resulting in these increments to technical knowledge is to neglect a critical determinant of the timing of innovation and the growth of productivity.

Finally of course, bringing an invention to the point of *technical* feasibility or workability is very different from establishing its *economic* superiority over existing techniques. This is a matter of continuing to improve its performance characteristics, often in very inconspicuous and unspectacular ways. Early workable versions of the diesel engine suffered from an excessive weight which rendered them economically inefficient, just as Trevithick's 1804 locomotive, in addition to generating an insufficient steam capacity, was also simply too heavy to be employed on the cast iron rails of the time. A commercially practicable helicopter required the availability of materials which could provide it with lightweight engines. Early versions of the jet engine suffered from unacceptably low performance characteristics until materials were made available to withstand high pressures and temperatures.[23] Precisely the same problems had, in a much earlier period, held up the introduction of the compound steam engine, which transformed ocean shipping, for almost a century. Although the compound steam engine had been patented in the late 18th century, its successful introduction awaited major breakthroughs in steel-making techniques which would make possible the production of high-quality boiler plates and boiler tubes that were so essential to high pressure and fuel economy.[24]

This discussion places me in the rather strange but unavoidable position of having to criticize economists for paying insufficient attention to the *economic* aspects of inventive activity. But it is a fact that much of the literature on innovation and diffusion simply *assumes* the existence of a profitable invention and then goes on to try to account for the lag in its adoption.[25] Attention then naturally focuses on such factors as characteristics of potential adopters, flaws in communications networks and imperfect information flows, ignorance, attitudes toward risk, uncertainty, social and cultural factors producing "Resistances to Change," etc.[26] Needless to say, these are highly proper subjects of analysis for the understanding of innovation and diffusion. However, studies of this kind often beg the fundamental question of whether or not a profitable invention actually exists. Clearly, if it does not, there is no lag to be accounted for. The potential profitability of an invention, in turn, involves a careful examination of the cumulation of small technical improvements of the invention over time, their implications for altered performance characteristics in economic terms and, as a result, a cost comparison of the new technology with the alternative technologies already available.

For example, synthetic rubber was known to be technically feasible for a long time. The basic scientific research had all been completed before the outbreak of the first World War. If we were to look at the lapse between the time when its technical feasibility had been established and the time it was introduced, and call that interval a "lag," the term would be grossly misleading. For what is ultimately of interest in introducing a new product is its economic feasibility. So long as natural rubber was available at low cost, as it was during the interwar years, the economic prospects for synthetic rubber were exceedingly dim. Synthetic rubber became economically significant when wartime circumstances sharply reduced the supply of natural rubber, raised natural rubber prices, and, as a result, drastically improved the economic prospects for the synthetic product.[27] The knowledge, over several earlier decades, that it was technically possible to produce synthetic rubber does not constitute evidence of a lag. Until the economic conditions generated by the second World War, all evidence suggested that synthetic rubber constituted an economically inferior technology. Similarly, the slow rate of introduction of atomic energy power generating plants into the United States over the past quarter century, although technical feasibility has been firmly established, has to be explained in terms of its inability to compete with the alternative fossil fuel technologies. That situation may change quickly in the immediate future. But it would clearly make no sense to speak of lags of two or three decades if by lags we intend to mean *a delay in the introduction of economically feasible technologies.* Clearly the substitution of a new technology for an established one needs to be understood as the resultant of the combination of forces driving down the supply schedule for the new technology and raising the supply schedule for the old one.

In its emphasis upon the discontinuities involved in the innovative process, and the willingness to reject or to depart from past practices, the Schumpeterian analysis has been misleading in another fundamental respect. Innovations hardly ever involve a total rejection of earlier practices but rather a selective rejection. Commercial success with technological innovations usually involves a careful discrimination among those aspects of past practices which need to be rejected and those which need to be continued. Although it has not usually been emphasized in the literature on entrepreneurship, the most successful entrepreneurs obviously had keen powers of discernment in this regard, particularly with respect to assessing the nature of the demand for the new product or service. For example, Thomas

Edison, in introducing the incandescent lamp, first made a very careful study of the gas industry. While he introduced a drastically new technical product, he also deliberately patterned many of his practices upon those of the old gas industry. Edison's *commercial* genius resided in an extremely shrewd awareness of those respects in which innovation called for continuity as well as discontinuity.[28]

Diffusion and development

The criticisms which I have leveled thus far against the artificial segregation of invention from innovation apply equally well to the segregation of invention from diffusion. Innovation is simply the beginning of the diffusion process. However, here again we have inherited from the Schumpeterian framework a sharp disjunction which emphasizes the high level of leadership and creativity involved in the first introduction of a new technique as compared to the mere imitative activity of subsequent adopters. Here also, as a result, the analysis of the diffusion process fails to focus upon continued technological and engineering alterations and adaptations, the cumulative effects of which decisively influence the volume and the timing of the product's sale. The diffusion process is typically dependent upon a stream of improvements in performance characteristics of an innovation, its progressive modification and adaptation to suit the specialized requirements of various submarkets, and the availability and introduction of other complementary inputs which decisively affect the economic usefulness of an original innovation. I have dealt with these and related aspects of the diffusion process elsewhere and therefore I will not attempt to reproduce the supporting evidence here.[29] However, a couple of additional points need to be made.

It is possible to state of these various postinnovation activities that they involve levels of creativity and imagination of a lower order than the original innovation - perhaps that they are even matters of engineering routine. While this is often - perhaps even usually - true, it should in no way reduce our interest in these activities insofar as our concern is with an economic problem and economic consequences. These activities appear to be central to the pace of the diffusion process. It is economically absurd to consider the innovation of the automobile as having been accomplished when there were a few buffs riding around the countryside terrifying horses. Innovation is, economically speaking, not a single well-defined act but a series of acts closely linked to the inventive process. An innovation acquires economic significance only through an extensive process of redesign, modification, and a thousand small improvements which suit it for a mass market, for production by drastically new mass production techniques, and by the eventual availability of a whole range of complementary activities, ranging, in the case of the automobile, from a network of service stations to an extensive system of paved roads. These later provisions, even if they involve little of scientific novelty, or genuinely new forms of knowledge, constitute uses and applications of knowledge from which flow the productivity improvements of innovative activity.

These observations apply as well to what is specifically called the "Development" component of "research and development" (R&D). Although we have grown accustomed to this phrase over the past twenty years, relatively little attention has been given to the development component. Indeed, it is not usually realized that, in recent years, some two-thirds of R&D expenditures consisted of development spending, less than 15% was basic research, and slightly over 20% was applied research.[30]

Economists have exhibited a consistent disinterest in the development process. One understandable difficulty in coming to grips with it is that it has become a kind of omnibus term (like the "service" sector of the economy) and therefore includes many quite disparate things. It includes trivial product differentiation activities which may be important from a marketing point of view - altering lipstick shades or redesigning the rear ends of automobiles so that they more closely resemble the insides of pinball machines - but which do not pose serious technological problems. It may involve solutions to problems which, from a technological point of view, may be neither difficult nor interesting, but economically very important.[31] It may, on the other hand, involve the solution of fundamental technological problems without which an idea is totally lacking in possible commercial applications. The diffusion of an innovation is inseparably linked with these ongoing technological activities which shift, often imperceptibly, the private payoff to individual adopters.

The relative neglect of the development process may also account for the failure to provide more coherent explanations than presently exist for the significant performance differences among countries in their commercial exploitation of new technologies. As a recent OECD study has put it: "There is . . . one conclusion that appears irrefutable. U.S. firms have turned into commercially successful products the results of fundamental research and inventions originating in Europe. Few cases have been found of the reverse process." [32]

The reasons for this American superiority vis-à-vis western Europe are obviously not simple. They probably involve organizational and managerial skills of a fairly subtle kind, some of which we may not even yet identify with any precision.[33] All that I want to suggest, for the moment, is that an important part of the explanation may lie in that collection of not-very-well defined activities which we call development as opposed to research. Characteristically, however, much more attention is generally devoted to the research than to the development component, even though the latter is, for the U.S., by far the largest proportion of total R&D expenditures.[34] Equally interesting and suggestive in this regard is the fact that, if we examine R&D expenditures among OECD member countries in the early 1960's, we find that the U.S. had a higher proportion of expenditures on the development component than any other country.[35]

Implications for the study of technological innovation

I conclude, then, that our dominant conceptualization of innovation has, in many basic respects, served to obscure rather than to illuminate the process of technological innovation. It has done this by creating artificial conceptual disjunctions between innovative activity and other activities with which it is not only linked, but which in fact constitute major parts of the historical process of innovation itself. It has done this primarily by employing concepts which do not explicitly recognize the role of patterns of events at the technological level. As a result, (1) we confine our thinking about innovations to features and characteristics which are likely to be true only of major innovations, (2) we focus disproportionately upon discontinuities and neglect continuities in the innovative process, (3) we attach excessive importance to the role of scientific knowledge and insufficient importance to engineering and other "lower" forms of knowledge, and (4) we attach excessive significance to early stages in the process of invention and neglect the crucial later stages.

I suggest, therefore, that our present conceptualization serves as an intellectual barrier to a better understanding of the nature of economic growth itself. In seeking a closer identification with the values and commitments of the natural scientist, the economist has inadvertently cut himself off from the study of forms of knowledge which ought to be the stock-in-trade of anyone concerned with long-term economic growth. How and where does the economic system generate technologically useful information? What kinds of forces provide the inducements? What conditions are conducive to the successful accumulation of such knowledge? What conditions encourage their rapid application? Only by posing such "low-level" questions will it be possible to come to grips with the nature of technological change in its full richness and diversity. For technological change in its economic aspect involves dealing with innumerable small increments to the stock of knowledge which are, from a strictly scientific point of view, totally uninteresting. Indeed, much of the knowledge even resists formalized statement or codification. Although we have made a concession to the recognition of some elements of such knowledge forms by coining the barbarous solecism "know-how," economists have shown remarkably little curiosity over the subject.

It is the essence of technological knowledge that it deals not with the general or the universal, but with the specific and the particular. In agriculture, much of this knowledge is location-specific. What seed varieties will grow best in a specific geographic location with its unique combination of rainfall, diurnal rhythms, soil chemistry, topography, etc.? What characteristics can be genetically engineered into new seed varieties to improve future crop performance? In the exploitation of mineral resources, the enormous variations in the richness of mineral deposits pose an unending set of technological problems as we move down the slope from the exploitation of high quality resources to the exploitation of resources of progressively lower qualities. Techniques such as beneficiation which have made possible the exploitation of low-grade taconite ores, improvements in sulphate pulping technology which made it possible for the wood-pulp industry to exploit the fast-growing southern pine which was previously unusable, the improvement of boiler designs to prevent leakage at high pressures, the development of new alloys with higher melting points - all these are developments of little scientific interest, dealing as they do with the detailed characteristics of a material or process under very specific circumstances. Yet it is precisely such forms of technological knowledge which are directly responsible for generating improvements in productive efficiency. Economists who regard such problems with disdain and who fail to enquire into the conditions which influence the rate at which such specific forms of knowledge are accumulated and applied, will necessarily remain dismally ignorant of a major source of productivity growth.

Implications for the study of social science innovations

What does this discussion of technological innovation suggest about innovation in the social sciences, and in economics in particular? Innovation in the social sciences is a more complicated process than in the natural sciences because there is a reciprocal interaction between the corpus of thought of a mature social science discipline like economics and our present day preoccupations and problems. These "external" concerns are continually influencing and refocusing our disciplinary interests in ways which have no clear counterpart in the natural sciences.

I suggest that what often happens in economics is that, as concern mounts over a particular problem - business instability, population pressures, economic growth, inadequacy of natural resources, environmental pollution - an increasing number of professionals commit their time and energies to it. We then eventually realize that there were all sorts of treatments of the subject in the earlier literature. Because interest in the problem was smaller at an earlier time, the problem was not attacked in a thorough and sustained way. As we now examine the problem more systematically we come to develop a heightened appreciation for earlier scholars who, in some sense, anticipated later developments. We then proceed to read much of our more sophisticated present-day understanding back into the work of earlier writers whose analysis was inevitably more fragmentary and incomplete than the later achievement. It was this retrospective view which doubtless inspired Whitehead to say somewhere that everything of importance has been said before - but by someone who did not discover it. The serious point, it seems to me, is that it is systematization which really counts. Basic insights are often around for a long time without having any significant impact. From the standpoint of the discipline of economics *post-Leontief,* it is obvious that Quesnay's Tableau Economique was really a primitive input-output model.[36] But it cannot be stressed too strongly that this was not obvious to Quesnay or to any of his contemporaries. In a sense, one can say that Leontief "merely" articulated the sectoral interconnections in a systematic and internally consistent way - but without that articulation what we have is a brilliant insight but not an operational tool.

There is a real danger here which is perfectly analogous to the problem which I discussed earlier of attaching excessive significance to breakthroughs in the early stages of the inventive process and neglecting the later stages. Good ideas usually acquire significance only when they are refined and elaborated and have gone through what is often an exhaustive process of patient modification and revision. Only then do they become useful in an operational sense. It seems to me that we seriously underrate, not only the significance of these later stages in the process of social science innovations, but that we vastly underrate the sheer size and complexity of the task and therefore the magnitude of the intellectual accomplishment involved. National income accounting can indeed be traced back at least as far as Gregory King's *Natural and Political Observations and Conclusions upon the State and Condition of England* (1696), but it was only the monumental labors of Simon Kuznets on the basic conceptual problems and estimating techniques which provided an invaluable set of tools for measurement, analysis, and policy formulation. Only after Kuznets (who certainly did not originate the basic idea) had resolved so many of the thorny and persistent problems, did we have, for the first time, a workable and valuable social science innovation.

After the basic intellectual innovation of national income accounting was published in Kuznets's first book on the subject, the subsequent diffusion of the innovation was highly selective in nature, reflecting not just "internal" issues but the relevance of the innovation to particular real-world concerns. Thus, given the nature of social priorities when Kuznets's early work was published, attention was focused upon his measures of output which were most useful in dealing with employment questions. His treatment of measures of welfare received much less attention at the time. The current attempt by Nordhaus and Tobin and others to provide measures of welfare represents a reversion to another dimension of Kuznets's early work which was long neglected but now resurfaces as a result of our current concerns with environmental problems of pollution, congestion, etc.

Even within a much shorter time frame, it is important to note that conceptual innovations in the social sciences, like technological innovations, are likely to be highly imperfect in their early stages - until they have gone through a stage of what A. P. Usher, in discussing technological inventions, called "critical revision." This critical revision is clearly visible in the decade or so following Keynes's 1936 publication of the General Theory - probably the most conspicuous intellectual innovation in economics in the 20th century. In a serious sense the revolution could not be completed until a decade or so of constructive controversy, during which time significant revisions, clarifications, and modifications were made by such contributors as Hicks, Harrod, Ohlin, Modigliani, and Lange. So that, not only do major innovations have a line - sometimes a long line - of clearly identifiable antecedents, but there is a further critical stage, following the major innovation, which moves us from the major insight to eventual scientific "marketability" and thus widespread diffusion.

The analogy with technological innovation can be extended one step farther. I emphasized earlier that there are important relationships of complementarity in accounting for the diffusion of innovation. It is a very common phenomenon that the success of one innovation is peculiarly dependent upon the availability of other inputs, sometimes requiring additional innovation. Most commonly, in the social sciences, this complementary input to an innovation is, simply, relevant empirical data, without which further exploitation of the innovation, or additional hypothesis testing, is impossible. For example, a good deal of theorizing about the properties of the consumption function did not generate interesting results until the availability - again, largely the product of Kuznets's early national income estimates - of rich cross-sectional times series data which permitted the testing of alternative hypotheses. Again, the pioneering work of Frisch and Tinbergen in econometric model building would not have been nearly so consequential had it not been for the advent of a major technological innovation - the computer - which made it possible to solve very large systems of equations. Here we have a very close complementarity between innovations from two entirely separate realms, electronic technology, and economic theory. Finally, individual social science innovations may have a magnified impact because of their simultaneous or near-simultaneous occurrence with *other* social science innovations. It seems highly plausible that the impact of the Keynesian Revolution would have been much smaller had it not been for the fact that it occurred at about the same time - during the 1930's - that reliable national income estimates were first being undertaken. The great early impact of the

Keynesian Revolution was due, in considerable measure, to the fact that empirical data were becoming available which could be directly "plugged in" to Keynesian macroeconomic models.[37]

One of the most intriguing aspects of the innovation process in the discipline of economics has been the remarkable failure to devote much attention to this most central phenomenon of economic change until very recent years. In what Baumol has called the "Magnificent Dynamics" of classical economics, it played a distinctly subordinate role. In neoclassical economics technological innovation was one of the forces explicitly excluded from the framework of analysis. Marx, of course, correctly regarded technological change as being absolutely central to the analysis of capitalist development but he was regarded as beyond the pale and this aspect of his work exercised no discernible influence upon the main stream of economic thinking. Why, however, did Schumpeter's work not exercise a much greater influence in at least directing economists to the study of technological innovation?

There is a variety of possible reasons. One is that Schumpeter suffered from very bad timing. His first major work, *The Theory of Economic Development,* came at a time (1912) when the main neoclassical tradition was still very much preoccupied with spelling out the principles of optimal resource allocation within a static framework of analysis. His *Business Cycles* had the misfortune to come upon the scene in 1939, shortly after the publication of Keynes's *General Theory,* and consequently received little attention.[38] His *Capitalism, Socialism and Democracy* appeared in 1942 and was swallowed up by the Second World War.

Although I do not mean to suggest that Schumpeter's treatment of technological innovation exercised no influence, I do think it is fair to say that, when economists finally turned their attention to this subject in a serious way in the mid-1950's, they had been brought to it by other forces. For one thing, Schumpeter emphasized the nonrational components of innovation decisions, the dependence upon charismatic social leadership which he designated "entrepreneurship." As a result, his work does not seem to have inspired anyone to analyze carefully the microanalytical components of the innovation process. Moreover, his unfortunate insistence upon the exogenous nature of inventive activity encouraged the belief that it was not an activity which was readily amenable to the tools of economic analysis. The heightened interest in technological change in the 1950's seems to have been primarily a product of a growing concern with long-term economic growth, which in turn was closely, although not uniquely, linked to a concern for the economic prospects for poor, recently liberated colonial areas. The critical event, however, which awakened the economics profession from its dogmatic slumber, was the discovery of The Residual by Abramovitz and Solow in 1950's, and the realization that only a rather small fraction of the observed long-term growth in American output per capita could be accounted for by the mere growth in the supply of conventionally measured inputs.[39] The resulting concern with issues of total factor productivity,[40] to which a significant proportion of the resources of the profession has been devoted in the past fifteen years or so, represents an attempt to deal with internal "puzzles" which have arisen within the discipline itself, as opposed to a quite separate literature which has been explicitly concerned to develop a body of analysis directly relevant to the formulation of economic policy in poor countries. This latter stream of research, which has been so directly concerned with relevance to immediate policy problems, has been distinctly inferior in terms of the quality of its contributions, and has had a less fundamental impact, than the former stream.

Finally, I would like to suggest that the economics profession has not made greater progress in the analysis of technological change for two additional reasons. The first is so simple and obvious that it is usually overlooked, and that is that technological change is an extremely complicated social process, inherently very difficult to model. The second, closely related point, is that it is a phenomenon having dimensions which do not fall conveniently within the boundaries of any single academic discipline. Research upon the subject must, necessarily, be interdisciplinary in nature. The exhortation to undertake interdisciplinary research, we all know, is a familiar one - just as we also know how infrequently it has been undertaken successfully. The crossing of disciplinary boundaries is likely to be a hazardous operation, not only intellectually but, perhaps even more important, professionaly as well, even when the outcome is successful. It will be accomplished, not in response to more exhortation, but only by the eventual realization that certain problems of major significance to practitioners of one discipline will resist satisfactory resolution unless there is a willingness - and a capacity - for crossing such lines whenever the intellectual chase demands it. Although the economist may aspire to the sublime and rarefied heights - and academic prestige - of the natural scientist, I suggest that an unkind Providence has so arranged the world that he can pursue some of his most urgent disciplinary problems only by a greater readiness to rub shoulders with engineers and technologists.

5

Neglected dimensions in
the analysis of economic change

The purpose of this paper is essentially exploratory. Its central analytical focus will be upon certain dimensions of the process of economic development which are, as yet, only very imperfectly understood. It is hoped that the formulation presented below will encourage students of disciplines outside of economics to undertake research which may, eventually, make an important contribution to our understanding of what economic growth is all about.

I

A convenient starting point will be to refer to the results of some recent research which has attempted to quantify the contributions of various factors to American economic growth during the past ninety years or so. Conventional economic theory suggests that the output of the economy at any point in time is a function of factor inputs and that, therefore, it ought to be possible to relate *changes* in output over time to systematic changes in factor inputs. An important segment of economic theory is devoted to exploring these relationships between variations in inputs and the associated varia-tions in outputs - these relationships being summarized in the term 'production function'.

A plausible inference of this approach is that one can explain economic growth as resulting from an acceleration in the economy's rate of capital formation. According to this view the rise in human productivity which is a central aspect of the development process is attributable to a speeding up of the rate at which the economy makes additions to its capital stock, that is to say, that the rise in output per man is essentially a function of a "capital-deepening" process. One of the most widely popularized theories of the "take-off" into economic growth in fact identifies this take-off with a sudden rise in the rate of saving and net capital formation (Rostow 1956).

The results of a recent study by Moses Abramovitz, however, raise serious questions concerning the adequacy of such an approach. Abramovitz set out to determine how much of the rise in per capita incomes in the United States between the decades 1869-78 and 1944-53 could be attributed to changes in the capital and labor inputs, as these inputs are conventionally defined and measured. Although his technique was necessarily based on some rather heroic simplifying assumptions, his results were nevertheless startling. It appears that changes in capital and labor inputs can account for only a very small fraction of the quadrupling of per capita incomes which took place during the period which he examined. During a period in which output per capita increased four times over the base decade, a weighted index of capital and labor inputs per capita rose by only 14 per cent.[1] If we look upon economic growth then, as a mere piling up of additional factor inputs, we have scarcely made a dent in the magnitude of the rise in per capita incomes. As Abramovitz so aptly puts it, what his calculations provide is a measure of our ignorance, which is large indeed.

The startling nature of these results has triggered off numerous attempts to measure the growth in output which can be attributed to technological change, i.e., to *shifts* in the production function as opposed to mere movements along an existing production function. This constitutes a signficant and long-overdue change in the direction of analytical effort, since economists, with few exceptions, have long tended either to exclude technological change from their theoretical work or to treat it as a purely exogenous variable.

The deficiencies of the conventional approach to economic growth which are so effectively highlighted by the Abramovitz study bring us to the central problems posed in this paper. The economy's output may be raised not only by increasing the supply of inputs (movements along an existing production function) or by technological change (shifts in the production function) but also by numerous kinds of alterations in the *qualities* of the inputs of a sort which typically escape the scrutiny of the economic theorist.[2] It is apparent that economic development is associated with important qualitative changes in the human agent as a factor of production. These improvements take such forms as changes in knowledge, technical skills, organizational and managerial abilities, levels of economic aspiration, responsiveness to economic incentives, capacity to undertake and to adapt to innovation, etc. What is as yet only very imperfectly understood is the nature of the mechanisms by which these alterations take place.

It is suggested here that a major neglected source of these improvements has derived from participation in economic activity itself. That is, the quality of the human agent as a factor of production is decisively affected in a variety of ways by the nature of his production and consumption[3] activities, which in turn appear to change in fairly systematic fashion as a result of economic development itself. The manner in which different patterns of economic activity (including both production and consumption) affect the human agent in any time period t may be of critical importance in determining the output of the economy in subsequent time periods t+1, t+2, t . . . n.

These feedback[4] phenomena upon the human agent have been neglected in part because of the long disinterest of the economist with questions dealing with economic growth and their preoccupation with problems of an essentially short-run nature. From the time of the so-called marginal revolution of the 1870's until recent years the central tradition in economics has been primarliy concerned with problems revolving around the conditions of optimum allocation of a fixed stock of resources.[5] As Jevons stated it in 1871: "The problem of Economics may, as it seems to me, be stated thus: Given, a certain population, with various needs and powers of production, in possession of certain lands and other sources of material: required, the mode of employing their labour which will maximise the utility of the produce." (Jevons 1911:267).

It is not our purpose here to minimize the importance of the problem of maximization of output from a fixed stock of resources which is, of course, fundamental. The point is, rather, that when we turn our attention to the problem of the *growth* of output over time, we are compelled to consider new problems of a sort which are not illuminated by static analysis. Economic growth is, in many important respects, a learning process, a process whereby the human factor acquires new skills, aptitudes, capabilities and aspirations. And the pattern of resource use which may maximize output from a given stock of resources may or may not generate the qualitative changes in the human agent which are most conducive to the growth of output in subsequent time periods.

Neoclassical economics fails to capture much of the explanation for the growth in productivity because of the failure to consider a variety of feedback mechanisms. We fail to consider, for example, the impact upon productivity of certain kinds of economic activities as opposed to others - such as manufacturing vs. agriculture. Different kinds of economic activities have different kinds of effects upon the productivity of the human agent. Moreover, we regard the household, as a unit of ultimate consumption, with excessive sanctity, and thereby fail to raise the question of how different consumption patterns (including leisure time activities) within the household react upon its members in their capacities as (present or future) producers. Posing questions of feedback in this fashion opens the door wide to the ministrations of the sociologist, and suggests a possible fruitful area of interdisciplinary research.

The theory of international trade centering upon the theory of comparative advantage is subject to the same strictures as the comparative statics of neoclassical economics, of which it is indeed merely a special case. This theory demonstrates how, given a fixed factor endowment in each country, total world output at any time is maximized through regional specialization and trade. A previously closed economy which is opened up to trade will undertake to reallocate its resources between production for domestic purposes and for export purposes in a manner based upon differences in initial factor endowment and therefore relative costs and prices. It is at least conceivable that the reluctance of many countries to conform to the role of primary product exporters which the theory of comparative advantage frequently appears to impose on them, may reflect an intuitive perception that such activities fail to generate the secondary feedback effects which are conducive to economic growth. (Myint 1954 and 1954-55).

II

The role of feedbacks from economic activity in determining the future growth of the economy was a problem to which the classical economists attached great importance. They were very much concerned with the individual's role in productive activity and the manner in which such activity created pressures and inducements which in turn formed his character, led to the acquisition of skills and inventive ability, created a system of values, determined his relative preferences for work and leisure, etc. A thorough examination of this embryonic economic sociology is eminently worth undertaking, but would reach monographic length, and cannot be attempted here. Some brief references, primarily to the works of David Hume and Adam Smith, must suffice to give some indication of the content and direction of this analysis.

The early classical economists - especially Hume and Smith - were very much interested in the process of economic evolution. In particular they were concerned with the manner in which a predominantly agrarian society transforms itself into one where commerce and industry flourish. Both agreed that there were inherent limits to the development of an agricultural economy - especially one dominated by a large landowning class - and that a predominantly agricultural economy was likely to remain a backward and stagnant one. Hume and Smith were agreed that men possess a certain predisposition to indolence[6] and that an agricultural society does not furnish sufficient motive to overcome this indolence. As Hume states:

Where manufactures and mechanic arts are not cultivated, the bulk of the people must apply themselves to agriculture; and if their skill and industry increase, there must arise a great superfluity from their labour beyond what suffices to maintain them. They have no temptation, therefore, to increase their skill and industry; since they cannot exchange that superfluity for any commodities, which may serve either to their pleasure or vanity. A habit of indolence naturally prevails. The greater part of the land lies uncultivated. What is cultivated, yields not its utmost for want of skill and assiduity in the farmers. (Hume 1955:10).

Not only do indolence and lethargy prevail in such a society in general; the habits and manners of the class of large landed proprietors which frequently dominate such a society are such that the potential economic surplus which might be devoted to productive investment is squandered in a display of idle dissipation and profligacy.

... as the spending of a settled revenue is a way of life entirely without occupation; men have so much need of somewhat to fix and engage them, that pleasures, such as they are, will be the pursuit of the greater part of the landholders, and the prodigals among them will always be more numerous than the misers. In a state, therefore, where there is nothing but a landed interest, as there is little frugality, the borrowers must be very numerous, and the rate of interest must hold proportion to it. The difference depends not on the quantity of money, but on the habits and manners which prevail.[7]

Rental incomes in a predominantly agricultural society are squandered in the maintenance of a large class of footmen and retainers, quite simply, because alternative forms of goods are, in large measure, not available. (Lewis 1955:227).

While an agricultural society without a growing commercial sector will be merely one where stagnation is self-perpetuating, the growth of a commercial and industrial sector is likely to generate numerous positive feedbacks conducive to continued growth. Not the least of these is the capacity and inducement to introduce progressive methods into agriculture itself. "When a nation abounds in manufactures and mechanic arts, the proprietors of land, as well as the farmers, study agriculture as a science, and redouble their industry and attention."[8]

More generally, the growth of commerce and industry introduces the public to a wide assortment of new consumer goods which provide a major incentive to industry and effort, thus overcoming the languor and indifference characteristic of a purely agricultural society and precluding the possibility of a backward sloping supply of labor at relatively low levels of income.

The most natural way, surely, of encouraging husbandry, is, first, to excite other kinds of industry, and thereby afford the labourer a ready market for his commodities, and a return of such goods as may contribute to his pleasure and enjoyment. This method is infallible and universal. (Hume 1955:146. Cf. Ricardo 1952:102-03).

It is in part this "demonstration effect" which accounts for the importance attached to foreign trade in an early stage of a country's development.

... this perhaps is the chief advantage which arises from a commerce with strangers. It rouses men from their indolence; and presenting the gayer and more opulent part of the nation with objects of luxury, which they never before dreamed of, raises in them a desire of a more splendid way of life than what their ancestors enjoyed.[9]

Equally important, Hume argues, is the fact that commerce begets both industriousness and frugality, qualities which are indispensable to the achievement of economic growth. For a commercial society provides a combination of inducements and opportunities which are not available in an agricultural society and thereby transforms the human agent from the casual pursuit of pleasure and diversion to the active pursuit of business profits.

There is no craving or demand of the human mind more constant and insatiable than that for exercise and employment; and this desire seems the foundation of most of our passions and pursuits. Deprive a man of all business and serious occupation, he runs restless from one amusement to another; and the weight and oppression, which he feels from idleness, is so great, that he forgets the ruin which must follow him from his immoderate expenses. Give him a more harmless way of employing his mind or body, he is satisfied, and feels no longer that insatiable thirst after pleasure. But if the employment you give him be lucrative, especially if the profit be attached to every particular exertion of industry, he has gain so often in his eye, that he acquires, by degrees, a passion for it, and knows no such pleasure as that of seeing the daily increase of his fortune. And this is the reason why trade increases frugality, and why, among merchants, there is the same overplus of misers above prodigals, as, among the possessors of land, there is the contrary.

Commerce increases industry, by conveying it readily from one member of the state to another, and allowing none of it to perish or become useless. It increases frugality, by giving occupation to men, and employing them in the arts of gain, which soon engage their affection, and remove all relish for pleasure and expense. It is an infallible consequence of all industrious professions, to beget frugality, and make the love of gain prevail over the love of pleasure. (Hume 1955:53).

Smith is in essential agreement with Hume on the character-forming impact of commercial activity on the commercial classes. "The habits, besides, of order, economy and attention, to which mercantile business naturally forms a merchant, render him much fitter to execute, with profit and success, any project of improvement." (Smith 1937:385). Moreover, "Whenever commerce is introduced into any country probity and punctuality always accompany it. These virtues in a rude and barbarous country are almost unknown." (Smith 1956:253).

Smith goes beyond Hume, however, by adding more specific dimensions to the impact of wealth, class and occupation upon the human actor.

Although the *desire* for affluence is one of the fundamental propelling forces of all mankind, the *attainment* of great wealth is likely to corrupt the effectiveness of its possessor. This is so because ". . . a man of large revenue, whatever may be his profession, thinks he ought to live like other men of large revenues; and to spend a great part of his time in festivity, in vanity, and in dissipation." [10] Moreover, if profit opportunities are so structured - for example, through monopoly or special privilege - that high profits are easily earned, this too destroys the effectiveness of the capitalist by releasing him from the disciplining forces of the competitive market place and corroding those characteristics which constitute the chief economic virtues of this class. "The high rate of profit seems everywhere to destroy that parsimony which in other circumstances is natural to the character of the merchant. When profits are high, that sober virtue seems to be superfluous, and expensive luxury to suit better the affluence of his situation." (Smith 1937:578. Cf. Malthus 1951:192).

It is for similar reasons that large landowners are likely to possess personal attributes inimical to growth:

It seldom happens . . . that a great proprietor is a great improver. . . . To improve land with profit, like all other commercial projects, requires an exact attention to small savings and small gains, of which a man born to a great fortune, even though naturally frugal, is very seldom capable. The situation of such a person naturally disposes him to attend rather to ornament which pleases his fancy, than to profit for which he has so little occasion. (Smith 1937:363-64).

The behavioural characteristics most conducive to a growth in agricultural productivity, Smith argues, are likely to be produced by a system of small proprietorships with security of tenure.

A small proprietor . . . who knows every part of his little territory, who views it all with the affection which property, especially small property, naturally inspires, and who upon that account takes pleasure not only in cultivating but in adorning it, is generally of all improvers the most industrious, the most intelligent, and the most successful. (Smith 1937:392. See also pp. 368-69).

One final point on the impact of productive activity upon the human agent, as conceived in classical economics. It is well known that Adam Smith placed enormous emphasis upon the role of the progressive division of labour as the main engine of economic growth. (Smith 1937: Book I, Chapter I). Less attention has been devoted to the fact that Smith also argued that this same division of labour, if left to itself, exerted a devastating effect upon the minds and character of the great mass of the population, i.e., the "labouring poor."

In the progress of the division of labour, the employment of the far greater part of those who live by labour, that is, of the great body of the people, comes to be confined to a few very simple operations, frequently to one or two. But the understandings of the greater part of men are necessarily formed by their ordinary employments. The man whose whole life is spent in performing a few simple operations, of which the effects too are, perhaps, always the same, or very nearly the same, has no occasion to exert his understanding, or to exercise his invention in finding out expedients for removing difficulties which never occur. He naturally loses, therefore, the habit of such exertion, and generally becomes as stupid and ignorant as it is possible for a human creature to become. . . . His dexterity at his own particular trade seems . . . to be acquired at the expense of his intellectual, social, and martial virtues. But in every improved and civilized society this is the state into which the labouring poor, that is, the great body of the people, must necessarily fall, unless government takes some pains to prevent it.[11]

Smith was content to argue that public education was necessary to offset these deleterious effects, but, if he took his own observations seriously, it is difficult to understand why they did not disturb him even more than appears to have been the case. It is hard to conceive of such workers playing the major role which Smith earlier attributes to them as sources of technological innovation - even within their own narrow, subdivided range of productive activities. (Smith 1937: Book I, Chapter I).

III

The classical economists were much concerned with the sort of feedback phenomena which we have discussed because they were deeply interested in problems of economic evolution and change. With the emergence of neoclassical economics in the latter part of the nineteenth century, interest in such phenomena was very largely submerged as the central theoretical interests of the economics profession were directed to the problem of optimum allocation of a fixed amount of resources. A serious assault upon the analysis of growth, which is once again a major preoccupation of economists, would appear to require that feedback phenomena again be explored, but this time in a more rigorous and systematic way.

We have now arrived at a critical juncture in our analysis. Neoclassical economics examines the conditions under which output from a given volume of resources will be maximized. But it does not examine the manner in which different patterns of resource use - production and consumption - affect the future quality of the human input as a productive agent. And, whereas for the former problem (output maximization) one may legitimately regard all economic activities as being on an equal footing, the critical growing points for subsequent economic growth may well be determined by the manner in which current economic activity modifies the character of the human input. This range of feedbacks deserves careful exploration. It may further serve to illuminate the widely held belief that, once development has proceeded to a certain point, it appears to acquire a momentum of its own and become, as it were, self-generating.

There now exists considerable empirical evidence suggesting that rising per capita incomes are associated with certain systematic alterations in the composition of resource use as well as the composition of output. The inadequacies of the data, the conceptual problems involved in interpreting them (Kuznets 1953: Chapters 6 and 7), and the obvious diversity in the historical experiences of different countries and regions all caution against hasty and premature generalization. In particular, much of the data available are cross-sectional by country, rather than long time series for each country, and the attempt to draw inferences about historical patterns from cross-sectional data is fraught with hazards. Some long-term secular trends, however, are fairly clear. Countries which have attained high levels of per capita income have experienced rather drastic changes in the relative importance of different industrial sectors - a sharp decline in agriculture, a growth in the manufacturing sector, and a later more pronounced expansion of transportation, communication, and the service sectors generally (retailing, finance, government, etc). Associated with this have been declines in unskilled workers and proprietors and managers (largely in the farm sector) and increases in clerical, professional and technical workers. These input changes in large part reflect the differential impact of technical change and the changing composition of output associated with rising incomes, such as the rising proportion of government expenditures in total output and the major compositional changes in consumer expenditures - growth in the relative importance of durable goods and in the provision of services, such as education, medical care and recreation. Superimposed on these changes is the decline in the household as a nucleus of productive activity and the increasing importance of the market nexus, an increase in leisure time, and a massive shift in the population from a primarily rural to a primarily urban environment. Phenomena such as these transform the human agent in ways about which we know relatively little, but which are clearly linked up with growth as a continuous (and perhaps in some ways self-reinforcing) process.

With respect to the impact of productive activities, one important area of exploration is the differential consequences of different sorts of employments. It is, for example, a widely held view that specialization in agriculture and primary production generally is less beneficial to the human agent than participation in manufacturing. (Singer 1950). This is so, it is held, because manufacturing activity creates skills, secondary educational effects, and external economies which, in their diffusion, are responsible for widespread future increases in productivity.

There may be a good deal to be said in favour of such a point of view, but on reading the arguments of its proponents, it is difficult to avoid the feeling that the generalizations are too sweeping and that the invidious comparison of agriculture in general with manufacturing in general is conducted on far too gross a level.

In considering the alleged failure of agriculture to generate beneficial secondary effects, it is important that we isolate those forces which are inherent in agricultural production from other forces which happen to be associated with agriculture in any particular geographic or historical context. For instance, relative factor endowment and therefore relative factor prices may be expected to be critical in determining the choice of productive techniques. A highly labour-abundant economy, *ceteris paribus,* will adopt labour-intensive techniques not only in agriculture but in industry and the service sector as well. This may be the dominant consideration in explaining the limited secondary benefits generated throughout much of the tropical and semi-tropical regions of the world which have concentrated on primary product extraction and export. Many of the human skills and aptitudes which are vital for growth are acquired in the production and employment of a (generally capital-intensive) machine technology, but a labor-abundant society will have minimum incentive to introduce such technology either in agriculture *or* industry. A highly elastic supply of labor at or near the prevailing wage has undoubtedly been a pervasive force in densely populated regions in accounting for both the productive methods employed as well as for the failure of wages to rise with the expansion of activity in the primary sector resulting from the "opening up" of these regions to world trade. (Lewis 1954).

It is highly important, furthermore, that we should distinguish clearly between underdevelopment of resources and backwardness of population, especially since it is a frequent practice to regard low income countries as possessing both characteristics. For, historically, many of the low income tropical countries experienced a highly intensive exploitation of their natural resources, but in a manner which has left the human agent virtually untouched. As Hla Myint has expressed it:

... in spite of the striking specialization of the inanimate productive equipment and of the individuals from the economically advanced groups of people who manage and control them, there is really very little specialization, beyond a natural adaptability to the tropical climate, among the backward peoples in their roles as unskilled labourers or peasant producers. Thus the typical unskilled labour supplied by the backward people is an undifferentiated mass of cheap manpower which might be used in any type of plantation or in any type of extractive industry within the tropics and sometimes even beyond it. ... Thus all the specialization required for the export market seems to have been done by the other co-operating factors, the whole production structure being built around the supply of cheap undifferentiated labour. (Myint 1954:153).

This failure to acquire economically relevant skills has frequently been further reinforced by factors not immediately resulting from the preoccupation with primary products itself. For, we might ask, since the growth of such activities necessarily requires a further growth of skilled supervisory and technical personnel, as well as a variety of middlemen and other business and marketing functions, why have native populations typically failed to acquire the skills which these important roles would seem to offer?

The answer to this question is complex and urgently requires further study, but it appears to be associated, at least in part, with the compelling pull, certainly during the early stages, of village or tribal communities, such that the native worker remains essentially a migrant, participating only on a temporary basis in the plantation or mine economy. (Berg 1961). Because of this status, labor turnover rates are high and the worker does not remain long enough to acquire the skills and disciplines which are required for upward mobility. This is further reinforced by the impediments to upward mobility and their corrosive effects upon incentives and aspirations, of discriminatory practices, whether official or unofficial. Under these circumstances it may not be primary production as such, but the specific role of supplier of unskilled labor which is imposed upon the native work force, which accounts for much of the failure to acquire skills of future economic usefulness. (Myint 1954:152-59. Cf. also Myint 1954-55).

A closely related set of considerations is the specific institutional context within which primary production takes place. Land tenure arrangements and marketing and credit systems may easily structure the framework of incentives and opportunities in such a way as to predetermine the nature of the impact of agricultural activity upon the human agent. (United Nations 1951 and Eckstein 1955).

A further point of major importance is the specific incidence of different kinds of crops. Agricultural products differ drastically in the kinds of knowledge, technical competence and even social systems required for their successful cultivation, and the secondary effects which they generate may also differ significantly. Kindleberger has noted the contrast between sugar and tobacco in countries like Cuba:

Sugar is produced on a sizeable scale with capital and unskilled labour in a highly seasonal burst of work, followed by a "dead season" of four or five months. Tobacco, on the other hand, calls for skilled labour, working all year round; a worker has a chance to develop his creative powers. In sugar, the land is owned by large companies, and it is impossible for a worker to move up the economic scale through acquiring land. (Kindleberger 1958:30. Cf. also Manners and Steward 1953).

It is clear - and important - that the kinds of skills generated by agriculture depend very much upon the kind of agriculture one has in mind. Some crops require an unskilled labor input performing nothing but simple, routinized, repetitive tasks - e.g., cotton. [12] Where this is the case the plantation system has been a logical development, since it is possible to centralize decision-making and to supervise the activities of large numbers of unskilled workers.

But midwestern U.S. agriculture has provided a radically different experience. The pattern of agricultural activity in the American midwest was of such a nature that it developed a high degree of commercial and technical sophistication on the part of the labor inputs. Much of the explanation lies in the fact that this was an agriculture centered on livestock husbandry which required a highly efficient and sophisticated system of managerial decision-making. Midwestern farming has been, to a considerable extent, an example of a complex system of vertical integration on the part of the individual producing unit - the individual farm typically produces the food-cereal products which constitute the basic food input of its livestock population.

The midwestern farm is often a fairly elaborate enterprise where the decision-maker must be close to the detailed day to day operations of the farm and which require a familiarity with market phenomena and a wide range of technical skills.[13] Midwestern farming has therefore produced effective managers and people well-versed in mechanical skills who have successfully transferred these skills to other sectors of the economy during the prolonged secular decline of the agricultural sector in the American economy.

One may suggest, then, that the system of small family proprietorships developed in the midwest was a logical outgrowth of the complex technical and managerial requirements of an agriculture centering on livestock husbandry. By contrast monoculture patterns such as prevailed in the American south and in many tropical countries generated a wholly different pattern in part because of the unskilled labour which it required.

The ramifications of these suggested contrasts extend even further. Many of the spectacular failures of the Soviet collectivization of agriculture can be attributed to factors such as have been suggested here. The contrast between Soviet and American agricultural performance reflects, to be sure, numerous differences in soil, climate, factor endowment, etc. Yet it is no accident (as Marxists themselves are fond of saying) that their comparative failures have been greatest in livestock husbandry.[14] These failures are, in large part, attributable to the attempt to impose a highly centralized management structure upon the production of agricultural products which, to be successfully raised, require a careful balance of strong personal interest and incentives with decentralized, "on-the-spot" managerial decisions.[15]

Further aspects of the possible impact of productive activity upon the human agent are illuminated when we focus our attention explicitly upon the fact that successful economic development necessarily involves continuous *alterations* both in the pattern of resource use in response to the changing composition of aggregate demand and in the kinds of technology employed in the productive process. Perhaps one of the most serious problems involved in primary product specialization as it has been carried out in many low income countries is that such skills as are involved are skills which are very specific to the production of one or a few commodities. These skills typically are attached to the exploitation of particular primary products and frequently cannot be employed in doing other things. They possess, as it were, little transfer value to other uses and therefore provide little capacity for shifting effectively to new products in the event of the decline in the demand for the product. Technologically speaking, they are "dead ends." Such economies, therefore, are likely to be highly vulnerable to changing world demand and to technological change in the advanced countries which often involves the development of synthetic substitutes or techniques which make it possible to reduce primary input requirements per unit of final output. It is this limited capacity to adapt to change, rather than primary product production *per se,* which is likely to prove fatal.

Here again it is important that our assertions should not be too sweeping and categorical and that further research be directed toward isolating the more specific factors and mechanisms which limit the capacity for adaptation. Not all primary production, as we have already argued, is equally deficient in generating technical skills, nor is all manufacturing activity equally successful in producing them. For example, one particular sector of the economy stands out in its role as a source of new technology appropriate to a country's factor endowment and in its ability to facilitate the adaptation to changing output: the capital goods producing sector. This sector - aside from construction - is usually undeveloped, or even nonexistent, in primary producing countries, and its undeveloped state would appear to constitute a handicap of enormous proportions. It is probable that one of the most important factors contributing to the viability and flexibility of industrial economies is the existence of a well-developed capital goods sector possessing the technical knowledge, skills and facilities for producing machinery to accommodate the changing requirements of productive activity *plus* the ability and the incentive for raising productivity of machinery production itself - thereby reducing its cost and encouraging its further adoption. Herein may lie the most important feedback of all which is central to explaining the differences in behavior between industrial and primary producing economies. Industrial societies, through the role of their highly developed capital goods producing industries, have, in effect, internalized in their industrial structure a technological capacity which undertakes technological change and adaptation almost as a matter of course and routine. Underdeveloped economies, of course, import much of their capital goods from abroad, but this expedient deprives them of a learning experience in the production, improvement and adaptation of machinery which may be vital to economic growth.

Even if the position taken here is valid, however, its policy implications are by no means obvious. For, by and large, the currently industrial countries developed under historical conditions of comparative labor scarcity and their capital goods sectors grew, in part, because the comparatively high price of labor provided a continuous inducement to the adoption of capital-intensive methods of production.[16] By contrast, entrepreneurs in the labor-abundant economies of most of the underdeveloped world possess a strong inducement to perpetuate labor-intensive techniques, as well as to undertake other forms of economic activity which may make little further contribution to the growth process. (Rosenberg, September 1960).

The abundance of labour in primary producing countries is likely to generate growth-inhibiting forces in other ways as well, via what may be referred to as "negative feedbacks." Extreme labor abundance not only induces the adoption of labor-intensive techniques but also leads to the development of attitudes and social institutions which, in turn, create a preoccupation with work-spreading arrangements and with aspects of productive activity which are simply irrelevant to growth and may easily constitute obstacles to it. Such an environment has led to make-work techniques and arrangements for spreading work among the largest number of people which are such a familiar feature of densely populated low-income countries. The virtual social obligation for the wealthy Indian to employ a large number of functionally specialized domestic servants is a case in point. (Cf. Moore and Feldman 1960:49-60). Furthermore, such environments are highly conducive to the proliferation of traditions and standards of craftsmanship - virtuoso performances of painstaking care and effort reflecting many years of patient apprenticeship and training - which are, at best, simply irrelevant to the skills required for adaptation and growth. At worst, such attitudes and interests discourage the exploration of new methods and techniques and constitute a powerful obstacle to innovation.

In contrast with the "negative feedbacks" which appear to play such an important role in primary producing low-income countries, it appears that, during some stages in their development at least, high-income industrial countries enjoy beneficial feedbacks from a rapid rate of growth of "knowledge-producing" occupations. Professor Fritz Machlup has recently completed a study of what he calls knowledge-producing occupations. Although there are many serious objections which may be raised concerning his definitions and criteria for classification, his conclusions nevertheless command considerable interest.

These are the trends read from the statistical series: (1) The knowledge-producing occupations have grown over the last 60 years much faster than occupations requiring manual labour. (2) The share of knowledge-producing occupations in the total labour force tripled between 1900 and 1959. (3) The share of these occupations in total employment has increased even more. (4) While in the first part of this century growth was fastest in clerical occupations, the lead was then taken by managerial and executive occupations, and more recently by professional and technical personnel. (5) The share of knowledge-producing occupations in total income has increased during the last decade. (6) The share of professional and technical personnel in total income has increased during the last two decades. (Machlup 1962:396).

We conclude on the basis of the preceding discussion that the impact of productive activity on the quality of the economy's future input is of decisive importance in determining the future growth of the economy.

IV

The final set of considerations to which we shall address ourselves can be dealt with more briefly, because they are admittedly even more speculative and less familiar than the subjects of the preceding discussion. Our general argument here is that a whole range of feedbacks are produced not only by the pattern of productive activity but by the pattern of consumer goods and services and, moreover, by the composition of the economy's entire range of final output - not only consumer goods and services but also investment goods and goods and services produced in the government sector. Indeed, it is necessary to go further and to insist that, for many of the problems connected with the growth process, it is becoming increasingly artificial even to maintain a rigid distinction between production and consumption. An important implication of this position is that we can no longer maintain a sharp separation between the firm and the household, regarding the former as the center of productive activity and the latter simply as the place where the output of the economy is passively consumed. (Cairncross 1958). For what goes on in the household may provide the key to important productivity changes, and it may well be that the limited success of the economist thus far in accounting for the extent of the rise in per capita incomes lies with an excessive preoccupation with the economic activity within the firm as the sole source of productivity improvements.

In one important respect economic analysis is currently attempting to incorporate some of the phenomena referred to here. It is now generally recognized that economics has, in the past, operated with a highly restricted concept of capital formation, confining itself to tangible capital of the kind purchased by business firms for direct use in the productive process.[17] Much of society's "investment" (if the term is used in the only meaningful sense of any current use of resources which increases future output) consists in investment in human capital; and a significant portion of the apparent discrepancy between the growth in output per capita and the growth in measured inputs, referred to earlier, is attributable to the exclusion of all capital which becomes embodied, so to speak, in the human agent. Such investments become increasingly important as an economy achieves higher levels of per capita income and it is apparent that the failure to include the expenditure of resources upon such activities as formal education and on-the-job training has imparted a major downward bias to our measure of capital formation and to our measures of growth-inducing forces generally.[18]

It is suggested that the notion of investment in human capital, although it constitutes an important corrective to an excessively restrictive concept of capital formation, is nevertheless merely a special species of a much larger genus. What now requires exploration are the multitude of ways in which resource use in the government and household sectors result in outputs which, through their feedback effects, modify the character of the human agent in ways which are further conducive to economic growth.

An adequate discussion of the role of the government sector would take us far beyond the limits of the present paper, and cannot be attempted here, but a few brief comments are in order. As Kuznets has persuasively argued, many of the expenditures of government in an industrial, urbanized society should be regarded as cost outlays necessary to the smooth functioning of such a complex society, and not as part of the flow of net final product. (Kuznets 1953: Chapters

6 and 7). On the other hand, a functional analysis of government expenditures will also reveal major growth-inducing activities financed (although not necessarily performed) within that sector. For example, the percentage of gross national product devoted to all research and development activities in the United States has risen abruptly over the past decade, rising from 1.4 per cent to 2.8 per cent of gross national product between the years 1953-54 and 1960-61, and most of this has been financed by the federal government. In 1960-61 the federal government financed 65 per cent, or $9 billion of the total outlay of $14 billion expended on research and development. Moreover, in the same period the federal government provided $745 million of the $1.3 billion spent upon basic research. (National Science Foundation 1963:134-42). Furthermore, government expenditures in the important areas of health and education show a secular rising trend and cross-sectional data indicate that such expenditures constitute a higher percentage of gross national product in high-income countries than in low-income countries. (Kuznets 1962:10).

The growing importance of the role of the government sector is also confirmed by cross-sectional data in addition to historical evidences of its growth in currently high-income countries. The available data indicate a high positive correlation between the size of a country's per capita income and government expenditures as a percentage of gross national product. The changing volume and composition of government expenditures may thus include a number of strategic, growth-inducing forces which increase in relative importance with the process of economic development itself.[19]

Increases in per capita consumption within the household may play an important role in economic growth not only through improving standards of nutrition, housing and health but also in the acquisition of skills and aptitudes; in improving the human agent's receptivity to certain learning processes and thereby raising his capacity to produce new knowledge, to innovate, and to adapt to change in his economic roles; and in changing the character of his motivations and aspirations which, in turn, modify his behavior in the economic arena. In short, the household must be examined as a "producer" of skills, aptitudes and aspirations.

The sharp increase in expenditures on consumer durables, which is associated with rising per capita incomes, is an important case in point. With rising per capita incomes the household itself becomes increasingly mechanized as it builds up a stock of consumer durables associated with the performance of household tasks, with amusement and recreation for the leisure time which is increasingly available, and with the provision of private transport facilities - at lower levels the bicycle and motor scooter and at higher levels the automobile. [20] Members of the household are, quite literally, immersed in an environment of considerable technological complexity. At these higher stages of per capita income the feedback from the consumption sphere to that of production skills and aptitudes may become extremely significant. For such surroundings both stimulate and, in some measure, compel a high degree of sophistication in the intricacies of modern technology and, in so doing, provide a socially costless diffusion of productive abilities. Consider furthermore the fearful array of children's toys of a mechanical and electronic nature, to say nothing of the specifically "educational" toys, games and devices directed at the children's market, or the "do-it-yourself" kits with which it is now even possible for youngsters to build simple model computers. And, of course, the impact of the automobile has been strictly *sui generis*. It has been a source of fascination to generations of adolescents and adults and has played an immensely important role in the distribution of mechanical skills among the American population.[21]

Our final point encompasses feedbacks which may originate in both the consumption and production spheres. It is the practice of economists to assume that consumer wants are autonomous and determined independently of the process through which they are satisfied. The consumer is visualized as possessing an ordered structure of preferences and entering into market relations with the purpose of maximizing his satisfactions subject to a budgetary constraint. By making this assumption it is possible to develop an important set of analytical relationships which illuminate the problem of optimum resource allocation. For purposes of analyzing the process of economic growth, however, this assumption is seriously deficient, since a major component of the growth process is a radical transformation of attitudes toward consumption and saving, and toward work and leisure. The changing structure of consumer wants and preferences, in other words, is itself a strategic variable in the growth process, as the classical economists recognized, since it is an important determinant of individual behavior and shapes the nature of his responses to economic incentives and opportunities.

Yet our ignorance on this subject is almost total, and a systematic examination of the ways in which wants are shaped and modified as a result of participation in economic activities is urgently required. For, as Marshall insisted many years ago:

... while wants are the rulers of life among the lower animals, it is to changes in the forms of efforts and activities that we must turn when in search for the keynotes of the history of mankind ... although it is man's wants in the earliest stages of his development that give rise to his activities, yet afterwards each new step upwards is to be regarded as the development of new activities giving rise to new wants, rather than of new wants giving rise to new activities. (Marshall 1948:85 and 89).

References

Moses Abramovitz, "Resource and Output Trends in the U.S. Since 1870," *American Economic Review Papers and Proceedings,* May 1956, pp. 1-23.

Francis Bator, *The Question of Government Spending* (Harper and Brothers, New York, 1960).

Elliot Berg, "Backward-Sloping Labour Supply Functions in Dual Economies - The African Case," *The Quarterly Journal of Economics,* August 1961, pp. 468-92.

A.K. Cairncross, "Economic Schizophrenia," *Scottish Journal of Political Economy,* February 1958, pp. 15-21.

Robert Campbell, *Soviet Economic Power* (Houghton Mifflin Company, Cambridge, 1960).

Hoang Van Chi, "Collectivization and Rice Production," *The China Quarterly,* January - March 1962, pp. 94-104.

Alexander Eckstein, "Land Reform and Economic Development," *World Politics,* July 1955, pp. 650-62.

Solomon Fabricant, *The Trend of Government Activity Since 1900* (National Bureau of Economic Research, New York, 1952).

H.J. Habakkuk, *American and British Technology in the 19th Century* (Cambridge University Press, Cambridge, 1962).

David Hume, *Writings on Economics* (University of Wisconsin Press, Madison, 1955) edited and introduced by Eugene Rotwein.

W. Stanley Jevons, *The Theory of Political Economy* (MacMillan and Company, London, 1911), Fourth Edition.

The Journal of Political Economy, October 1962, Part II, special supplement on "Investment in Human Beings."

Charles Kindleberger, *Economic Development* (The McGraw-Hill Book Company, New York, 1958).

Simon Kuznets, *Economic Change* (W.W. Norton and Company, New York, 1953).

Simon Kuznets, "Quantitative Aspects of the Economic Growth of Nations: VII. The Share and Structure of Consumption," *Economic Development and Cultural Change,* January 1962, Part II, pp. 1-92.

W.A. Lewis, "Economic Development with Unlimited Supplies of Labour," *The Manchester School,* May 1954, pp. 139-91.

W.A. Lewis, *The Theory of Economic Growth* (Richard D. Irwin, Inc., Homewood, Illinois, 1955).

Fritz Machlup, *The Production and Distribution of Knowledge in the United States* (Princeton University Press, Princeton, 1962).

Thomas Robert Malthus, *Principles of Political Economy* (Reprinted by Augustus M. Kelley, Inc., New York, 1951), Second Edition.

R.A. Manners and J.A. Steward, "The Cultural Study of Contemporary Societies: Puerto Rico," *American Journal of Sociology,* September 1953, pp. 123-30.

Alfred Marshall, *Principles of Economics* (MacMillan and Company, London, 1948), Eighth Edition.

Alison Martin and W. Arthur Lewis, "Patterns of Public Revenue and Expenditure," *The Manchester School,* September 1956, pp. 203-44.

John Stuart Mill, *Principles of Political Economy* (Longmans, Green and Company, London, 1909).

Wilbert Moore and Arnold Feldman, *Labor Commitment and Social Change in Developing Areas* (Social Science Research Council, New York, 1960).

Hla Myint, "An Interpretation of Economic Backwardness," *Oxford Economic Papers,* June 1954, pp. 132-63.

Hla Myint, "The Gains from International Trade and the Backward Countries," *The Review of Economic Studies,* 1954-55, Vol. XXII (2), No. 58, pp. 129-42.

National Science Foundation, *Twelfth Annual Report,* 1962 (United States Government Printing Office, Washington, D.C., 1963).

Harry Oshima, "Share of Government in Gross National Product," *American Economic Review,* June 1957, pp. 381-90.

David Ricardo, *Works* (Cambridge University Press, Cambridge, 1952), Vol. VIII, edited by Piero Sraffa.

Nathan Rosenberg, "Capital Formation in Underdeveloped Countries," *American Economic Review,* September 1960, pp. 706-15.

Nathan Rosenberg, "Some Institutional Aspects of the *Wealth of Nations,*" *The Journal of Political Economy,* December 1960, pp. 557-570.

W.W. Rostow, "The Take-Off into Self-Sustained Growth," *Economic Journal,* March 1956, pp. 25-48.

Walt W. Rostow, *The Stages of Economic Growth* (Cambridge University Press, Cambridge, 1960).

Theodore Schultz, "Investment in Man: An Economist's View," *The Social Science Review,* June 1959, pp. 109-17.

Theodore Schultz, "Capital Formation by Education," *The Journal of Political Economy,* December 1960, pp. 571-83.

Theodore Schultz, "Investment in Human Capital," *American Economic Review,* March 1961, pp. 1-17.

Hans Singer, "The Distribution of Gains Between Investing and Borrowing Countries," *American Economic Review Papers and Proceedings,* May 1950, pp. 473-85.

Adam Smith, *Lectures on Justice, Police, Revenue and Arms* (Reprinted by Kelley and Millman, Inc., New York, 1956).

Adam Smith, *The Wealth of Nations* (Random House, New York, 1937).

Robert Solow, "Technical Change and the Aggregate Production Function," *Review of Economics and Statistics,* August 1957, pp. 312-20.

United Nations, Department of Economic Affairs, *Land Reform: Defects in Agrarian Structure as Obstacles to Economic Development* (New York, 1951).

6

The direction of technological change: inducement mechanisms and focusing devices*

Introduction

One of the things which is perfectly obvious about societies which have achieved high degrees of industrialization is that they have acquired unusual skills in problem-solving activities. Industrial societies have learned how to solve certain kinds of problems, and understanding this creative capacity is basic to an understanding of the growth process. What is less obvious, however, is that the developed countries never solve more than a small fraction of the problems they are capable of solving. Rather, they solve some fraction of the problems which happen to be formulated and actively pursued. This suggests that our understanding of the process of technological change may be advanced by exploring the manner in which problems are formulated at the firm level.

This paper is intended as a sort of historical reconnaissance mission. It represents an attempt to establish certain generalizations concerning the problem-solving process in industrializing countries in the past two centuries. Our interest is in the forces which provide inducements to technical change, and in examining these forces - what Hirschman has called "inducement mechanisms"[1] - we will not confine ourselves to the more conventional framework of economic reasoning. Indeed, the present paper has been prompted in some measure by the extreme agnosticism to which one is led on the subject of technological change by recent theorizing. It used to be thought possible to explain the factor-saving bias, which inventions took, in purely economic terms. It will be recalled that Hicks stated, in his *Theory of Wages:*

The real reason for the predominance of labour-saving inventions is surely that which was hinted at in our discussion of substitution. A change in the relative

*The author has benefited from discussions with Edward Ames, Albert Hirschman, William Parker, Charles Plott, Eugene Smolensky, and Paul Strassmann.

prices of the factors of production is itself a spur to invention, and to invention of a particular kind - directed to economizing the use of a factor which has become relatively expensive. The general tendency to a more rapid increase of capital than labour which has marked European history during the last few centuries has naturally provided a stimulus to labour-saving invention.[2]

It now turns out that, at least within the framework of a purely competitive model, we cannot even say this much. For, as has been pointed out in several places, Hick's position that changes in factor prices lead to innovations involves a confusion between technological change and factor substitution. The current position, as expressed by Fellner, Salter, Samuelson, and others, is that under competitive conditions an individual firm is simply not interested in the particular factor-saving bias of technical improvements. The argument is that a firm always has an incentive to reduce any portion of its costs. The market mechanism provides no incentive to look for inventions which have any particular factor-saving bias. Indeed, the position is that in competitive equilibrium it does not even make sense to speak of "dear" labor or "cheap" labor. After all, when each factor is being paid the value of its marginal product, then all factors are equally "cheap" and equally "dear" in the eyes of a competitive firm.

Salter's position may be quoted as representative of a larger genus. Speaking of Hicks's theory of induced inventions, he stated: "If . . . the theory implies that dearer labor stimulates the search for new knowledge aimed specifically at saving labor, then it is open to serious objections. The entrepreneur is interested in reducing costs in total, not particular costs such as labor costs or capital costs. When labor costs rise any advance that reduces total cost is welcome, and whether this is achieved by saving labor or capital is irrelevant. There is no reason to assume that attention should be concentrated on labor-saving techniques."[3]

This is, in a sense, very disquieting - not only because economic historians have long used differences in factor prices (and differences in rates of change of factor prices) as an explanation for the particular factor-saving bias of technological change. More importantly, the individuals directly involved in the process have, on numerous occasions, explained their own intent and motivation in these terms. Their contemporaries, especially in the nineteenth century, employed similar explanations with almost monotonous regularity.

Perhaps what is needed is to break out of an excessively restrictive competitive framework and to approach the problem of technological change from different vantage points. The present paper will therefore look for clues to the process of technological change in several areas of the historical experience of the past century and a half or so. It will pose the question: have there been forces at work in recent history which have in fact pushed exploratory activity in specific directions?[4] Nothing will be proven since historical examples by themselves prove nothing; hopefully, however, a beginning will be made toward a better understanding of some of the mechanisms which initiate and direct exploratory activities.

I

It will be argued first that, in looking for the origins of technological changes in the manufacturing sector, the technological level itself has been badly neglected. What will be said should not be confused with a crude form of technological determinism, where social, economic, and political changes are explained in terms of antecedent changes in technology. Rather, what is asserted is that technology is much more of a cumulative and self-generating process than the economist generally recognizes. Technological change, when approached from the point of view of economic theory, is likely to be treated as a realm which passively adjusts to the pressures and signals of economic forces, mediated through the market place and through factor prices in particular. The opposite danger, to which the following remarks are subject, is that they will be interpreted as saying that economic forces do not condition the direction in which technological changes move. Our position, then, is that the ultimate incentives are economic in nature; but economic incentives to reduce cost always exist in business operations, and precisely because such incentives are so diffuse and general they do not explain very much in terms of the *particular sequence and timing of innovative activity*. The trouble with the economic incentives to technical change, as an explanatory variable, is precisely that they *are* so pervasive. In the realm of pure theory, a decision maker bent upon maximizing profits under competitive conditions will pursue any possibility for reducing costs, regardless of which factor will be economized on. What forces, then, determine the directions in which a firm actually goes in exploring for new techniques? Since it cannot explore in all directions, what are the factors which induce it to strike out in a particular direction? Better yet, are there any factors at work which compel it to look in some directions rather than others?

In answering this question it may be necessary to modify the model of the maximizing firm to recognize explicitly real-world forces which constrain and otherwise influence its behavior. Within the realm of exploratory activities which may yield new, cost-reducing techniques, where does the firm place its resources? Some firms no doubt are able to take a very long time horizon and survey, with equanimity and fine impartiality, the whole spectrum of possibilities. But most firms - or at least most decision makers - are under pressure to undertake actions which promise a payoff in a relatively short time period and with at least most of the constraints imposed by the existing plant. They are confronted with the existing range of productive activities as an inevitable starting point. They are naturally led to search the technological horizon, as it were, within the framework of these current activities and to attack the most restrictive constraint. My primary point is that most mechanical productive processes throw off signals of a sort which are both compelling and fairly obvious; indeed, these processes when sufficiently complex and interdependent, involve an almost compulsive formulation of problems. These problems capture a large proportion of the time and energies of those engaged in a search for improved techniques.

Such signals have not, of course, been confined to the industrial sector of the economy. In agriculture, for example, the mechanization of reaping may be said to have been effectively "signalled" by the compelling and obvious need to harvest the wheat crop within a very limited number of days in order to prevent spoilage. [5] Nevertheless, it seems to be a historical generalization, with which this paper is in fact much concerned, that such signalling has been more conspicuously effective in the industrial sector.

It will be argued that complex technologies create internal compulsions and pressures which, in turn, initiate exploratory activity in particular directions. These pressures operate at the plant level and also often within the components of the final product itself. The improved designs of automobile engines have led - through the achievement of higher speeds - to the invention of improved braking systems. The need for such systems - or perhaps one should say the penalty for *failing to provide* such a system along with a more powerful engine - was painfully obvious to all automotive engineers (if not to all drivers). Similarly, to someone constructing a hi-fi system it is obvious that the benefits of a high-quality amplifier are lost if it is attached to a low-quality loudspeaker.[6] An understanding of these interdependencies may help us to understand better than we do the path and the process of technological change.

The notion of compulsive sequences will not be new to economic historians. Reference to the imbalances in the relation between machines is virtually de rigueur in any treatment of the English cotton textile industry in the eighteenth century (Kay's flying shuttle led to the need for speeding up spinning operations; the eventual innovations in spinning in turn created the shortage of weaving capacity which finally culminated in Cartwright's introduction of the power loom).[7]

In a sense the capital goods sector is always being bombarded with messages of the sort that say: "I expect to be able to earn a profit if I can produce a new device which will conform to certain specifications. But no machinery now exists which can produce such a device. Therefore you can earn a profit by devising and selling machines which will produce according to these specifications. Do so." The early prototype for this relationship was James Watt's protracted search for a cylinder bored to a minimum degree of accuracy in its diameter. The required degree of precision was eventually attained with the use of John Wilkinson's boring mill, and the commercial practicability of the steam engine really dates from the use of this mill in preparing cylinders.[8] This backward linkage, to use Hirschman's terminology, has been an enormously important source of technical change in the Western world, and it can be argued that a responsive machinery-producing industry has been the key to successful industrialization. My concern, however, is not primarily with the historical responsiveness of certain supply factors but with the manner in which the demand for new techniques emerges and is perceived.[9]

II

Let us consider some historical episodes. In the early 1890s, at the beginning of the bicycle craze, the machining of bicycle-wheel hubs posed a serious production problem. Forming tools, which previous to 1890 had been applied to metals of soft composition, such as the caps of salt and pepper boxes, were successfully applied to hardened metals in the shaping of the outside of the wheel hubs. However, this use of the forming tool created an imbalance between the operations carried on for the outside and the inside of the hub. Since the forming tool now worked more rapidly on the outside of the hub than the conventional drills worked on the inside, it was not possible to derive the fullest gains from the use of the forming tool. This imbalance was eventually corrected by the introduction of the oil-tube drill. This drill had an oil channel leading to or near the point, and it made possible the lubrication and cooling of cutting edges as well as the removal of chips, and therefore speeded up drilling operations. In doing so, the oil-tube drill corrected the initial imbalance and brought about a closer synchronization between the operations on the outside and the inside of the hub.[10]

A similar sort of technical imbalance between interdependent processes led to important improvements in the internal structure of the Bessemer converter. The early Bessemer converters required a considerable amount of expensive, auxiliary equipment. This equipment was utilized only a small percentage of the time, partly because the bottom of the converter wore out, due to the extreme temperatures of the process, after one to three heats. At that point the converter had to be allowed to cool so that a man could climb inside and repair the lining. Alexander Holley addressed himself directly to correcting this imbalance by devising ways of speeding up the use of the converter. His eventual solution was a removable shell for the bottom of the converter which made it unnecessary to let the converter cool and therefore led to a considerable saving of time and reduction in costs.[11]

According to Bessemer's own testimony, the search activity which led to the development of the Bessemer process of steel production was itself initiated by a problem Bessemer confronted which conforms precisely to what is being called here a technical imbalance. Bessemer stated in his autobiography that his attention was directed to the subject of artillery by the outbreak of the Crimean War in 1854. He developed a new gun of superior power, capable of firing a heavy, elongated projectile. To his dismay, however, he concluded that the gun could not be safely fired if it were constructed of the standard cast iron employed in artillery pieces. The Bessemer process was the product of Bessemer's search for a superior metal which would make his newly designed gun practicable by withstanding the severe strains imposed by the heavy weight of the projectiles.[12]

The introduction of high-speed steel into the machine tool industry constitutes a major example of technological imbalance. High-speed steel is a steel alloy (combined with tungsten, vanadium, and chromium) which raised enormously the red hardness of cutting tools[13] and therefore made it possible to remove metal by cutting at dramatically higher speeds and also by taking heavier cuts in the metal. High-speed steel was first developed by Frederick W. Taylor and his associates and exhibited in operation at the Paris Exposition of 1900. "At that exhibition they exhibited tools made of this steel, in use in a heavy and powerful lathe, taking heavy cuts at unheard of speeds - 80, 90 or 100 feet per minute, instead of the 18 to 22 feet per minute that previously had been the maximum for heavy cuts in hard material."[14] It turned out, however, that it was impossible to take advantage of higher cutting speeds with machine tools designed for the older carbon steel cutting tools because they could not withstand the stresses and strains or provide sufficiently high speeds in the other components of the machine tool. As a result, the availability of high-speed steel for the cutting tool quickly generated a complete redesign in machine tool components - the structural, transmission, and control elements. As one machine tool authority put it: "During the first decade of the 20th century we see high-speed steel revolutionizing the lathe - as it does all production machine tools. Beds and slides rapidly become heavier, feed works stronger, and the driving cones are designed for much wider belts than of old. The legs of big lathes grow shorter and shorter, and finally disappear as the beds grow down to the floor. On these big machines massive tool blocks take the place of tool posts, and multiple tooling comes into vogue."[15]

The final effect of this redesigning which was initiated by the use of high-speed steel in cutting tools was to transform machine tools into much heavier, faster, and more rigid instruments which, in turn, enlarged considerably the scope of their practical operations and facilitated their introduction into new uses. Much of the progress in machine tools resulted from the generation of imbalances between the machine itself and the cutting tool. Improvements in the cutting tool required machines of greater strength, rigidity, capacity to withstand stress, etc. Improvements in the design and operation of the machines, in turn, were useless without improvements in the properties of the cutting tool.[16]

In this exploratory activity, however, which was initiated to accommodate the new requirements of high-speed steel, inventions were made which went far beyond the need merely to make such accommodations. The need to adapt speeds and feeds to the enlarged cutting capacity of the tool was a major stimulus to the development of new speed-changing devices. For example, the cone pulley, a primitive device for altering the speed of the machine tool in accordance with the requirements of the work in hand, was replaced by much more sophisticated gear-change devices which enabled the operator to vary the speed merely by shifting a lever.[17]

This "over-shooting" of the mark is, of course, characteristic of exploratory activities. Such activities are undertaken in response to certain kinds of stimuli, and although the stimuli must presumably exceed some minimum magnitude in order to be perceived and acted upon, the "size" of the discovery need bear no systematic relationship to the "size" of the initial stimulus.

In the case of machine tools, the interdependence is of four component parts. In addition to the cutting tool itself there are:

1. Structural or frame elements whose function it is to carry or support work and tool.
2. Transmission elements which give the work or tool, or both, movements for shaping the work to be produced.
3. Control elements for both adjusting the structural elements relatively to each other and controlling the function of the transmission for moving either tool or work or both.[18]

This sequence - changes in one component of an interdependent system creating a stimulus for changes elsewhere in this system - has been a highly fruitful source of technical change in the machine tool industry. Awareness of imbalances between components has continually led to an exploration of possibilities for corrective action whose eventual result was major improvements in productivity. The imbalance between high-speed cutting tools and the lathe was exactly duplicated in the relationship between milling cutters and milling machines from the 1880s to the early years of the twentieth century. Thus Woodbury tells us:

De Leeuw's analysis had shown that the cutters of the time were not as strong as the machines that were driving them and therefore gave out long before the maximum power of the machine was reached. On this basis Cincinnati Milling Machine Company designed a new face mill, later marketed by Union Twist Drill Company, which increased the metal removal per horsepower by 50%. Further experiments on a larger scale were carried out by A.L. De Leeuw at Cincinnati

Milling Machine Company which led to the design of wide spaced cutters as more durable and more efficient in removing metal.[19]

A set of interrelated changes entirely analogous to those following upon the introduction of high-speed steel on the lathe may also be observed in the history of the grinding machine. There the sequence of technical imbalances and their systematic correction and "overcorrection" was triggered by the introduction of artificial abrasives in the construction of the grinding wheel. The sequence culminated in the conversion of the grinding machine from an instrument used only to perform finishing operations on components which had acquired their basic shapes on a lathe, to a machine performing heavy production operations on the automobile. Indeed, the grinding machine in this form became indispensable to the emerging automobile industry because it provided the only way, at the time, of precision machining of the strong, light alloy steels which played such a prominent part in automobile components.[20]

The history of technology is replete with examples of the beneficent effects of this sort of imbalance as an inducement for further innovation. To venture somewhat farther afield, the history of military technology provides numerous examples of improvements in weaponry resulting from the continuous rivalry between offensive and defensive weapons. The contemporary race between the development of missiles and antimissiles is merely the latest - and most destructive - stage of a compulsive rivalry which has been going on for many centuries. A classic case occurred with the development of high-powered, rifled ordnance in the 1850s, in large measure as a result of the work of Joseph Whitworth and William Armstrong. The great increase in the destructive power of offensive weapons led to the application of armored plate to ships to *protect* them against such weapons. It was out of this interplay that, in thirty or forty years, the modern warship was developed.[21] Even a hundred years ago, however, this rivalry was recognized as something which was already a time-honored process. An article in *Mechanics Magazine* in 1867 concluded its evaluation of the current experiments involving new armored plates by stating: "It appears to be the old story over again; no sooner is a gun or projectile produced which carries everything before it, than a target is devised by which it can be stopped."[22]

Technical imbalances leading to changes in complementary processes may be clearly observed even in medieval siege machinery. After successive improvements in the trebuchet, the medieval artillery piece, it became apparent in the thirteenth century that further improvements in accuracy could not be achieved unless the projectile itself was altered.

... the very fact that the power of a trebuchet could be so nicely regulated impelled Western military engineers to seek even greater exactitude in artillery attack. They quickly saw that until the weight of projectiles and their friction with the air could be kept uniform, artillery aim would still be variable. As a result, as early as 1244 stones for trebuchets were being cut in the royal arsenals of England calibrated to exact specifications established by an engineer: in other words, the cannon ball before the cannon.[23]

Part of the reason for the effectiveness of technological disequilibria in inducing innovations is that they involved compulsive sequences. The relationship among components was usually such that some imbalance *had* to be corrected before an initial innovation could be fully exploited. Such a situation therefore continually directed the attention of technically competent personnel to the solution of problems of obvious practical importance.[24]

III

The apparent recalcitrance of nineteenth-century English labor, especially skilled labor, in accepting the discipline and the terms of factory employment provided an inducement to technical change which was in many ways analogous to the technical disequilibria discussed so far. The most important point is that the threat of worker noncompliance - in the last resort, strikes - served as a powerful agent in focusing the attention of decision makers on obvious and major threats to their profit positions. Here again it may be said that employers had a general, diffuse economic incentive to reduce costs, regardless of which factor inputs were being economized on. But it was also obvious that the labor input was the one which posed the greatest threat, in that its services were likely to be withheld at times and in ways which constituted a serious threat to a firm's profit prospects. The threat of such withdrawals, then, was a powerful force in directing energies in a search for labor-saving machines.

The view was widely expressed in nineteenth-century England, especially in the period 1820-60, that strikes were a major reason for innovations. Contemporary observers who agreed on little else, who were as far apart in their ideological biases and commitments as, say, Karl Marx and Samuel Smiles, were in complete accord on this point. We have, moreover, the evidence of numerous inventors themselves, who testified that they undertook the search process which led to a particular invention as the result of a strike or the threat of a strike.

In the *Poverty of Philosophy,* Marx stated:

In England, strikes have regularly given rise to the invention and application of new machines. Machines were, it may be said, the weapon employed by the capitalists to quell the revolt of specialized labour. The *self-acting mule,* the greatest invention of modern industry, put out of action the spinners who were in revolt. If combinations and strikes had no other effect than that of making the efforts of mechanical genius react against them, they would still exercise an immense influence on the development of industry.[25]

In *Capital,* Marx stated: "It would be possible to write quite a history of the inventions, made since 1830, for the sole purpose of supplying capital with weapons against the revolts of the working-class."[26] Here again he asserted that the most important of these inventions was Richard Roberts's self-acting mule, because it "opened up a new epoch in the automatic system."[27] Roberts's mule was invented in 1825 as the result of a strike on the part of the skilled and highly independent mule-spinners. A delegation of Manchester cotton manufacturers appealed to Roberts to help them in their desperate plight, and Roberts's inventive skill brought forth the self-acting mule. This famous episode was dwelt upon at some length, not only by Marx, but by Smiles[28] and Ure, who concluded his discussion by stating, in one of his inimitable didactic excursions, "that when capital enlists science in her service, the refractory hand of labour will always be taught docility."[29]

Another of Roberts's important inventions, the so-called Jacquard punching machine, owed its origin to somewhat similar circumstances.[30]

Marx had pointed out that "W. Fairbairn discovered several very important applications of machinery to the construction of machines, in consequence of strikes in his own workshops."[31] Fairbairn himself described how his riveting machine had originated with a

stoppage of a part of the works at Manchester by a strike of the boiler-makers. For some time previously we had been busily engaged in the construction of boilers, and nothing could have been more injurious than the stoppage of the works at such a time. I remonstrated with the men, but without effect, and perceiving no chance of coming to terms in any reasonable time, I determined to do without them, and effect by machinery what we had heretofore been in the habit of executing by manual labour.[32]

After some experimentation, with the assistance of one of his employees, Fairbairn devised a riveting machine for which a patent was procured in 1837.

The new machine effected a complete revolution in boiler-making and riveting, and has substituted the rapid and noiseless work of compression for the eternal din of the hammer; besides making the work infinitely superior in quality and strength.

The introduction of the riveting machine gave great facilities for the despatch of business. It fixed, with two men and a boy, as many rivets in one hour as could be done with three men and a boy in a day of twelve hours on the old plan; and such was the expedition and superior quality of the work, that in less than twelve months the machine-made boilers were preferred to those made by hand, in every part of the country where they were known.[33]

Babbage (and others) attributed the invention of the system of grooved rollers for rolling skelps for musket barrels to a strike of the gun-skelp forgers who were, in substantial measure, superseded by the invention. He points out also that the technique of welding the skelps into a finished gun barrel was perfected as a result of difficulties encountered with a combination of workers.[34]

The uncertainties which the supply of labor posed in the British gun-making industry eventually led to the first large-scale borrowing of American technology by the Old World. For the British government's demand for firearms was not only highly erratic but also, in time of national emergency, highly inelastic. Since military firearms, up to the 1850s, were produced entirely by skilled craftsmen whose supply was certainly highly inelastic in the short run, the entry of the government into the market almost invariably led to strikes or threats of strikes unless wages were raised substantially. This situation resulted in an almost complete break-down in supplies of firearms on the eve of the outbreak of the Crimean War in 1854. In its determination to free itself from what it regarded as an intolerable dependence on a small number of skilled workers, the British government sent a commission to the United States to examine American gunmaking machinery. This commission purchased large quantities of machines which were set up at the Enfield Arsenal. The purchase marked the introduction of American mass-production technology into Europe, where the system was long referred to as the "American System of Manufacturing."[35]

Many contemporaries cited the long engineers' strike of 1851 as an important direct cause of the introduction of new, labor-saving techniques.

Mr. Nasmyth, in his evidence before the Trades Unions Commissioners, described very graphically how the long strike of 1851 made him anxious to develop to the utmost the use of labour-saving machinery. "The great feature," he said, "of our modern mechanical improvement has been the introduction of self-acting tools. All that a mechanic has to do, and which any lad is able to do, is, not to labour, but to watch the beautiful functions of the machine. All that class of men, who depended upon mere dexterity, are set aside altogether. I had four boys to one mechanic. By these mechanical contrivances I reduced the number of men in my employ, 1,500 hands, fully one half. The result was that my profits were much increased."[36]

Finally, we may cite Smiles's observations on the impact of strikes in his discussion of Nasmyth's experiences.

Notwithstanding the losses and suffering occasioned by strikes, Mr. Nasmyth holds the opinion that they have on the whole produced much more good than evil. They have served to stimulate invention in an extraordinary degree. Some of the most important labour-saving processes now in common use are directly traceable to them. In the case of many of our most potent self-acting tools and machines manufacturers could not be induced to adopt them until compelled to do so by strikes. This was the case with the self-acting mule, the wool-combing machine, the planing machine, the slotting machine, Nasmyth's steam arm, and many others.[37]

It would be easy to multiply examples.[38] Such multiplication, however, would add nothing further to our point, which is simply to suggest that strikes or fear of strikes have served, historically, as a powerful agent for directing the search for new techniques in a particular direction. The preoccupation with substituting capital for labor (especially skilled labor) was more than just a matter of wage rates. Perhaps even more important was the great nuisance value of strikes. In part at least, entrepreneurial behavior must be understood in terms of an aversion to the uncertainties presented by strike possibilities whose disruptive effects were regarded as intolerable intrusions into the domain of managerial decision making and responsibility.[39]

IV

There is a third category of mechanisms which seems to have been important, historically, in providing successful inducements to technical change. Perhaps the most general way of expressing it is that an accustomed source of supply was, for some reason, cut off or drastically reduced, causing major disruption due to the unavailability either of alternative sources of supply or of satisfactory substitutes. The most common cause of such disruption, of course, has been the outbreak of war. In each case the imposition of a previously nonexistent contraint led to a search activity out of which a satisfactory or superior substitute, or more productive process, eventually emerged.

Thus, France's early commercial leadership in the synthetic alkali industry, developed upon the Leblanc process, owed a great deal to the fact that France was cut off from its supplies of Spanish barilla during the Napoleonic Wars.[40] An eminent observer, Robert Owen, said more generally in describing the effect of the Napoleonic Wars: "The want of hands and materials created a demand for and gave great encouragement to new mechanical inventions and chemical discoveries, to supersede manual labour in supplying the materials required for warlike purposes."[41]

The cutting off of American cotton supplies to Britain during the American Civil War and the resulting cotton famine apparently served as an important inducement to technical change in British textiles. Although these changes included, as one would expect, methods which, by reducing waste, economized on the cotton input, the changes were by no means confined to such economies.[42] During World War I, the Germans were deprived of their imports of Chilean nitrates by the English blockade, and in their efforts to find a substitute they perfected the nitrogen fixation process.[43] On the other hand, the cessation of trade with Germany also resulted in the cutting off of U.S. imports of German chemical products, particularly dyes. By the end of the war the United States had developed a dye industry of substantial proportions.[44] Finally, as is well known, the cutting off of the supply of southeast Asian natural rubber due to Japanese occupation early in the Second World War was responsible for advancing the research activity which culminated in the emergence of the American synthetic rubber industry.[45]

In all of these cases, of course, a major deterrent to most innovations - ignorance of the possible size and nature of the market - was nonexistent. The market for the commodity was well established, and the "message" to potential suppliers of substitutes was unmistakably clear. It was further dramatized by the abruptness with which supplies were terminated, the sharpness of the domestic rise in price, and, in some cases, the strategic importance of the commodity.

Although the instances cited have in common the impact of war in cutting off users from traditional suppliers, such need not be the case. The attempt of the Japanese, after occupying Formosa in 1895, to exploit that island's virtual monopoly of the supply of natural camphor was a powerful stimulus to the search for a synthetic process, which was eventually perfected.[46]

Even an act of legislation that imposes a constraint may lead to exploratory activities which eventually confer an advantage to those who were constrained. Thus, Keirstead states that "the Swedish law against stream pollution forced Swedish producers of chemical pulps to work out means of utilizing waste liquors. Waste liquors from the sulphate process are ordinarily recovered in Canada, but waste sulphite liquor is not. The recovery process of waste sulphite liquor in Sweden has consequently given the Swedish sulphate producers an advantage over their Canadian and American competitors."[47] Similarly, James Watt's invention of the epicyclic (or "sun and planet") gear for the conversion of reciprocating into continuous rotary motion owed much to his apprehension over possible litigation. Watt rejected the more obvious alternative of a crank mechanism for this purpose because of the possibility that it may have been covered by the recent patent by James Pickard of Birmingham.[48]

The building of the tubular railroad bridge across the Menai Straits in the 1840s is a further interesting example of the benificent - and quite unanticipated - effects of the introduction of a constraint. Due to restrictions imposed by the admiralty (which was concerned with the headroom and width that the bridge would leave for navigation) on the form and size of bridge which it would allow to span the straits, an entirely new technique of bridge building was invented. The solution of the problems involved in building the bridge subject to admiralty constraints resulted in fundamental advances in knowledge concerning the structural properties of iron - resistance of beams and plates, strength of girders, compression and tensile strengths, etc. This new knowledge had a wide range of immediate applications, not only in other bridges, but in the construction of cranes, ships, multistorey buildings, steam engines - indeed, wherever iron was used as a building material.[49]

All of this, admittedly, brings us perilously close to the proposition that "new constraints are good for you," which is in some ways analogous to a view occasionally expressed during the 1950s about heart attacks. Obviously, when stated this way, it is nonsense. Yet, if we qualify the proposition to state that "new constraints have sometimes proven beneficial," the proposition is obviously defensible[50] ("Sweet - sometimes - are the uses of adversity"). The critical question is, of course, when, and under what circumstances? To answer this question in a nontautological or nontrivial way we would have to understand both the nature of the creative process and the working of social systems much better than we do at present.[51]

What is clear, however, is that there have existed a variety of devices at different times and places which have served as powerful agents in formulating technical problems and in focusing attention upon them in a compelling way. That there was an economic counterpart to this focusing process in the form of sharply rising prices, reductions in cost, and expectations concerning future profits hardly needs to be said. But the ordinary messages of the marketplace are general and not sufficiently specific. The market rewards reductions in cost, but this is true of all reductions in cost, wherever attained. It does not specify the directions in which cost reductions should be sought. The mechanisms examined here share the property of forcefully focusing attention in specific directions. They called attention decisively to the existence of problems the solutions to which were within the capacity of society at the time, and which had the effect of either increasing profits or preventing a decline that was anticipated with a high degree of probability.

These and other focusing devices deserve more careful examination in any serious exploration of the sources of technological change. It would be very interesting to examine, for example, the learning experiences and other creative opportunities associated with accidents and other disasters. Our ancestors, who had not yet taken to the air or experienced the fearful carnage of a motorized age,[52] were nevertheless plagued by exploding boilers, collapsing bridges, and sinking ocean-going vessels. Although the cost was very high, much information was gleaned from these accidents,[53] but, perhaps more important, they powerfully dramatized weaknesses in existing techniques and thus provided a strong impulse to research efforts.

Disasters such as fires at sea were almost always followed by a rush of patent applications for devices which purportedly would prevent their recurrence.[54] Nobel's invention of dynamite, a comparatively safe explosive, owed much to the unfortunate tendency of his shipments of nitroglycerin to explode in transit aboard railroad trains or, in some spectacular instances, in midocean.[55] The nearly disastrous bursting of a hydraulic cylinder due to the enormous pressure to which it was subjected during the raising of the Britannia tubular bridge across the Menai Straits led directly to major advances in knowledge concerning the crystallography of metals. The new knowledge thus gained resulted in important improvements in design over a range of objects wherein cylinders were subjected to severe internal pressures, for example, cast-iron guns.[56]

All the cases dealt with have a common denominator in the expectation of profit, but in all cases also there are forces pointing emphatically in certain directions. This is the underlying unity. The inducement mechanism is the prospect of making extra-high profits, or the unbearable prospect of losing a marvelous and tangible opportunity to do so, just as in the cases of strikes and wars the mechanism is the unbearable prospect of being put out of business or having one's attempt at a more orderly planning process reduced to chaos. The kinds of agonizing uncertainties created by the imminent danger of a strike cannot be adequately reduced, as they often are in modern decision theory, to a probabilistic treatment. Probabilistic statements, by definition, do not apply to single events but only to classes of events. The employer, faced by a strike next week, cannot necessarily console himself with the thought that his situation is an unlikely one, and concentrate his attention on the mathematical expectation of the future. If the strike takes place, there may *be* no future - at least for his firm. It may well even be that the apparently *irrational* determination of so many nineteenth-century firms to introduce labor-saving innovations (irrational from the point of view of modern theory) really arose from their determination to engage in *rational* long-term planning - a procedure which was impossible as long as strike possibilities by indispensable skilled personnel hung like a sword of Damocles over their heads.[57]

It is possible, furthermore, that threats of deterioration or actual deterioration from some previous state are more powerful attention-focusing devices than are vague possibilities for improvement. There may be psychological reasons why a worsening state of affairs, or its prospect, galvanizes those affected into a more positive and decisive response than do potential movements to improved states. The same sort of asymmetry which Duesenberry postulated for consumer units confronted with the need to adjust to a downward revision in their incomes may hold for decision makers who control the allocation of resources for exploring the technological horizon.[58] Such asymmetrical behavior may possibly be treated more appropriately within a "satisficing" model of entrepreneurial behavior and response, where alternative technologies are explored only when a firm's profit position falls below some minimum acceptable level.[59] In any case, it is clear that threats to an established position have often served as powerful inducements to technical change.

Finally, an important common denominator running through these historical examples is the persistence with which firms attack what, at any given time, they regard as the most restrictive constraint on their operations. This suggests that it may be possible to formulate a microeconomic approach to technical change in terms of a bottle-neck analysis. If we would like to understand the kinds of problems to which technically competent personnel are likely to devote their attention, we must come to grips with their inevitable preoccupation with day-to-day problems posed by the existing technology. We might here invoke what March and Simon call "Gresham's Law of planning" which, succinctly stated, amounts to the proposition that "Daily routine drives out planning."[60] If we pay more attention to the cues thrown out by this daily routine, we may gain a clearer understanding of the process of technical change.

7

Karl Marx on
the economic role of science*

It is not the articles made, but how they are made, and by what instruments, that enables us to distinguish different economical epochs. [Marx 1906, p. 200]

The purpose of this paper is to examine certain aspects of Marx's treatment of rising resource productivity and technological change under capitalism. Many of the most interesting aspects of Marx's treatment of technological change have been ignored, perhaps because of the strong polemical orientation which readers from all shades of the political spectrum seem to bring to their reading of Marx. As a result, much has been written about the impact of the machine upon the worker and his family, the phenomenon of alienation, the relationship between technological change, real wages, employment, etc. At the same time, a great deal of what Marx had to say concerning some 300 years of European capitalist development has received relatively little attention. This applies to his views dealing with the complex interrelations between science, technology, and economic development.

It is a well-known feature of the Marxian analysis of capitalism that Marx views the system as bringing about unprecedented increases in human productivity and in man's mastery over nature. Marx and Engels told their readers, in *The Communist Manifesto*, that "the bourgeoisie, during its rule of scarce one hundred years, has created more massive and more colossal productive forces than have all preceding generations together. Subjection of Nature's forces to man, machinery, application of chemistry to industry and agriculture, steam-navigation, railways, electric telegraphs, clearing of whole

*The author is grateful to Professors Stanley Engerman and Eugene Smolensky for critical comments on an earlier draft.

126

continents for cultivation, canalisation of rivers, whole populations conjured out of the ground - what earlier century had even a presentiment that such productive forces slumbered in the lap of social labour?" (Marx and Engels 1951, 1:37). No single question, therefore, would seem to be more important to the whole Marxian analysis of capitalist development than the question: Why is capitalism such an immensely productive system by comparison with all earlier forms of economic organization? The question, obviously, has been put before, and certain portions of Marx's answer are in fact abundantly plain. In particular, the social and economic structure of capitalism is one which creates enormous incentives for the genera-tion of technological change. Marx and Engels insist that the bourgeoisie is unique as a ruling class because, unlike all earlier ruling classes whose economic interests were indissolubly linked to the maintenance of the status quo, the very essence of bourgeois rule is technological dynamism.[1] Capitalism generates unique incentives for the introduction of new, cost-reducing technologies.

The question which I am particularly interested in examining is the role which is played, within the Marxian framework, by science and scientific progress in the dynamic growth of capitalism. For surely the growth in resource productivity can never have been *solely* a function of the development of capitalist institutions. It is easy to see the existence of such institutions as a necessary condition but hardly as a sufficient condition for such growth. Surely the technological vitality of an emergent capitalism was closely linked up with the state of scientific knowledge and with industry's capacity to exploit such knowledge.

Marx's (and Engels's) position, briefly stated, is to affirm that science is, indeed, a fundamental factor accounting for the growth in resource productivity and man's enlarged capacity to manipulate his natural environment for the attainment of human purposes. How-ever, the statement requires two immediate and highly significant qualifications, which will constitute our major concern in this paper: (1) science does not, according to Marx, function in history as an independent variable; and (2) science has come to play a critical role as a systematic contributor to increasing productivity only at a very recent (from Marx's perspective) point in history. The ability of science to perform this role had necessarily to await the fulfillment of certain objective conditions. What these conditions were has not been understood adequately.

I

Marx's treatment of scientific progress is consistent with his broader historical materialism. Just as the economic sphere and the requirements of the productive process shape man's political and social institutions, so do they also shape his scientific activity at all stages of history. Science does not grow or develop in response to forces internal to science or the scientific community. It is not an autonomous sphere of human activity. Rather, science needs to be understood as a social activity which is responsive to economic forces. It is man's changing needs as they become articulated in the sphere of production which determine the direction of scientific progress. Indeed, this is generally true of all human problem-solving activity, of which science is a part. As Marx states in the introduction to his *Critique of Political Economy:* "Mankind always takes up only such problems as it can solve; since, looking at the matter more closely, we will always find the problem itself arises only when the material conditions necessary for its solution already exist or are at least in the process of formulation" (Marx 1904, pp. 12-13).

Marx views specific scientific disciplines as developing in response to problems arising in the sphere of production. The materialistic conception of history and society involves the rejection of the notion that man's intellectual pursuits can be accorded a status independent of material concerns. It emphasizes the necessity of systematically relating the realm of thinking and ideas to man's material concerns. Thus, the scientific enterprise itself needs to be examined in that perspective. "Feuerbach speaks in particular of the perception of natural science; he mentions secrets which are disclosed only to the eye of the physicist and chemist: but where would natural science be without industry and commerce? Even this 'pure' natural science is provided with an aim, as with its material, only through trade and industry, through the sensuous activity of men" (Marx and Engels 1947, p. 36). Egyptian astronomy had developed out of the compelling need to predict the rise and fall of the Nile, upon which Egyptian agriculture was so vitally dependent (Marx 1906, p. 564, n. 1). The increasing (if still "sporadic") resort to machinery in the seventeenth century was, says Marx, "of the greatest importance, because it supplied the great mathematicians of that time with a practical basis and stimulant to the creation of the science of mechanics."[2] The difficulties encountered with gearing as waterpower was being harnessed to larger millstones was "one of the circumstances that led to a more accurate investigation of the laws of friction."[3]

These themes are repeated by Engels, who asserts that "from the very beginning the origin and development of the sciences has been determined by production."[4] In accounting for the rise of science during the Renaissance, his first explanation again drew upon the requirements of industry.

If, after the dark night of the Middle Ages was over, the sciences suddenly arose anew with undreamt-of force, developing at a miraculous rate, once again we owe this miracle to production. In the first place, following the crusades, industry developed enormously and brought to light a quantity of new mechanical (weaving, clock-making, milling), chemical (dyeing, metallurgy, alcohol), and physical (spectacles) facts, and this not only gave enormous material for observation, but also itself provided quite other means for experimenting than previously existed, and allowed the construction of *new* instruments; it can be said that really systematic experimental science now became possible for the first time.[5]

Moreover, in a letter written in 1895, Engels stated: "If, as you say, technique largely depends on the state of science, science depends far more still on the *state* and the *requirements* of technique. If society has a technical need, that helps science forward more than ten universities. The whole of hydrostatics (Torricelli, etc.) was called forth by the necessity for regulating the mountain streams of Italy in the sixteenth and seventeenth centuries. We have only known about electricity since its technical applicability was discovered" (Marx and Engels 1951, 2: 457, Letter from Engels to H. Starkenburg, January 25, 1895; emphasis Engels's).

This statement is probably the most explicit and direct assertion in the writings of Marx and Engels that factors affecting the demand for science are overwhelmingly more important than factors affecting its supply. Scientific knowledge is acquired when a social need for that knowledge has been established. Science is, however, not an initiating force in the dynamics of social change. Developments in this sphere are a response to forces originating elsewhere. Thus, Marx and Engels appear to be presenting a purely demand-determined explanation of the social role of science. Scientific enterprise supplies that which industry demands, and therefore the changing direction of the thrust of science needs to be understood in terms of the changing requirements of industry.

II

In this section I will argue that, while the demand-oriented component of the argument just presented is indeed a major part of the Marxian view, there are also vital but less conspicuous elements in Marx's argument which have been ignored. Without these additional and more neglected elements one cannot explain a central thesis which emerges out of Marx's view: namely, that it is only at a particular time in human history that science is enlisted in a crucial way in the productive process. It is only at a very recent point in history, Marx argues, that the marriage of science and industry occurs. Moreover, this marriage does *not* coincide with the historical emergence of capitalism. In fact, Marx is quite explicit that the union of science and industry comes only centuries after the arrival of modern capitalism and the emergence of sophisticated bodies of theoretical science. If arguments based upon the existence of capitalist incentives and demand forces generally were a sufficient explanation, the full-scale industrial exploitation of science would have come at a much earlier stage in Western history. But it did not. Why?

Stripped to its essentials, Marx's answer is that the handicraft and manufacturing stages of production lacked the technological basis which would *permit* the application of scientific knowledge to the solution of problems of industrial production.[6] This essential technological basis emerged only with modern industry. The immense and growing productivity of nineteenth century British industry was really, in Marx's view, the resultant of three converging sets of forces: (1) the unique incentive system and capacity for accumulation provided by capitalist institutions, (2) the availability of bodies of scientific knowledge[7] which were directly relevant for problem-solving activities in industry, and (3) a technology possessing certain special characteristics. It is this last category which is least understood and to which we therefore now turn.

Historically, capitalist relationships were introduced in an unobtrusive way, by the mere quantitative expansion in the number of wage-laborers employed by an individual owner of capital (Marx 1906, p. 367). The independent handicraftsman, operating with a few journeymen and apprentices, gradually shifted into the role of a capitalist as his relationship with these men assumed the form of a permanent system of wage payments and as the number of such laborers increased.[8] The system of manufacture, therefore, while introducing social relationships drastically different from the handicraft system of the medieval guilds which preceded it,[9] initially employed the same technology.[10]

From Marx's mid-nineteenth-century vantage point, the system of manufacture had actually been the dominant one throughout most of the history of capitalism - from "roughly speaking . . . the middle of the 16th to the last third of the 18th century" (Marx 1906, p. 369; see also p. 787). Manufacture involved a significant regrouping of workers and a redefinition of the responsibilities of each. Whereas a medieval handicraftsman would himself perform a succession of operations upon a product, the manufacturing system divided up the operation into a succession of steps, each one of which was allocated to a separate workman.[11]

The essence of the manufacturing system, therefore, is a growing specialization on the part of the individual worker. While this in turn has psychological and social consequences of the greatest importance for the worker with which Marx was very much concerned,[12] it continued to share with the earlier handicraft system an essential feature. That is to say, although the product now passed through a succession of hands, and although this reorganization raised the productivity of labor, it nevertheless perpetuated the industrial system's reliance upon human skills and capacities.[13] Whereas the critical skill was formerly that of the guild craftsman, it is now the unremitting repetition of a narrowly defined activity on the part of the detail laborer. More precisely, the productive process now pressed against the constraints imposed by the limited strength, speed, precision, and, indeed, the limited number of limbs, of the human animal.

So long as the worker continues to occupy strategic places in the productive process, that process is limited by all of his human frailties. And, of course, the individual capitalist is, in many ways, continually pressing the worker against those limits. But the point which Marx is making here is of much broader significance: The application of science to the productive process involves dealing with impersonal laws of nature and freeing itself from all dependence upon the organic. It involves calculations concerning the behavior of natural phenomena. It involves the exploitation of reliable physical relationships which have been established by scientific disciplines. It involves a degree of predictability of a purely objective sort, from which the uncertainties and subjectivities of human behavior have been systematically excluded. Science, in short, can only incorporate its findings in impersonal machinery. It cannot be incorporated in human beings with their individual volitions, idiosyncracies, and refractory temperaments. The manufacturing period shared with the earlier handicraft system the essential feature that it was a tool-using economy where the tools were subject to human manipulation and guidance. It is this element of human control, the continued reliance upon the limited range of activities of the human hand, and not the nature of the power source, Marx insists, which is decisive in distinguishing a machine from a tool.

The machine proper is . . . a mechanism that, after being set in motion, performs with its tools the same operations that were formerly done by the workman with similar tools. Whether the motive power is derived from man, or from some other machine, makes no difference in this respect. From the moment that the tool proper is taken from man, and fitted into a mechanism, a machine takes the place of a mere implement. The difference strikes one at once, even in those cases where man himself continues to be the prime mover. The number of implements that he himself can use simultaneously, is limited by the number of his own natural instruments of production, by the number of his bodily organs. . . . The number of tools that a machine can bring into play simultaneously, is from the very first emancipated from the organic limits that hedge in the tools of a handicraftsman.[14]

III

What, then, is the distinctive technological feature of modern industry? It is that, for the first time, the design of the productive process is carried out on a basis where the characteristics of the worker and his physical endowment are no longer central to the organization and arrangement of capital. Rather, capital is being designed in accordance with a completely different logic, a logic which explicitly incorporates principles of science and engineering.[15] The subjectivity of a technology adapted, out of necessity, to the capacities (or, better, the debilities) of the worker is rejected in favor of the objectivity of machinery which has been designed in accordance with its own laws and the laws of science.

In Manufacture it is the workmen who, with their manual implements, must, either singly or in groups, carry on each particular detail process. If, on the one hand, the workman becomes adapted to the process, on the other, the process was previously made suitable to the workman. This subjective principle of the division of labour no longer exists in production by machinery, Here, the process as a whole is examined objectively, in itself, that is to say, without regard to the question of its execution by human hands, it is analysed into its constituent phases; and the problem, how to execute each detail process, and bind them all into a whole, is solved by the aid of machines, chemistry, etc.[16]

The shift from the hand-operated to the machine-operated process is a momentous one, for the simple reason that machine processes are susceptible to continuous and indefinite improvement, whereas hand processes are not.[17] The factory system makes possible the virtual routinization of productivity improvement.[18] By breaking down the productive process into objectively identifiable component parts, it creates a structure of activities which is readily amenable to rigorous analysis. "The principle, carried out in the factory system, of analysing the process of production into its constituent phases, and of solving the problems thus proposed by the application of mechanics, of chemistry, and of the whole range of the natural sciences, becomes the determining principle everywhere."[19] Thus, historical development has brought technology to a point where it has become, for the first time, an object of scientific analysis and improvement.

A characteristic feature is, that, even down into the eighteenth century, the different trades were called "mysteries" (mystères); into their secrets none but those duly initiated could penetrate. Modern Industry rent the veil that concealed from men their own social process of production, and that turned the various spontaneously divided branches of production into so many riddles, not only to outsiders, but even to the initiated. The principle which it pursued, of resolving each process into its constituent movements, without regard to their possible execution by the hand of man, created the new modern science of technology. The varied, apparently unconnected, and petrified forms of the industrial processes now resolved themselves into so many conscious and systematic applications of natural science to the attainment of given useful effects. Technology also discovered the few main fundamental forms of motion, which, despite the diversity of the instruments used, are necessarily taken by every productive action of the human body; just as the science of mechanics sees in the most complicated machinery nothing but the continual repetition of the simple mechanical powers.

Modern Industry never looks upon and treats the existing form of a process as final. The technical basis of that industry is therefore revolutionary, while all earlier modes of production were essentially conservative.[20]

In its most advanced form, therefore, "modern industry . . . makes science a productive force distinct from labour and presses it into the service of capital" (Marx 1906, p. 397).

Before capitalism could reach this stage of self-sustaining technological dynamism, however, another critical condition needed to be fulfilled. Machinery cannot fully liberate the economy from the output ceiling imposed by dependence upon human skills and capacities so long as these things continue to be essential in the production of the machines themselves. In the early stages of modern industry, machines were, inevitably, produced by direct reliance upon human skills and capacities. The manufacturing system responded to the demand for the new inventions by creating new worker specializations.[21] While this sufficed in the early stages of the development of modern industry, improvements in machine design and performance and increasing size eventually came up increasingly against the limitations of the human machine maker.

Modern Industry was crippled in its complete development, so long as its characteristic instrument of production, the machine, owed its existence to personal strength and personal skill, and depended on the muscular development, the keenness of sight, and the cunning of hand, with which the detail workmen in manufactures and the manual labourers in handicrafts, wielded their dwarfish implements. Thus, apart from the dearness of the machines made in this way, a circumstance that is ever present to the mind of the capitalist, the expansion of industries carried on by means of machinery, and the invasion by machinery of fresh branches of production, were dependent on the growth of a class of workmen, who, owing to the almost artistic nature of their employment, could increase their numbers only gradually, and not by leaps and bounds. But besides this, at a certain stage of its development, Modern Industry became technologically incompatible with the basis furnished for it by handicraft and Manufacture. The increasing size of the prime movers, of the transmitting mechanism, and of the machines proper, the greater complication, multiformity and regularity of the details of these machines, as they more and more departed from the model of those originally made by manual labour, and acquired a form, untrammelled except by the conditions under which they worked, the perfecting of the automatic system, and the use, every day more unavoidable, of a more refractory material, such as iron instead of wood - the solution of all these problems, which sprang up by the force of circumstances, everywhere met with a stumbling-block in the personal restrictions which even the collective labourer of Manufacture could not break through, except to a limited extent. Such machines as the modern hydraulic press, the modern powerloom, and the modern carding engine, could never have been furnished by Manufacture.[22]

The vital step, therefore, is the establishment of the technological conditions which would make it possible to use machinery in the construction of machines, thus bypassing the central constraint of the old manufacturing system. "Modern Industry had therefore itself to take in hand the machine, its characteristic instrument of production, and to construct machines by machines. It was not till it did this, that it built up for itself a fitting technical foundation, and stood on its own feet. Machinery, simultaneously with the increasing use of it, in the first decades of this century, appropriated, by degrees, the fabrication of machines proper."[23] Marx singles out, not only the new power sources which offered gigantic quantities of energy subject to careful human regulation, but also that indispensable addition to the equipment at the disposal of the machine maker, the slide rest. This simple but ingenious device of Henry Maudsley replaces, as Marx perceptively notes, not any particular tool, "but the hand itself" (Marx 1906, p. 408). In this sense it is a strategic technological breakthrough, fully comparable in importance to the steam engine.

The improvements in the machinery-producing sector constitute a quantum leap in the technological arsenal at man's disposal. They make it possible to escape the physical limitations of a tool-using culture. They do this, ironically as Marx points out, by providing machines which reproduce the actions of a hand-operated tool, but do so on a "cyclopean scale."[24]

IV

Thus, I would interpret the Marxian position to be that it is the changing requirements of industry and the altering perception of economic needs which provide the stimulus to the *pursuit* of specific forms of scientific knowledge. But I would also conclude that the Marxian position cannot be adequately described as a demand-induced approach without doing a severe injustice to the subtlety of Marx's historical analysis.[25] For the ability to apply science to the productive sphere turns upon industry's changing *capacity* to utilize such knowledge, a capacity which Marx explicitly recognizes has been subjected to great changes over the course of recent history. Indeed, Marx himself, as I have tried to establish, devoted considerable effort to the elucidation of the factors which have shaped society's altering capacity to absorb the fruits of scientific knowledge.[26]

Nor did Marx argue that the historical sequence in which scientific disciplines actually developed was also directly determined by economic needs. For example, in discussing the relative pace of development in industry and agriculture, he states that productivity growth in agriculture had, historically, to await the development of certain scientific disciplines, and therefore came later, whereas industry progressed more rapidly than agriculture at least in large part because the scientific knowledge upon which industry relied had developed earlier. "Mechanics, the really scientific basis of large-scale industry, had reached a certain degree of perfection during the eighteenth century. The development of chemistry, geology and physiology, the sciences that *directly* form the specific basis of agriculture rather than of industry, does not take place till the nineteenth century and especially the later decades."[27]

This strongly suggests at least some degree of independence and autonomy on the part of science in shaping the sequence of industrial change, in spite of the fact that, as we saw earlier, Marx and Engels usually emphasize the cause-effect relationships which run from industry to science. If the growth in agricultural productivity is dependent upon progress in specific subdisciplines of science, and if the existence of profitable commercial opportunities in agriculture cannot "induce" the production of the requisite knowledge, then factors internal to the realm of science must be conceded to play a role independent of economic needs.

Moreover, it is especially curious to find that Engels is content to state, as quoted earlier, that "from the very beginning the origin and development of the sciences has been determined by production" (Engels 1954, p. 247). For Engels himself, in the *Dialectics of Nature,* had also presented a classification scheme for the sciences which emphasized a hierarchy of increasing complexity based upon the forms of motion of the matter being analyzed. Increasing complexity is identified with the movement from the inorganic to the organic, from mechanics to physics to chemistry to biology. [28] Engels even goes so far as to speak of an "inherent sequence," [29] which he clearly believes has structured the *historical* sequence in which nature's secrets have been progressively uncovered. But, if one accepts this intuitively plausible view, then surely there is much more to "the origin and development of the sciences" than can be accounted for by the specific demands being generated in the productive sphere. Surely the historical fact that the biological sciences came to the assistance of agriculture long after the mechanical sciences were being utilized by industry is a sequence originating, not in economic needs, but in the differing degrees of complexity of these scientific disciplines. Engels's formulations particularly seem to overemphasize the importance of demand-induced incentives to the neglect of supply side considerations, even though he is obviously sensitive to these supply variables in other contexts.

In Engels's defense one must recall, of course, the unfinished, indeed often merely fragmentary condition of his *Dialectics of Nature*.[30] It is entirely possible that, had he the opportunity, he would have resolved these apparent inconsistencies. But it is expecting far too much to look to either Marx or Engels for the resolution of these deep and thorny problems. We are still, today, a long way from being able to incorporate the history of science in an orderly manner into our understanding of the economic development of the Western world.[31]

Conclusion

There are several possible meanings which can be attached to the statement that "the origin and development of the sciences has been determined by production."

1. Science depends upon industry for financial support.

2. The expectation of high financial returns is what motivates individuals (and society) to pursue a particular scientific problem.

3. The needs of industry serve as a powerful agent in calling attention to certain problems (Pasteur's studies of fermentation and silkworm epidemics).

4. The normal pursuit of productive activities throws up physical evidence of great importance to certain disciplines (metallurgy and chemistry, canal building and geology). As a result, industrial activities have, as a byproduct of their operation, provided the flow of raw observations upon which sciences have built and generalized.

5. The history of individual sciences, including an account of their varying rates of progress at different periods in history, can be adequately provided by an understanding of the changing economic needs of society.

I believe that Marx and Engels subscribed to propositions 1-4 without qualification. I believe they often *sounded* as if they subscribed to the fifth proposition. However, I think the preceding

138   The generation of new technologies

discussion has established that they subscribed to the fifth proposition only subject to certain qualifications - qualifications which strike me as being, collectively, more interesting than the original proposition.

References

Bernal, J.D. *Science and Industry in the Nineteenth Century.* London: Routledge & Kegan Paul, 1953.

_____ . *Science in History.* 4 vols. Cambridge, Mass.: M.I.T. Press, 1971.

Bober, M.M. *Karl Marx's Interpretation of History.* New York: Norton, 1965.

Engels, Frederick. *Socialism: Utopian and Scientific.* Chicago: Kerr, 1910.

_____ . *Herr Eugen Duhring's Revolution in Science.* New York: International, 1939.

_____ . *The Dialectics of Nature.* Moscow: Foreign Languages, 1954.

Marx, Karl. *A Contribution to the Critique of Political Economy.* Chicago: Kerr, 1904.

_____ . *Capital.* Vol. 1. Chicago: Kerr, 1906.

_____ . *Capital.* Vol. 3. Moscow: Foreign Languages, 1959.

_____ . *Theories of Surplus Value.* 3 pts. Moscow: Progress, 1963.

_____ . *The Poverty of Philosophy.* Moscow: Foreign Languages, n.d.

Marx, Karl, and Engels, Frederick. *The German Ideology.* New York: International, 1947.

_____ . *Selected Works.* 2 vols. Moscow: Foreign Languages, 1951.

Sweezy, Paul. "Karl Marx and the Industrial Revolution." *Events, Ideology and Economic Theory,* edited by Robert Eagly. Detroit: Wayne State Univ. Press, 1968.

Zvorikine, A. "Technology and the Laws of its Development." *The Technological Order,* edited by Carl F. Stover. Detroit: Wayne State Univ. Press, 1963.

Part 3

Diffusion and adaptation of technology

8

Capital goods, technology, and economic growth*

One of the things which we all "know" about American economic history is that the relative scarcity of labor in the United States has led to the development of our well-known, much-admired labor-saving technology. But why, in underdeveloped countries, with abundant supplies of labor and scarce capital, has not the scarcity of capital led to the development of capital-saving techniques?

It is at once apparent that there is a confusion with respect to what we all "know" about American economic history - or, at least, that there is a highly important distinction which is typically glossed over. Scarcity of labor has, in the United States, resulted in the adoption of labor-saving techniques, just as, in underdeveloped countries, the abundance of labour has led to the adoption of labor-intensive, capital-saving techniques. So far there is complete symmetry in our treatment. Differences with respect to factor endowment, and therefore with respect to factor prices, dictate different optimum techniques along an existing production function.

What is frequently asserted with respect to American economic development is an additional and much more important proposition, namely that labor scarcity has in fact led to the development of a new, labor-saving technology, to *shifts* in the production function, and not merely to movements along an existing production function in accordance with factor endowment and prices. But, on reflection, it appears reasonable to ask why this should be the case only when it is labor which is scarce, and not capital. If the relative scarcity of a

*The author has benefited from helpful comments by his colleagues, Edward Ames and June Flanders, and from a reading of H. J. Habakkuk's recent book, *American and British Technology in the 19th Century,* Cambridge University Press, 1962, which raises some of the problems discussed in this paper, although in a different context. Whereas Habakkuk is concerned with the British and American economies in the nineteenth century, we shall be primarily interested in the problems of underdeveloped countries.

Our special interest in this paper is in the production of machinery. Throughout, therefore, "capital goods" should be taken to refer to producers' durable equipment and to exclude the output of the construction goods industry.

factor of production has been responsible for innovations which economize on the use of that factor in the United States, why have poor countries not had similar experiences with respect to their scarcest factor of production? For the common observation is not so much that technical change in underdeveloped countries has any particular sort of bias, but rather that it is entirely or virtually nonexistent.

Here, then, is the lack of symmetry between factor endowment and technical change which calls for explanation. It is a generally accepted proposition that the scarcity of a particular factor of production - labor - has led to a dynamic technology in the United States. However, in economies where another factor of production has been relatively scarce - capital - the result has been technological stagnation. Why didn't the underdeveloped countries develop their own - capital-saving - technology? If the following explanation, which Hicks offers to account for the labor-saving bias of Western technology, has any validity, why has there not been a parallel capital-saving path of technological innovation in the poor, capital-scarce countries of the underdeveloped world?

The real reason for the predominance of labour-saving inventions is surely that which was hinted at in our discussion of substitution. A change in the relative prices of the factors of production is itself a spur to invention, and to invention of a particular kind - directed to economising the use of a factor which has become relatively expensive. The general tendency to a more rapid increase of capital than labour which has marked European history during the last few centuries has naturally provided a stimulus to labour-saving invention.[1]

It is suggested here that an important aspect of the nature and impact of technological change is illuminated if we focus attention more explicitly on the role of the capital goods sector and, more particularly, on the relationship between the capital goods and consumer goods sectors as technological change occurs. The capital goods sector obviously plays a crucial role in the process of technological innovation. All innovations - whether they involve the introduction of a new product or provide a cheaper way of producing an existing product - require that the capital goods sector shall produce a new product (machine) according to certain specifications. We may usefully look upon the capital goods sector as one which is, in effect, primarily engaged in custom work. That is, firms in this industry are typically highly specialized in the sense that each firm produces a relatively narrow range of output (at least when the aggregate demand for capital goods is sufficiently large) in response to specifications laid down by a wide range of customers in the consumer goods and other capital goods industries.

The efficient operation of this sector, in turn, is dependent upon the achievement of a sufficiently high level of demand for capital goods. We revert here to Smith's time-honored dictum (the full implications of which are not yet completely appreciated) that "the division of labour is limited by the extent of the market." However, it is extremely important to realize that the strictures imposed by limited market size are much more serious in the case of capital goods than in the case of consumer goods. An economy may be sufficiently large to make possible all the economies of specialization available to the producers of consumer goods without being nearly large enough to generate optimum conditions for the producers of capital goods. There exists, in other words, a discontinuity with respect to minimum size requirements between the capital goods and consumer goods industries.[2]

It should also be pointed out that economists, in attempting to account for improvements in efficiency, have been far too much preoccupied with bigness (economies of scale) at the firm level, and have devoted insufficient attention to changing patterns of specialization within industries (or sectors) which do not involve bigness in the size of the individual firm. The importance of the growth in markets is not necessarily bigness but rather an increased division of labor among firms in the specific sense of a narrowing down of the product range and the ability to concentrate on a limited range of products possessing certain specified properties, performing specific functions, and meeting highly specialized requirements. This is strikingly evident in the machine-tool industry which has never attained to bigness at the individual firm level, has consisted of large numbers of firms, and has been - since the last few decades of the nineteenth century - highly specialized by firm in the sense defined above.[3]

For this reason we wish to distinguish between capital goods industries, whose output constitutes replacement of or additions to the economy's stock of physical capital, and producer goods industries, an imprecise term generally used to designate not only the capital goods industries but also all intermediate goods which are used as inputs by firms.[4] We suggest that capital goods producers (machinery producers), who typically produce a heterogeneous output, usually enjoy economies of specialization, while many producers of intermediate goods, whose output is typically homogeneous (chemicals, iron and steel, metals generally) enjoy economies of scale. Economies of scale, then, is a more comprehensive concept: firms which achieve economies of scale are also specialized, but firms may also achieve economies of specialization of a sort which do not involve significant economies of scale. The economies of specialization referred to derive not from the production of a completely homogeneous product but from the concentration upon a relatively narrow (heterogeneous) product range which in turn requires a relatively homogeneous collection of resources in their production. The point is that the typical machine-producing firm produces small batches of output drawn up to specifications reflecting the unique requirements of the user, but each such batch differs only slightly, and all draw upon a homogeneous collection of resources - each firm possessing plant facilities, designing abilities, and other technological "know-how" which is geared to the effective solution of a very limited range of production problems. Thus, in the American economy, not only do individual machine-tool firms concentrate on a very limited range of tools (single firms producing only milling machines or boring machines or turret lathes) but frequently also they produce only various modifications of a single basic machine type for firms in a single industry. The truly mass-production industries, such as automobiles, are served by an extraordinary complex of relatively small specialist firms, each constructing very limited numbers and ranges of tooling devices for specific mass-production processes - dies, jigs, fixtures, gauges, moulds, etc.[5] The obvious advantage of this arrangement is that there is an important learning process involved in machine production, and a high degree of specialization is conducive not only to an effective learning process but to an effective *application* of that which is learned. This highly developed facility in the designing and production of specialized machinery is, perhaps, the most important single characteristic of a well-organized capital goods industry and constitutes an external economy of enormous importance to other sectors of the economy. But for such a pattern of specialization among firms to develop, capital goods producers must be confronted with an extremely large demand for their output.

Thus, even in a domestic market as large as the British a recent study has concluded that the market was insufficiently large to generate an environment conducive to the emergence of technically-progressive machinery-producing specialist firms.

In certain industries the British market has not been big enough to encourage the growth of specialist producers of equipment - who themselves might have created new possibilities of progress. We have found examples of this relating to paper, bread, rubber, plastics, fine chemicals, aircraft, and scientific instruments. There are some cases where one or a few specialist producers exist, but progress in design is slow: in a bigger market there would be more specialist producers, more competition in design, and a better chance that good designs would be produced.[6]

Now, if we really take seriously what we say about the beneficent forces of competition, the situation confronting underdeveloped economies must be dreary indeed, since they may be unable, in certain cases, because of limited demand, to support even a single modern firm, to say nothing of a competitive industry. The competitive pressures which normally act as a spur to innovation and change and which compel individual firms, on pain of extinction, either to explore new techniques or to adopt superior techniques which have been developed elsewhere, will be virtually nonexistent. If, in J.M. Clark's phrase, we conceive of "competition as a dynamic process," the sources of such dynamism may unfortunately not be available in important sectors of underdeveloped economies. The absence of salubrious competitive pressures thus reinforces the handicap resulting from the inability to achieve an optimum degree of specialization by firm.

Within this context a further point about our own industrial development may be made, pertaining to the especially crucial role of the "transportation revolution" in making possible the growth of our capital goods sector. It was surely not an historical accident that specialized machine producers emerged on the national scene in the thirty or forty years after 1840 - coinciding exactly with the laying down of a national railway network. Until roughly 1840 machinery production was not only relatively unspecialized - each producer typically undertaking a wide range of output - but it was also, because of the very high cost of transporting machinery, a highly localized operation - each producer typically producing for a very limited geographic radius. The growing specialization in machine production after 1840, the emergence of large numbers of producers each of whom typically concentrated on a very narrow range of machines, was closely linked up with the transportation improvements and consequent reduction in freight costs during the period. Highly specialized machinery could not be produced for a severely restricted geographic market for the reason already discussed with respect to underdeveloped countries - insufficiency of market demand. The growth in the size of the market to individual producers of machines, resulting from the reduction in freight costs, was therefore peculiarly important to the process of specialization in the production of capital goods.

We are now in a position to examine the nature of the handicap which confronts underdeveloped economies and also to appreciate more fully the role of the capital goods sector in the process of economic growth. Economists who have been concerned in recent years with the prevailing technology of underdeveloped countries[7] have concentrated their attentions primarily on the question why underdeveloped countries have not developed a modern labor-intensive technology appropriate to their factor endowment. This is, of course, a highly important question.[8] But the manner in which the question is formulated is such that it bypasses an important part of the factor adjustment process for the economy as a whole - specifically the unique role played by the capital goods sector. What is important is not just the development of capital-saving innovations - although this is certainly very important. What is also important is improving the efficiency with which the existing types of capital goods are produced. Underdeveloped countries have been deficient on both accounts, but the latter deficiency has received practically no attention. They have therefore missed a major source of capital-saving for the economy as a whole.

Historically, a major source of capital-saving innovation has been improvements in the efficiency of capital goods production. The important analytical point is that any cost reduction in the capital goods sector - whether it is immediately labor-saving or capital-saving in its factor-proportion bias - is a capital-saving innovation to the economy as a whole. Recognition of the necessarily capital-saving nature of innovations in the capital goods sector goes back at least to vol. iii of Marx's *Capital* (chaps. iv and v), but has appeared in the theoretical literature dealing with technical progress only very recently.[9]

Many of the major innovations in Western technology have emerged in the capital goods sector of the economy. But underdeveloped countries with little or no organized domestic capital goods sector simply have not had the opportunity to make capital-saving innovations because they have not had the capital goods industry necessary for them. Under these circumstances, such countries have typically imported their capital goods from abroad, but this has meant that they have not developed the technological base of skills, knowledge, facilities, and organization upon which further technical progress so largely depends.

A capital-saving stage is an inevitable but later stage of the sequence by which an industrial economy accommodates itself to an innovation. When a new machine is introduced there exists, by definition, no established system of organization to produce the machine. As Marx stated: "There were mules and steam-engines before there were any labourers whose exclusive occupation it was to make mules and steam-engines; just as men wore clothes before there were such people as tailors."[10]   Currently developed economies, then, have gone through the following historical sequence: with the growth in the demand for machinery the capital goods industry became gradually more and more highly specialized and subdivided in order to undertake the production of machines, the cost of producing machines was thereby sharply reduced, and as a result capital-saving for the economy as a whole was achieved.

Some remarks by Stigler in a somewhat different context may be cited on this point:

If one considers the full life of industries, the dominance of vertical disintegration is surely to be expected. Young industries are often strangers to the established economic system. They require new kinds or qualities of materials and hence make their own; they must overcome technical problems in the use of their products and cannot wait for potential users to overcome them; they must persuade customers to abandon other commodities and find no specialized merchants to undertake this task. These young industries must design their specialized equipment and often manufacture it, and they must undertake to recruit (historically, often to import) skilled labor. When the industry has attained a certain size and prospects, many of these tasks are sufficiently important to be turned over to specialists.[11]

If then, we consider the process of innovation over time, there is a high probability that labor-saving innovations are likely to be followed by capital-saving innovations, and this may provide an important key to understanding the dynamics of technical change. If we start with a new innovation which is labor-saving and capital-intensive, the new machine itself will, almost inevitably, be produced inefficiently in the early stages. This is not only because it is experimental or because there are "bugs" in the early stages, although this is often certainly the case, but also because the capital goods (machinery-producing) sector is, itself, not tooled-up or equipped for producing the machine at low per unit cost. The introduction of the new product requires a process of adaptation and adjustment in the capital goods industry which did not initially exist. There is, as it were, a period of technical gestation during which time the resources of the capital goods industry accommodate themselves to the specific requirements of the new product. This entire "breaking-in" process is a capital-saving process for the economy as a whole, and its final result is an upward shift to a new production function.[12]

It may well be that, historically, this has been the most important path which capital-saving has taken. If so, it carries the implication that it may be very hard for an underdeveloped country to start right off upon a capital-saving path. Perhaps it is necessary first to take the labor-saving path - at least until one has built up a substantial stock of capital and a capital goods industry catering for a market which exceeds some critical minimum size.

Here we may point to a somewhat different reason from the one typically emphasized for the importance of accelerating the rate of domestic capital formation in achieving economic development. A high rate of capital accumulation may be crucial in that it is a precondition for the growth in the absolute size of the capital goods sector. Such an enlargement is essential if this sector is to achieve the minimum size which it requires in order to achieve the high degree of specialization which is so critical to its effective operation. In this sense it may be possible to argue that a high rate of investment is an important determinant of rapid technological change. If this is so, then underdeveloped countries are doubly handicapped: low rates of capital formation perpetuate low capital/labor ratios and therefore low levels of labor productivity; and the failure to achieve a well-developed capital goods sector means a failure to provide the basis of technical skills and knowledge necessary to the development of capital-saving techniques and therefore a reinforcement of their state of technical backwardness. The kinds of skills which are needed to develop a technology more appropriate to their own peculiar factor endowments are, themselves, undeveloped.

In this sense - the absence of the appropriate pool of mechanical skills, knowledge, and facilities - there may be an important asymmetry, reflecting factor endowment. *Not* with respect to the selection among existing alternatives (labor-abundant economies select labor-intensive alternatives, capital-abundant economies select capital-intensive ones) but with respect to the preconditions for technical innovation and progress. Labor-scarce economies are likely to generate labor-saving technical progress, because such economies are likely to develop the pool of mechanical skills necessary for innovation. A critical aspect of the labor-saving path, then, is the "production" of skills and familiarity with technical processes upon the part of the human agent which are essential preconditions for technical change of any sort. However, labor-abundant economies are not likely to generate a stream of capital-saving innovations because labor-abundant economies, largely because of the stagnation and backwardness of their capital goods industries, are not likely to provide the necessary skills and aptitudes conducive to technical progress in the first place.[13]

A closely related point may be made. Although labor scarcity does not, by itself, lead to innovation, it does lead to the adoption and utilization of techniques at the capital-intensive end of the spectrum. And this, in turn, may be expected to have important consequences. Specifically, it leads to the establishment of a sizable capital goods industry. As a result, such an economy (via economies of specialization, acquisition of knowledge and skills, and familiarity with the technology of capital goods production) may become an efficient capital goods producer and achieve all the conditions, in its capital goods sector, which are indispensable to innovation (and therefore capital-saving) in that sector.

Via this somewhat circuitous route, then, with its intermediate steps concerning the role of the capital goods sector, we arrive at a position in essential agreement with Hicks's conclusion that "the general tendency to a more rapid increase of capital than labour which has marked European history during the last few centuries has naturally provided a stimulus to labour-saving invention." But the different *route* by which this agreement has been reached - the special role of the capital goods sector - has important implications to the analysis of economic backwardness. For we have here at least a partial explanation of the perpetuation of low rates of capital formation which, it is generally agreed, is such a central feature of nondeveloping economies. Although there is general consensus that nondeveloping economies are characterized by very small annual increments to their capital stock, there is little agreement on the reasons for this deficiency. The reader of the literature is treated to a curious amalgam of economic and sociological explanations, with supposedly "irrational" preferences playing an important role.[14] Our analysis, however, suggests that, in underdeveloped countries, the investment decision is likely to be heavily weighted by an unfavourable relative price structure which acts as a serious impediment to investment activity. The investment decision, after all, involves computation of a prospective rate of return which is determined by the present price of capital goods and the anticipated future price of consumer goods. But it should be clear that the relative inefficiency of the capital goods industries in underdeveloped countries and therefore the high price of capital goods is responsible for yielding low or even negative rates of return on a wide range of prospective investments. A major handicap of underdeveloped countries, then, is located in their inability to produce investment goods at prices sufficiently low to assure a reasonable rate of return on prospective investments.[15] Reasoning symmetrically, one of the most significant propelling forces in the growth of currently high-income countries has been the technological dynamism of their capital goods industries which has maintained the marginal efficiency of capital at a high level. Empirical evidence lending support to this point has recently been adduced by Kuznets, who found that the ratio of capital goods prices (most particularly producers' durable equipment) to consumer goods prices was substantially higher in less developed countries than in the more advanced countries.[16] Kuznets's data conform with what we should expect to find on the basis of our analysis of the role of the capital goods industries and reinforce our conclusion that underdevelopment can be explained as a basically economic phenomenon.

9

Economic development and the transfer of technology: some historical perspectives

I

Back in an earlier, more naïve day, we managed to allow ourselves to believe that there was a purely technological solution - a cheap "technological fix" - to the problems of poverty and economic backwardness which beset most of the human race. In the Point Four of his 1949 inaugural address President Truman spoke optimistically of the incalculable benefits which technical assistance could bring to improving the desperate plight of the poor throughout the world. "We must embark," he said, "on a bold new program for making the benefits of our scientific advances and industrial progress available for the improvement and growth of underdeveloped areas."

Unfortunately we exaggerated from the outset what could be accomplished solely by making Western techniques available. In some measure the present mood of disillusion with foreign aid in general is due to the unrealistically high expectations which we once attached to such programs (including their effectiveness as weapons of politics and diplomacy). We now realize that there are a host of difficulties - institutional and otherwise - which hamper the successful adoption of foreign technology. I would like to contribute some observations of my own on the obstacles in the way of a successful transfer of technology. I will also, before I am done, go one pessimistic step further, and suggest that there are serious problems, not just in facilitating the transfer but in the very nature of technology which we presently have to offer and its possibly limited relevance (especially in agriculture) to the problems of poor countries. But it should be emphasized at the outset that the main thrust of this paper is to pose problems for the international transfer of technology which emerge from an examination of earlier - primarily 19th-century - mechanisms through which such transfers have taken place. I do not propose to offer authoritative conclusions or recommendations but rather - and much more modestly - to use historical experience to sharpen and to enlarge our appreciation for the difficulties which may still lie ahead.

I propose, first, to discuss the role of the capital goods industries in the 19th century as centers for the creation and diffusion of new techniques. I start from the proposition that technical change is not, and never has been, a random phenomenon. In both the United States and the United Kingdom in the 19th century, technological change became institutionalized in a very special way - that is, in the emergence of a group of specialized firms which were uniquely oriented toward the solution of certain kinds of technical problems. The rapid rate of technological change was completely inseparable from these capital goods firms. In fact, I would regard the emergence of such firms as the fundamental institutional innovation of the 19th century from the point of view of the industrialization process. I am particularly interested in the factors which affected the success and failure of these industries in the context of a developing industrial economy. In general, I shall argue that these industries performed certain critical functions which accounted for the speed of the growth process in the 19th century. But I am anxious to free myself of the charge of suggesting that poor countries today will have to tread exactly the same path. The most important reason why poor countries may *not* have to tread the same path as their industrial predecessors is precisely that industrial countries have already done so. One of the advantages of not taking the lead in economic development is that, once an objective has been reached and clearly demarcated, other and easier routes to attain that objective may become obvious. Or, to put the point a little differently, there is no reason to believe that the optimal path in the development of a new technology is the same as the optimal path for transferring and adapting that technology, once it has been developed.[1] In fact, I want to insist on this point. Economic growth has never been a process of mere replication. Although one can identify certain broad phenomena and functions common to all rapidly growing economies (changes in the composition of output and labor force, urbanization) the actual paths to development show very wide institutional variations. Certainly the latecomers on the development scene (e.g., Germany, Japan, Russia) followed very different paths and sequences than did pioneering England.[2]

Perhaps it should also be said that the policy implications of what follows are not at all clear - at least not to me. It is not my intention to press the analysis of technology transfer in the last third of the 20th century onto a 19th-century Procrustean bed. Rather, I want to draw upon the experience of the 19th and early 20th centuries as a way of suggesting what are likely to be problems of considerable interest and importance.

II

I have been very much interested in problems of the transmission of technology in the 19th century, not only from one country to another but from one industry to another. The problems involved are not entirely unrelated. Simply because transmitting technologies from advanced to underdeveloped countries presents many unique problems, we should not ignore the fact that transmitting technology from one use to another always presents certain elements of novelty.[3]

In both the United States and United Kingdom, to a very surprising degree, innovations in the area of machine technology were transmitted from existing uses to new uses by a very personal mechanism. Impersonal forms of communication - such as trade journals - were often important in generating interest but hardly ever provided enough of the highly specific information required for the successful transmission of a new productive technique.

There are remarkable similarities, in the United States and the United Kingdom, in the pattern of transmission. A dominant pattern in both countries went as follows: A firm develops a new technique or process in response to a problem in a particular industry - say firearms manufacture. It later becomes apparent that this technique is applicable to typewriters. The technique is "transferred" to the typewriter industry by the firm which developed the technique actually undertaking the production of typewriters (Remington). In other words, a very common mode of technology transfer *between industries* took place as a result of firms adding to, or switching, their product lines.

An alternative way of stating this is to say that, in the area of machine technology, firms became increasingly specialized by *process* rather than product. In both countries we find firms undertaking a combination or a sequence of products which can be understood only in terms of their common underlying technical processes. We can find many examples of firms in each country undergoing the same product sequences over time - for example, firearms, sewing machines, bicycles - or undertaking the same product combinations. If we examine historically the operation of multiproduct firms in both countries, we find that they frequently arrived at the combination through the development of a specialized skill at a particular process - for example, hydraulics (Armstrong, Tangyes, Fawcetts) - or an unusual expertise in the property of a material - for example, William Fairbairn in Manchester and William Sellers in Philadelphia both undertook activities where the structural properties of iron were critical to success - bridges, iron ships, multistory dwellings, cranes, etc. The center for the transmission of relevant knowledge and techniques from one industry to another, and for the application of known techniques to new uses, was, to a very considerable degree, the individual firm.

Where the transfer of technology involved places geographically distant from one another, the reliance upon the migration of trained personnel (at least temporarily) was very strong. This was true even within the United States. It would be possible to trace the diffusion of machine tool technology in America in the 19th century in terms of a genealogical table showing the movement of a very small number of highly skilled mechanics from the places where they received their training (apprenticeship and early "on-the-job" learning) to the firms where they eventually brought their valuable skills; often these were new firms established by the mechanic himself.[4]

I have been impressed by the extent to which the transfer of technological skills - even between two countries so apparently "close together" as Britain and the United States in the mid-19th century - was dependent upon the transfer of skilled personnel.[5] This was demonstrated by several striking episodes. For example, when the British government purchased a large quantity of American gun-making machinery in the 1850s and introduced it into the Enfield Arsenal, it was found to be absolutely essential to employ a large number of American machinists and supervisory personnel. At about the same time, Samuel Colt, impressed with the backwardness of British firearms technology as compared with his own plant in Hartford, set up a factory in London for the purpose of producing firearms with American machinery and British workers. Although Colt had been fantastically successful in the United States - where he ran the largest private armory in the world - his London plant had to be closed down, apparently because of the difficulties encountered in employing British workers with unfamiliar American machinery.[6]

There is much evidence that the transmission of industrial technology from England to the Continent in the first half of the 19th century was also heavily dependent upon the same sort of personal mechanism. David Landes has suggested, for example, that in spite of legal prohibitions until 1825, there were at least 2,000 skilled British workers on the Continent providing indispensable assistance in the adoption of the newly developed techniques.

The best of the British technicians to go abroad were usually entrepreneurs in their own right, or eventually became industrialists with the assistance of continental associates or government subventions. Many of them came to be leaders of their respective trades: one thinks of the Waddingtons (cotton), Job Dixon (machine-building), and James Jackson (steel) in France; James Cockerill (machine construction) and William Mulvany (mining) in Germany; Thomas Wilson (cotton) in Holland; Norman Douglas (cotton) and Edward Thomas (iron and engineering) in the Austrian empire; above all, John Cockerill in Belgium, an aggressive, shrewd businessman of supple ethical standards, who took all manufacturing as his province and with the assistance of first the Dutch and then the Belgium governments made a career of exploiting the innovations of others.[7]

These episodes suggest an important point with respect to the transfer of technology. The notion of a production function as a "set of blueprints" comes off very badly if it is taken to mean a body of techniques which is available independently of the human inputs who utilize it. The point has been well expressed by Svennilson:

It would be far too crude to assume, as often seems to be the case, that there is a common fund of technical knowledge, which is available to anybody to use by applying his individual skill. We must take into account that only a part, and mainly the broad lines, of technical knowledge is codified by non-personal means of intellectual communication or communicated by teaching outside the production process itself. The technical knowledge of persons who have been trained in actual operations has a wider scope, especially as regards the application of more broad knowledge. The "common fund," thus, covers only part of the technical knowledge to which individuals or groups of persons apply their personal skill.[8]

If, with Svennilson, we define "know-how" as "the capacity to use technical knowledge," then it is apparent that such know-how is essential for the successful utilization of the technical information incorporated in the economist's production function. And such know-how has been transmitted personally in the past simply because it is a noncodified kind of skill which has been largely acquired by direct exposure to and participation in the work process.[9] A somewhat amusing, yet apposite, instance of the role of know-how has recently been cited by Lucius Ellsworth, who points out that although we possess a complete set of the appropriate craft tools, no one in the United States today *knows how* to use coopers' tools for making wooden barrels - a once-honored and widely practiced craft.[10]

The extent of specialization in the production of machinery was of critical importance in the development of industrial skills. To appreciate this importance we must think in dynamic terms as well as in terms of static allocative efficiency. For there is a crucial learning process involved in machinery production, and a high degree of specialization is conducive both to a more effective learning process and to a more effective *application* of that which is learned. This highly developed facility in the designing and production of specialized machinery is, perhaps, the most important single characteristic of a well-developed capital goods industry, constituting an external economy of great importance to other sectors of the economy.[11]

It is a common practice to look upon industrialization as involving not only growing specialization but also growing complexity and differentiation. While this is certainly true in the sense that there takes place a proliferation of new skills, facilities, commodities, and services, it also overlooks some very important facts. The most important for present purposes is that industrialization was characterized by the introduction of a relatively small number of broadly similar productive processes to a large number of industries.[12] This follows from the familiar fact that industrialization in the 19th century involved the growing adoption of a metal-using technology employing decentralized sources of power.

The use of machinery in the cutting of metal into precise shapes involves, to begin with, a relatively small number of operations (and therefore machine types): turning, boring, drilling, milling, planing, grinding, polishing, etc. Moreover, all machines performing such operations confront a similar collection of technical problems, dealing with such matters as power transmission, control devices, feed mechanisms, friction reduction, and a broad array of problems connected with the properties of metals (such as ability to withstand stresses and heat resistance). It is because these processes and problems became common to the production of a wide range of disparate commodities that industries which were apparently unrelated from the point of view of the nature and uses of the final product became very closely related (technologically convergent) on a technological basis - eg., firearms, sewing machines, and bicycles.

Because of this technological convergence, the machine tool industry, in particular, played a unique role both in the initial solution of technological problems and in the rapid transmission and application of newly learned techniques to other uses. I have found it useful to look at the machine tool industry in the 19th and early 20th centuries as a center for the acquisition and diffusion of new skills and techniques in a "machinofacture" type of economy. Its chief importance lay in its strategic role in the learning process associated with industrialization. This role was a dual one: (1) new skills and techniques were developed or perfected there in response to the demands of specific customers, and (2) once they were acquired, the machine tool industry was the main transmission center for the transfer of new skills and techniques to the entire machine-using sector of the economy.

The questions which currently trouble me are: How do we institutionalize such activities now? What current substitutes are available for the performance of the vital activities of the production and diffusion of industrial skills - activities which were largely performed by the capital goods industries in the 19th century? The experience of successfully industrializing countries in the 19th century indicates that the learning experiences in the design and use of machinery were vital sources of technological dynamism, flexibility, and vitality. Countries which rely upon the importation of a foreign technology are thereby largely cut off from this experience, but other institutional alternatives may well be available. The contemporary rise of the multinational firm in such areas as chemicals, chemical process plants, plastics, and electronic capital goods may reflect, more than anything else, an attempt to fill this gap: to provide institutional and organizational mechanisms which will facilitate the transfer of know-how. This occurs largely by institutional innovations which make the services of the personnel with the appropriate knowledge and skills readily available, on short notice, anywhere in the world.[13]

III

Anglo-American experience with the introduction and diffusion of technology in the 19th century points strongly to the importance of the composition of consumer demand and the malleability of public tastes. The willingness of the public to accept a homogeneous final product was a decisive factor in the transition from a highly labor-intensive handicraft technology to one involving a sequence of highly specialized machines. Across a whole range of commodities we find evidence that British consumers imposed their tastes on the producer in a manner which seriously constrained him with respect to the exploitation of machine technology. English observers often noted with some astonishment that American products were designed to accommodate, not the consumer, but the machine. Lloyd noted, for example, of the American cutlery trade, that "where mechanical devices cannot be adjusted to the production of the traditional product, the product must be modified to the demands of the machine. Hence, the standard American table knife is a rigid, metal shape, handle and blade forged in one piece, the whole being finished by electroplating - an implement eminently suited to factory production."[14]   Even with respect to an object so ostensibly utilitarian as a gun, the British civilian market was dominated by peculiarities of taste which essentially precluded machine techniques. English civilians were in the habit of having their guns made to order, like their clothing, and treating their gunsmith in much the same manner as their tailor. In fact, we hear a Colonel Hawkins instructing his upper-class readers in the 1830s that "the length, bend and casting of a stock [gunstock] must, of course, be fitted to the shooter, who should have his measure for them as carefully entered in a gunmaker's books, as that for a suit of clothes on those of his tailor."[15]   On the other hand, Siegfried Giedion has pointed out that "in the United States, where department stores had slowly been developing since the forties, ready-made clothes, in contrast to Europe, were produced from the start."[16]   Giedion's point is doubly interesting in the present context because, in fact, the United States, although beginning industrialization later than England, took an early lead in the manufacture of both guns and ready-made clothing (as well as in new techniques of merchandising).

There is, it seems to me, an intimate interrelationship between composition of demand and homogeneity of product, on the one hand, and the range of technological possibilities open to society on the other. Certainly the producer must have some minimum degree of freedom to design his product in order to make it suitable for mass production techniques. There is very little doubt that, in this respect, the American producer had several degrees of freedom more than did his English counterpart. This difference had a great deal to do with the fact that mass production technology essentially originated in the United States rather than England, with early American leadership in products which were particularly well suited for mass production - firearms, sewing machines, typewriters, and later automobiles - and with the slow acceptance of many of these techniques in England.

But I am anxious not to leave the impression that Anglo-American differences can all be explained in terms of differences in consumer tastes, although I do think these were important. They seem to have been part of a much more widespread phenomenon in Britain as compared with the United States of "customer initiative" as opposed to "producer initiative." If we examine the relation between producers of capital equipment and their purchasers in both countries, we also find analogous differences. That is, in America the producer of capital goods took the initiative in matters of machine design and successfully suppressed variations in product design which served no clearly defined purpose. He brought about, in other words, a high degree of standardization in the machinery, which very much simplified his own production problems and in turn reduced the price of capital goods. Producer initiative was a very important factor in developing patterns of efficient specialization in American capital goods production.

In England, on the other hand, the capital goods producer remained, to a surprising degree, what Landes has aptly called a "custom tailor working in metal." The explanation for this difference in roles is not at all clear. In the case of many kinds of capital goods, the variations in machine design were quite unrelated to any functional justification. Buyers of machinery were in the habit of drawing up blueprints with highly detailed specifications which the machine producer had to agree to provide as a condition of fulfillment of the contract. Many of the specifications were often nothing more than the result of historical accident. In England the initiative in matters of machine design was emphatically in the hands of the buyer, and this had serious implications for the machine-producing sector both as a transmission center for new techniques and as a producer of low-cost machinery.[17]

The problems encountered by the machinery-producing sector

were further intensified by the role of that strange British institution, the consulting engineer. These engineers were imbued with a professional tradition which often led to an obsession with technical perfection in a purely engineering sense, and they imposed their own tastes and idiosyncrasies upon product design. In America, by contrast, the engineer and engineering skills were more effectively subordinated to business discipline and commercial criteria and did not dominate them.[18] The result was to perpetuate, in Great Britain, a preoccupation with purely technical aspects of the final *product* rather than with the productive *process*. This is reflected in the history of the automobile, where, of course, British engineers developed an early reputation for high-priced, high-quality engineering. In 1912, when Henry Ford was preparing to demonstrate to the world the great possibilities of standardized, high volume production, an influential British trade journal commented: "It is highly to the credit of our English makers that they choose rather to maintain their reputation for high grade work than cheapen that reputation by the use of the inferior material and workmanship they would be obliged to employ to compete with American manufacturers of cheap cars."[19]

The earlier design of locomotives was a classic case of a rampant and pointless individualism. By 1850, after the early period of experimentation (when there was an important justification for variations in design),[20] the basic features of a locomotive stabilized.[21] But innumerable minor variations of this basic design persisted. "Probably five distinct classes of locomotives would afford a variety sufficiently accommodating to suit the varied traffic of railways," an expert wrote in 1855, "whereas I suppose the varieties of locomotives in actual operation in this country and elsewhere are very nearly five hundred."[22] There is much authoritative evidence that, by comparison with American practice, a needless proliferation of designs and specifications prevailed in Great Britain throughout a broad range of iron and steel products and engineering works generally.[23]

This absence of standardization vastly complicated the process of adopting the technology and organization of mass production. It seriously inhibited the growth of specialization by firms which was so characteristic of the American machinery industry in the second half of the 19th century. For example, British railroad companies built their rolling stock in their own workshops. Private locomotive producers worked essentially for the export trade, and therefore large numbers of firms in Britain were engaged in small-batch production of a wide range of railroad equipment. During the 1920s one locomotive manufacturer was making equipment for twenty-four different gauges, varying from 18 inches to 5 feet 6 inches, and locomotive sizes ranging from 6 to 60 tons.[24] Nothing comparable to the Baldwin Locomotive Company of Philadelphia ever developed, under these circumstances, in Britain. And what was characteristic of British locomotive production was broadly true of all British engineering, where general engineering firms predominated - that is, firms which undertook to produce a range of products which would have seemed incredibly diverse by American standards. British engineering and equipment-producing firms were typically engaged in "batch" or "small-order" production. At the same time, of course, the absence of standardization further increased capital costs by raising the costs of repair and necessitating larger inventories.

Similarly, one of the most characteristic features of automobile production in Britain was the extent to which each firm produced its own components. In part this was due to the failure to standardize, but cause and effect are very hard to disentangle in these matters. There is much evidence suggesting that, in the early days of the automobile industry, British firms found it impossible to make satisfactory arrangements for the production of components by other firms and therefore fell back upon the expedient of producing such components themselves. When William Morris (later Lord Nuffield) made up his mind to purchase the main components of his automobiles from specialist producers (1913), he found that they could not meet his requirements, and he finally had to import components from America.[25]

All this was a very serious handicap because it meant that individual firms were responsible for producing a range of products which was beyond their organizational and technical competence.[26] From the perspective of American industrialization, where highly elaborate patterns of interfirm specialization developed, this was one of the most striking features of British industry. Not only did railways produce their own locomotives and automobile companies their own components, but users of capital equipment such as machine tools often made the tools themselves - for example, sewing machine companies and textile machinery firms - long after such production had been taken over by specialist firms in the United States.[27]

IV

Although I have already suggested several factors which may account for differences in the degree of specialization in the United States and Britain, I am far from convinced that such factors as differences in taste, standardization, and producer initiative constitute a complete explanation. To put the point in its most general way: American firms showed a much greater talent than British firms for coordinating successfully their relationships with other firms upon whom they were dependent for the supply of essential inputs. American firms learned to integrate their own operations with those of their main suppliers in a way which enabled them to confine their own productive operations to a limited number of specializations. Clearly the British failure to subcontract more efficiently was closely related to the factors which we have already discussed. But the failure seems also to have included an inability of British firms to rely more confidently upon the fulfillment of contracts with the degree of precision in engineering specifications which is so important to the assembly of components into a complex final product. F. W. Lanchester, in looking back upon his early experiences in the auto industry, where he attempted to persuade craftsmen to work to standard instructions so as to achieve interchangeability, commented: "In those days, when a body-builder was asked to work to drawings, gauges or templates, he gave a sullen look such as one might expect from a Royal Academician if asked to colour an engineering drawing."[28] The difficulties of successful subcontracting of high-precision components in the face of such attitudes can be readily imagined.

Furthermore, I wonder if the failure might not also involve the absence of some more subtle managerial or administrative talents which may have been available in greater abundance in the United States. For example, after discussing vertical integration as a possible solution to the problems of cooperation and communication among chemical firms, contractors, and component makers, Freeman has pointed out:

The trend in the United States has been in the opposite direction. The oil and chemical firms make increasing use of specialist contractors and specialist component-makers. The contractors tend to divest themselves of manufacturing and fabricating activities. Studies of the aircraft industry and large weapon systems have found that prime United States contractors make far more use of specialist sub-contractors and external economies than European aircraft firms. When the prime contractor (whether in aircraft or chemical plant) uses large numbers of specialist sub-contractors, this means that he must acquire considerable skills in systems management and must be capable of enforcing very high technical standards on sub-contractors.[29]

In another study, of electronic capital goods, Freeman emphasizes that the innovative process is now highly dependent upon a successful collaboration between producers of the final product and specialist makers of components.

The development process in the electronic capital goods industry consists largely of devising methods of assembling components in new ways, or incorporating new components to make a new design, or developing new components to meet new design requirements. (This is not quite so simple as it may sound. There are more than a hundred different components in a TV set, more than 100 thousand components in a large computer, and more than a million in a big electronic telephone exchange.) Consequently, there must be close collaboration between end-product makers and component makers.[30]

There seem to be some kinds of talents involved here which economists have not yet identified very precisely but which may be important determinants of success in the utilization of certain kinds of technology. The Japanese, it is worth noting, have been extremely successful (especially since the 1920s) at subcontracting, and this ability is almost certainly closely connected with the important role which small firms continue to play in Japanese industry.[31] The Japanese, as Strassmann points out, have been highly successful in the "inter-firm coordination of schedules and standards in subcontracts."[32] On the other hand, the Russians have been singularly unsuccessful in introducing subcontracting arrangements. This failure has been closely connected with many of the organization failures of Russian central planning.[33]

V

The existence of a well-developed domestic capital goods industry is important to the transmission of technology for a reason which is not yet sufficiently explicit. I have discussed the role of these industries so far in terms of their *capacities,* but we must recognize also their *motivations.* It is the producers of capital goods who have the financial incentive and therefore provide the pressures (marketing, demonstration) to persuade firms to adopt the innovation (which they produce). Creating a capital goods industry is, in effect, a major way of *institutionalizing* internal pressures for the adoption of new technology. In America the producers of capital goods have always played a major role in persuading and educating machinery users about the superiority and feasibility of new techniques.

This is an extremely important activity in overcoming the inevitable combination of inertia, ignorance, and genuine uncertainty which surrounds an untried product. The introduction of the diesel locomotive by General Motors is a classic case in point. In the United States both the railroad companies and the locomotive producers were extremely skeptical of the diesel engine and resisted its introduction. It took great promotional effort on the part of GM, which developed the diesel, to induce the railroads even to consider and experiment with the innovation. This kind of promotional activity, on the part of capital goods industries with a strong personal motive to gain acceptance for their product, seems to have been a critical factor in the American experience. Here again it has to be asked: what mechanisms or institutions can be substituted for the motivational pressure, provided in the past by domestic capital goods producers, on the part of poor countries which rely on distant foreign producers for their capital equipment?

In general the relationship between machinery producers and their customers was more successful in the United States than in England or elsewhere in Europe. In the confrontations that took place between these two groups, there seems to have been an interchange of information and communication of needs to which the machinery producers responded in a highly creative way. They learned to deal with the requirements of their customers at the same time that the machinery user learned to rely heavily on the judgment and initiative of the machinery supplier. It was, in part, the relative harmony and mutual confidence of these relationships which made it possible for machinery makers to eliminate customer preferences that were technically nonessential or irrelevant and therefore to design more highly standardized machinery. This was a process which involved an intimate knowledge of customer activities and needs and which presupposed frequent face-to-face confrontations and exchange of information (at which the British were much less successful than the Americans).[34]

Here the extent of machinery standardization intrudes itself again as a significant variable. I think it is reasonable to argue that, once the technical characteristics of a new machine have become stabilized, the amount of initiative which it is possible for capital goods suppliers to exert with their customers will depend on the extent of standardization. With extreme heterogeneity (British railroads) the role of the capital goods producer tends to become passive and adaptive rather than active and initiatory. Initiative will then reside with the user of equipment, and it will be difficult for anyone to anticipate and cater to his needs (tailor syndrome). This may have a profound effect not only on the ability of the capital goods industries to generate innovations but also on the efficiency with which they are able to produce their machines. This certainly seems to have been an important factor in accounting for differences in performance between British and American capital goods industries.

In the 19th century, a major source of capital saving for the economy as a whole came through improvements in the efficiency of capital goods production. The important analytical point is that any cost reduction in the capital goods sector - whether it is labor saving or capital saving in its immediate factor-proportion bias - is a capital-saving innovation to the economy as a whole. Such a capital-saving stage has been an inevitable but later stage of the sequence by which industrial economies have accommodated themselves to an innovation. When a new machine was introduced, there existed, by definition, no established system of organization for producing the machine. As Marx once stated: "There were mules [spinning mules] and steam-engines before there were any laborers whose exclusive occupation it was to make mules and steam-engines; just as men wore clothes before there were such people as tailors." [35] The fact of the matter is that a very high fraction of new inventions, new products, or new processes, once conceived, are of no economic relevance until the capital goods industries have successfully solved the technical and mechanical problems or developed the new machines which the inventions require. This creative process on the part of the capital goods industries has been badly slighted, and a recognition of this process may help us understand the widely observed failure of poor countries to develop techniques with the factor-saving bias which they need. In the currently developed economies, the growing efficiency of the capital goods sector was a major source of capital saving for the economy, *both* because a large proportion of all innovations originated in that sector and because it developed a capacity for producing capital goods at lower and lower cost.[36]

From this perspective, some of my concerns about the prospects for poor countries which rely on the importation of foreign capital equipment are obvious. Of course it is of enormous benefit to them to be able to import this equipment, even where the equipment is not optimally factor-biased. But if new techniques are regularly transferred from industrial countries, how will the learning process in the design and the production of capital goods take place? Reliance on borrowed technology perpetuates a posture of dependency and passivity. It deprives a country of the development of precisely those skills which are needed if she is to design and construct capital goods that are properly adapted to her own needs. What, then, are the prospects for underdeveloped countries ever becoming efficient producers of capital goods and, in particular, developing a technology with factor-saving biases more appropriate to their own factor endowments? In the past, as I have argued, the appropriate skills were acquired through an intimate association between the user and the producer of capital goods. In the absence of these experiences, what substitute mechanisms or institutions can be established to provide the necessary skills?

The problem takes on additional importance when we adopt a more realistic conception of the nature of technical change than we have inherited from Schumpeter. Schumpeter accustomed economists to thinking of technical change as involving major breaks, giant discontinuities or disruptions with the past. This rather melodramatic conception fitted in well with his charismatic approach to entrepreneurship. But technological change is also (and perhaps even more importantly) a continuous stream of innumerable minor adjustments, modifications, and adaptations by skilled personnel, and the technical vitality of an economy employing a machine technology is critically affected by its capacity to make these adaptations.[37] The necessary skills, in the past, were developed and diffused in large measure by the capital goods sector. The skills are, inevitably, embodied in the human agent and not in the machine, and unless these skills are somehow made available, the prospects for the viability of a machine technology may not be very good.

As part of our Schumpeterian heritage, we still look upon the transformation of an "invention" into an "innovation" as the work of entrepreneurs. But from a *technological* perspective, it is much more the work of the capital goods industry. This is most obviously the case where the invention is a cost-reducing *process* and does not have to be marketed to final consumers. But in making new products and processes practicable, there is a long adjustment process during which the invention is improved, bugs ironed out, the technique modified to suit the specific needs of users, and the "tooling up" and numerous adaptations made so that the new product (process) can not only be produced but can be produced at low cost. The idea that an invention reaches a stage of commercial profitability *first* and is then "introduced" is, as a matter of fact, simple-minded. It is during a (frequently protracted) shakedown period in early introduction that it becomes obviously worthwhile to *bother* making the improvements. Improvements in the *production* of a new product occur *during* the commercial introduction.

Alternatively put, there has been a tendency to think of a long precommercial period when an invention is treated as somehow shaped and modified by exogenous factors until it is ready for commercial introduction. This is not only unrealistic; it is a view which has also been responsible for the neglect of the critical role of capital goods firms in the innovation process.

We are now coming to realize that modern technology has a long umbilical cord. Innumerable unsuccessful foreign aid projects in the past twenty years - including Russian-sponsored as well as American-sponsored projects - have confirmed that when modern technology is carried to points remote from its source, without adequate support-ive services, it will often shrivel and die. This is partly because the technology emerged in a particular context, often in response to highly narrow and specific problems, such as may have been defined by a particular natural resource deposit. But, more important, the technology functions well only when it is maintained and nourished by an environment offering it a range of services which are essential to its continued operation. These would include the ability to diagnose correctly the causes of machine breakdown or other sources of inferior performance, the availability of facilities and personnel to perform repair work and to provide routine maintenance and repairs, and the provision of spare parts. In the absence of the kinds of skills produced and embodied in a capital goods industry, repair and maintenance costs in the use of machinery are likely to remain high.

A major reason for a domestic capital goods industry, therefore, is that the ability to utilize complex machinery effectively - whatever the country of origin of the machinery - depends upon the kinds of skills which such an industry uniquely makes available.

Moreover, physical proximity between the producer and user of machinery seems to have been indispensable in the past for reasons which we do not really understand but which seem to be rooted basically in the problem of communications. Successful technological change seems to involve a kind of interaction that can best be provided by direct, personal contact. Successful instances of technological change in the past have involved a subtle and complex network of contacts and communication between people, a sharing of interests in similar problems, and a direct confrontation between the user of a machine, who appreciates problems in connection with its use, and the producer of machinery, who is thoroughly versed in problems of machinery production and who is alert to possibilities of reducing machinery (and therefore capital) costs.

VI

With all of the difficulties attaching to the transfer of industrial technology, such technology is nevertheless much easier to transfer than agricultural technology because industrial technology is at least very much self-contained. It tends to be a relatively closed system. Even its dependence upon human inputs can be reduced by making the technology more capital-intensive. But where ecological relationships are involved, as they are in agriculture, everything is quite different. Here there are important interactions between the human enterprise and specific features of the natural environment. Here natural phenomena participate in a much more active way, and productive activity must be more highly responsive to even minor variations and peculiarities of the environment. Agricultural activity is much more closely enmeshed with the natural environment, and this, in turn, has important consequences. It means that solutions to agricultural problems and the ability to make appropriate adaptations to local conditions will hinge upon kinds of knowledge which are not ordinarily possessed by the cultivator: biology, botany, biochemistry, genetics, etc. Major breakthroughs in agriculture are likely to come from the application to local circumstances of scientific disciplines which do not ordinarily flourish in poor countries. I would therefore argue that a major goal of institution building in poor countries should be to equip them with the kind of research abilities and facilities that will enable them to produce the new knowledge required for their own peculiar agricultural resources.

Agriculture's close involvement with nature has another important consequence for the acceptance of new techniques: because of the importance of even minor variations in rainfall, sunshine, soil content, topography, plant diseases, etc., there is a much higher degree of uncertainty concerning the application of new agricultural techniques. Institution building in agricultural environments will have to take this element of uncertainty prominently into account. Institutional arrangements which serve to reduce this uncertainty or to protect the cultivator against a loss that could be personally disastrous may prove to be extremely important in speeding the adoption of new techniques.

In examining the sources of productivity improvement in both industry and agriculture, we must pay attention to the fact that the performance of individual industries will frequently depend not only on resources within that industry but on the availability and the effectiveness of industries which stand in an important complementary relationship with it. The technological inputs (including knowledge) which crucially affect the success of industry A are produced by industries B, C, and D. Much of the discussion of the prospects and possibilities for technical improvement in poor countries has suffered from ignoring such interindustry relationships. We have to take account of these relationships and learn how to exploit them.

This point is particularly pertinent to the behavior of agriculture. The history of countries with highly productive agricultural sectors indicates clearly that the major sources of improvements in agricultural productivity have been generated *outside* the agricultural sector. Unless an economy is well equipped with these complementary sources, agriculture is not likely to experience rapid improvements in efficiency. In the American experience the sources of productivity growth in agriculture have come from the machinery-producing sector, which developed a mechanical technology appropriate to agriculture; from research at agricultural experiment stations and other educational institutions; from the fertilizer industry; and increasingly in the 20th century from the study of genetics and from chemistry (including soil chemistry). In fact, in American agriculture the industries which supply agriculture with certain of its inputs have played a role in the introduction of new technologies entirely analogous to that of the capital goods industries in relation to the manufacturing sector.[38] It is the absence of these complementary sectors and resources which is often so damaging to underdeveloped agriculture.[39] If, for example, the fertilizer industry can produce fertilizer only at a very high cost, then all the extension work and exhortation in the world will not induce the peasant to use it - he will reject its use on perfectly rational cost-benefit grounds. What is needed is research oriented toward reducing the cost of the inputs which agriculture receives from the industrial sector. An extension service will not perform an important social function until it has valuable information to extend.[40]

In broad terms, the history of Japanese agricultural development seems to confirm the importance of these complementary relationships. Japan represents perhaps the most remarkable success story of the 20th century - a story that began 100 years ago. The great improvements in Japanese agricultural productivity, which provided the basis for the growth of the rest of the economy, rested firmly upon inputs provided from outside the agricultural sector. The growth of her agricultural productivity was due to a long series of dynamic interactions between the needs of the farm sector, on the one hand, and the combined ingenuity of industry and the educational establishments to supply these needs, on the other. The outcome of these interactions included the provision of low-cost commercial fertilizer, new farm implements, high-quality education and extension work, and, at a later stage, the beginnings of mechanization. Although it is relatively easy to point out the things which were done well in Japan's spectacular success story, the really critical question is how to incorporate such functions in a set of institutions which will perform successfully in a drastically different environment.[41]

VII

The exact institutional arrangements which will make for success will always be difficult to specify a priori. I would like to suggest that this is so mainly because success or failure always depends on the environment in which any particular institution is immersed - that is, on the complex of *other* institutions as well as widely shared values and traditions which critically affect its operation.

If history teaches us anything in this regard, it is that a wide diversity of institutional forms have proved to be successful under differing conditions. Similarly, history indicates also the very variable experience of the *same* institution when placed in different contexts. Thus, while on the American scene it has been customary to celebrate the virtues of the family farm, the family farm in France has been a favorite whipping boy, held responsible for the continuation of outmoded farm practices, for excessively small farm units, and for an insufficient use of capital. And, while the American owner-operated farm has indeed been a remarkable success, it has been critically dependent for this success upon a wide range of organizational arrangements which absolutely defy simple ideological classification.[42]

The kinds of institutions which will function successfully in a particular environment and in the pursuit of particular objectives is something which it is not easy to generalize about a priori. [43] Certainly this is confirmed by our own recent history and experience, which suggests that a wide diversity of organizational tactics have proven to be successful. Consider the changing mix of government participation in various economic activities in recent years. Not only do we have public production of the TVA variety, but we also have public regulation and control of private utilities, public subsidies to private enterprise, public highways financed on a user-cost basis, public and private institutions of higher education living happily together (or almost so), public support of medical research which is financed partly from private sources, and public sponsorship of specialized activities contracted to private organizations - the Atomic Energy Commission and even the RAND Corporation. If we look specifically at the realm of technology, we find that varying mixes of the public and private sectors have been responsible for a broad range of technological breakthroughs in areas such as the development of atomic energy and its application to peaceful uses, medical research, jet propulsion, electronics, and agriculture. I point this out not to suggest that there is anything optimal about the public-private mix in the American economy but rather to indicate the wide range of institutional devices and combinations which have proved to be successful. Indeed, one of the most spectacularly successful kinds of institutions in our "capitalist" economy in recent years has been the nonprofit corporation, which has been a major source of both new knowledge and new technology.

Clearly there is no single "best-way-of-doing-things" to which we have rigidly adhered in all sectors of our economic life. If the American experience suggests anything, it is that there has been no single institutional formula for success. It is reasonable to expect that the success formula will be both different and equally variable in poor countries. Without question we ought to attach the highest research priority to finding out what combinations are likely to prove successful in the special environments of individual countries. This is an area where history can provide some guidance and valuable insights but certainly no authoritative answers.

10

Selection and adaptation in the transfer of technology: steam and iron in America 1800-1870

The acquisition of new technologies by noninitiating countries is a subject of major importance and, at the same time, one which has been conspicuously neglected. Historians of industrialization in the past have focused attention forcefully upon the identification of first beginnings and the examination of the claims to priority of different individuals. These questions, indeed, are of basic importance in the history of *invention*. This preoccupation, however, has led to the neglect of the diffusion process,[1] even *within* the initiating country. Indeed, the problem has often been assumed out of existence by virtually identifying the date of first technological success with economic impact. This has frequently occurred because, in the past, technological history and economic history inhabited quite different worlds - or at least their authors did. Thus, in his classic account of the steam engine Dickinson, in describing the *consequences* of the improvement which James Watt had introduced, states that, by the end of the eighteenth century, "The textile industry had been transformed from handicraft into a machine industry."[2] One must regretfully record that Dickinson, who wrote with such commanding authority on technological matters was, quite simply, a half century off the mark in his economic history.[3]

The point at issue is fundamental because, to put it bluntly, the social and economic consequences of technological changes are a function of the rate of their diffusion and not the date of their first use. Since an invention, looked upon as the outcome of an economic activity, has the attractive property of having to be "produced" only once (unlike tons of steel or bottles of wine), from the point of view of *world* history the critical social process requiring examination is that of diffusion.

In examining this diffusion process, economic considerations are paramount. Even in a world of perfect, instantaneous and free communications, with no barriers whatever to the transmission of knowledge, the export of goods or the movement of people[4], we should not expect a smooth, uniform diffusion of new techniques. Since countries differ, sometimes drastically, in the availability of factors of production (including, it should be noted, the age structure as well as the size of their existing stocks of capital goods), techniques which are efficient (i.e., minimize costs for a given volume of output) in one environment may not be efficient in another. Therefore, between two countries as differently endowed as Britain and the U.S. in the early nineteenth century, we should expect the borrowing country to borrow in a highly selective fashion: i.e., to borrow some techniques rapidly, others more slowly, and perhaps yet others not at all. Underlying the selective nature of the transfer and diffusion, then, is an economic mechanism, based upon factor proportions and factor prices, which determined the expected profitability of different techniques in a new environment.

Such a pattern of selectivity, it is argued, would constitute rational economic behavior and would reflect differences in the expected profitability of the "menu" of techniques which are available for borrowing by the noninitiating country. The timing of the transfers, and therefore the length of the observed historical lags, would also be affected by a second factor related to differences in resource endowment. That is, new techniques frequently require considerable *modification* before they can function successfully in a new environment. This process of modification often involves a high order of skill and ability, which is typically underestimated or ignored. Yet the *capacity* to achieve these modifications and adaptations is critical to the successful transfer of a technology - a transfer which is too frequently thought of as merely a matter of transporting a piece of hardware from one location to another. Much of this paper will be organized around these twin themes of selectivity and modification and their implications for understanding the transfer of industrial technology to the United States.

The steam engine

There are many possible ways of thinking about technology. At one level it may be observed as a set of principles and techniques which has become embodied in a particular collection of machines, tools, or commodities. In its concrete form, however, the technology is always more specific than the knowledge which it embodies, because it has been shaped and modified in response to a very particular set of needs, goals or resource constraints. It is helpful in examining the acquisition of steam engine technology in the U.S. to think of it in the more general and diffuse sense as a working knowledge and capacity for applying the power of steam to perform useful work. At this stage the steam engine represented a potential; it was a power source which could be used to do many possible things; most obviously, it could turn the machinery of a mill or factory or it could propel a vehicle over land or water.

The specific uses to which this newly acquired technology was actually put in the U.S., and the timing and the location of these applications, reflected entrepreneurial expectations about the recognized or anticipated needs of this new society insofar as these needs were expected to find expression in the market place. That is to say, entrepreneurs engaged in introducing a newly developed foreign technology for domestic use attempted to do so in a manner best calculated to promote their future profits. In so doing they were necessarily highly sensitive to the size and structure of domestic demand, to their perception of relative costs, and to their expectation about likely future changes in these magnitudes.

Within this economic matrix, then, we need to put aside the fact that the steam engine had many unique and interesting aspects from the point of view of technological history and regard it as "merely" a new source of motive or traction power.

Motive power. As a source of motive power the steam engine competed with water, animal power and, in some places, wind and, of course, human muscle power. The willingness to adopt the new power source depended, first of all, upon the demand for power and, secondly, upon the availability and the cost of alternative forms of power. These considerations explain why, within the U.S., steam power for motive purposes was only slowly adopted. Industry in the early 19th century was heavily concentrated in New England. Although the demand for power was great and the capacity to produce steam engines was certainly well established by the 1830's, an excellent and cheaper substitute in the form of water power was provided along the fall line by the numerous, fast-flowing rivers and streams of New England. The growing reliance upon steam power was, in fact, closely connected with the westward movement of population and industry into geographic regions which offered fewer sources of water power and where, as a result, the economic significance of steam power was greater.

It is important to note, then, that mechanization in the U.S. did not initially involve any massive shift to new power sources. U.S. census reports indicate that American manufacturing relied upon water as its main source of power until well into the second half of the 19th century. Indeed, as late as 1869 steam power accounted for barely over one-half of primary-power capacity in manufacturing - 51.8% as compared to 48.2% for water.[5] In general, it may be said that steam power tended to be adopted in locations where water power sources were scarce and where fuel was either abundant or cheaply transportable. One important consequence of improvements in transportation, therefore, was to expand the area within which steam power could be profitably employed.[6]

It is quite clear that the relatively slow adoption of the stationary steam engine even in the second third of the 19th century was *not* due to an inability of Americans to manufacture steam engines. In fact, data collected in the late 1830's demonstrate that by 1838 there were some 250 or so steam engine builders, mostly very small scale businesses, widely scattered throughout the U.S. A government report on steam engines in the U.S. (known to be somewhat incomplete) estimated that there were more than 1800 stationary steam engines in operation in 1838.[7] American mechanical skills by this date were obviously well developed, and there was no slavish reliance upon or imitation of British models and techniques. This is most apparent in the American use of the high-pressure steam engine. Although the high-pressure steam engine had been developed simultaneously by Richard Trevithick in Great Britain and Oliver Evans in the U.S. in the opening years of the 19th century, British practice subsequently strongly favored the low-pressure stationary engine while Americans overwhelmingly adopted the high-pressure stationary engine.[8] On the other hand, the American preference for water power over steam, especially in New England, was a consequence of the abundant supplies of running water, which Americans exploited by the construction of the cheaper forms of water wheels. These were, in a strictly engineering sense, highly inefficient. However, they were preferred to wheels which utilized a higher proportion of the available water power, but which involved a larger initial capital expenditure.[9] The situation changed after 1840 as the limited number of more attractive water sites even in New England was exhausted and demand for power continued to grow rapidly. Nevertheless, although the efficiency (as measured, e.g., in coal requirements per unit of horsepower) of the steam engine improved and its cost declined, there were also major improvements in the utilization of water power. This particularly took the form, after 1840, of the water turbine, an important invention of French origin.[10] The decisive significance of geography in influencing the selection of technology is indicated by the fact that, as late as 1869, by which time steam had just surpassed water on a national basis as a source of power, less than 30% of the power employed in New England manufacturing establishments was derived from steam.[11] New England industry for long remained heavily concentrated in communities which offered superior access to water power - Lowell, Lawrence, Hadley Falls, Holyoke, Chicopee, Springfield, Waterbury, Manchester.

Tractive power. Although the availability of a good substitute slowed the pace at which the stationary steam engine was acquired in the U.S., the situation was vastly different in the application of steam power to transportation. The U.S. comprises a land area of continental proportions.[12] So long as population was confined to a relatively narrow strip of land east of the Appalachians, the Atlantic Ocean and its bays, sounds and tidal rivers offered a reasonably adequate basis for the movement of goods. But the exploitation of the trans-Appalachian west was dependent upon innovations which would liberate commerce from the prohibitively high cost of land transport and the upstream haulage of goods. It is perhaps not too much to say that the major economic consequence of the acquisition of the steam engine in the New World before the Civil War lay in its application to new forms of transport - the steamboat and later the railroad - which (together with the canal) provided a network for the cheap movement of goods - especially bulky agricultural products. The contrast between the relatively slow adoption of the stationary steam engine and the rapid exploitation of the steam engine for transport purposes is, in fact, highly instructive. It is an excellent demonstration of the manner in which social needs, as expressed in the market place, and as they influence business profit expectations, shaped and directed the pattern of innovative activity. For within the context of the antebellum American economy the abundance of water power in the east and the huge size of the land area awaiting exploitation in the west shifted the attention of inventors and entrepreneurs away from the use of the steam engine as a stationary power source and into its rapid exploitation as a device for the conquest of space.

As a source of tractive power, the steam engine in the early 19th century had many substitutes - animals and animal-drawn devices, running water (down which keelboats and flatboats could be cheaply floated) and sailing vessels. But the geography - the scale and disposition of resources - was such that the vast area of the trans-Appalachian west could not be exploited without major reductions in transport costs from some new source. This was clearly perceived by a series of fertile and inventive minds - Oliver Evans, John Fitch, Robert Fulton, James Ramsey and John Stevens - who were aware of the great lag in the reduction of overland transport costs behind those of water, and for whom the steam engine represented, above all, a new and much-cheapened form of transportation. If America led the world in the development and exploitation of the steamboat by the second quarter of the 19th century, it was largely because the economic payoff to this innovation was correctly perceived to be very great in a large, undeveloped country, rich in natural resources, and with a vast system of natural internal waterways. By contrast with Great Britain and western Europe, which were comparatively well served by coastal waterways and canals, the steamboat in the immense Mississippi basin offered a form of transportation far superior at the time to the available alternatives. Although the early experiments with the steamboat were, inevitably, conducted on eastern waterways, men like Fitch, Evans and Fulton clearly understood that the great utility of the steamboat would be on the western rivers.[13]

The steamboat quickly achieved a great success on western rivers and, in this form, the steam engine dominated the economic life of the Mississippi valley for a generation (see Table 1).

Even the twelvefold increase in tonnage between 1820 and 1860 drastically understates the growth in transport services provided by the steamboat because of major increases in speed and in cargo capacity in relation to measured tonnage. Because of such improvements Hunter concludes that "... the facilities of steamboat transportation on the western rivers during a period of forty years multiplied not twelvefold but more than one hundred and twentyfold."[14]

Table 1. *Steamboats operating on the
Western rivers 1817-1860*

Year	Number	Tonnage
1817	17	3,290
1820	69	13,890
1823	75	12,501
1825	73	9,992
1830	187	29,481
1836	381	57,090
1840	536	83,592
1845	557	98,246
1850	740	141,834
1855	727	173,068
1860	735	162,735

Source: Hunter, op. cit., p. 33.

The extent to which American exploitation of steam was dominated by the steamboat before 1840 may be simply stated.
According to the *Report on Steam Engines,* almost 60% of all power
generated by steam in 1838 was accounted for by steamboat
engines.[15]

If one asks precisely what it was which was "acquired" from the
Old World by the Americans who created this vast inland fleet, the
answer is by no means obvious. Certainly there was the steam engine
itself, and the knowledge of its operating principles as well as
techniques of construction. Some of the steam engines installed in
the first steamboats were imported from Great Britain - indeed,
Fulton himself had installed a Boulton and Watt steam engine in the
Clermont. But America quickly acquired the skills to provide her
own engines, and produced them to her own design in large
quantities in the main centers of steamboat construction - Pittsburgh,
Cincinnati and Louisville.[16]

Basically the steamboat at the time of its inception was a clumsy
merger of the steam engine with an ordinary oceangoing vessel. With
its deep and rounded hulls, projecting keel, heavy frame and
low-profile superstructure of a sailing vessel, it was remarkably
ill-adapted to the shallow-water navigation of inland rivers. The
transformation of the basic design of the steamboat occurred very
quickly, and yet it is difficult to identify this transformation with
specific inventions. Rather, the steamboat underwent a series of
innumerable - indeed continuous - changes in structural design and

proportions which made of it the maneuverable, flat-bottom, shallow draft, high superstructured vessel which it had to be to negotiate western rivers with substantial cargoes.[17] The steamboats were powered by high pressure engines whose wastefulness in fuel consumption as compared with low-pressure engines was more than offset by low cost of construction and repair and relative compactness. Since cordwood was abundant and cheap along the banks of the Ohio and Mississippi Rivers, economy of fuel consumption was of less consequence than in the east, where there was a greater preference for the low pressure engine.[18]

The story of the steamboat, then, is a story of continuous modification and adaptation of a general concept - the possibility of propelling a large cargo vessel by means of a steam engine - to the highly specific characteristics of a particular environment. The end-product of this adaptiveness - one might even say "evolution" - was a unique instrument ideally suited to a particular set of economic and geographic circumstances.

Although space considerations preclude any detailed treatment of the acquisition of railroad technology, a few remarks are in order. In contrast to the steamboat where the U.S. asserted an early leadership, the U.S. relied much more heavily upon British leadership and prior development in the case of railroads.

The U.S. was much more directly dependent upon Britain in the application of the steam engine to land transport than it had been in the case of the steamboat. Britain was clearly the pioneer and major innovator in the case of the railroad, and America the follower. Yet it is notable how quickly the student acquired a competence and became independent, at least with respect to rolling stock. Only a limited number of locomotives were imported from Britain even in the earliest years. In the early 1830's locomotives began to be built in foundries, machine shops and textile machinery plants, and in Philadelphia the specialized Baldwin locomotive works was established in 1832 and the Norris firm two years later. Only 117 of the 450 locomotives in the U.S. at the end of 1839 had been imported from England, and most of these imports had occurred before 1836.[19]

The situation was very different with respect to rails, where Britain's more advanced iron-making plants and rolling mills conferred a price advantage on British rail makers great enough to overcome the additional costs of ocean shipping, commissions, etc.

American railroads remained heavily dependent upon imports of British rails for many years. During the western railway construction boom from 1849 to 1854, 80% of American railway iron requirements were met by imports.[20] American iron-making capacity expanded thereafter (with some initial emphasis on rerolling), in no small measure due to the insistent demands of the railroads. Although America supplied most of her own rails by the late 1850's, British imports continued to be a significant proportion of the total during peak construction years such as the late 1860's, early 1870's, and, to a lesser extent, early 1880's.[21]

In spite of this reliance upon British experience and supplies, American railroads showed a high degree of responsiveness to unique aspects of the American scene, and American engineers and builders demonstrated an imaginative capacity to modify and adapt the foreign model in significant ways. The relative cheapness of wood to iron in the U.S. led to extensive substitution of wood in uses where the British had employed iron. In numerous ways capital outlays in the early years were reduced by building the roads to specifications which British engineers would have considered intolerable: steep gradients and sharp curvatures were tolerated, and the construction of tunnels was avoided whenever possible.[22] One of the main results of such construction was to raise fuel input requirements of railroad operation. Given the general abundance of fuel and relative scarcity of capital, however, such a substitution of (cheap) resource inputs for (expensive) capital inputs was eminently "rational." The Americans also reduced initial construction costs by purchasing low-quality British rails.[23]

Moreover, as the relative prices of inputs altered, so did railroad practice. For example, up until the Civil War American locomotives were almost totally committed to wood as a fuel.

Within a score of years thereafter, a transformation so rapid had occurred that twenty times more coal than wood was being consumed annually, and more than a fourth of bituminous coal output was regularly absorbed by the railway sector. The underlying mechanism is almost a text book illustration of substitution in response to changing relative prices. To begin with, eastern railroads with large coal deposits along their lines, and hence both low coal prices and elastic supply, invested in research necessary to eliminate the troublesome technical problems that had limited the development of coal-burning locomotives. Once successful, the eastern railroads penalized by high wood prices and the western railroads favored by low coal prices led the parade to mineral fuel. . . .[24]

Finally, we may note that the Russians sought out an American engineer to supervise the construction of a railway from St. Petersburg to Moscow as early as the 1840's. Such a selection, somewhat startling at first glance, may have reflected a shrewd awareness that Russian and American conditions closely resembled one another, and that an American engineer would be more likely than an English one to modify the new railroad technology in accordance with the peculiar needs of the Russian environment.[25]

Iron

Improvements in metallurgy were fundamental to that process which we call the industrial revolution. The new machines, the remarkable feats of construction and civil engineering, and the revolution in transport would have been inconceivable without large quantities of cheap iron (and later, steel). In examining this transfer, we are confronted with the intriguing fact that, unlike the rapid imitation or even leadership which the U.S. quickly demonstrated with respect to the uses of the steam engine, the country lagged several decades behind the best practice techniques of the British iron industry. It is a striking fact that, whereas the mechanical technology of the steam engine fairly leaped across the barrier of the Atlantic Ocean, changes in metallurgical technology moved at a pace which seemed almost glacial by comparison.

This statement requires some qualification. The lag in the transfer of the new iron technology was much smaller at the stage of refining the pig iron into wrought iron (the complex of innovations involving puddling with use of mineral fuel and the shaping of the wrought iron by rolling instead of hammering) than at the stage of smelting of the iron ore in the blast furnace.[26]

The introduction of mineral fuel into the blast furnace in England was the outcome of a protracted search for a cheap substitute for wood fuel going back to the early 17th century. While we cannot be detained here by this fascinating piece of technological history, it should at least be pointed out that very little is known of the diffusion of Darby's coke-smelting process after his initial success, apparently in 1709.[27] The shift from charcoal to coke in the blast furnace was accelerated in the second half of the 18th century and was virtually completed by 1800.[28]

The situation in the U.S. stood in striking contrast to this. As late as 1840 almost 100% of all pig iron produced in the U.S. was still made with charcoal, and even as late as 1860 only a scant 13% of American pig iron was being smelted with the "modern" fuel - coke.[29]

In coming to grips with the American lag in the adoption of the new iron technology, we must recognize that metallurgy differs from more purely mechanical technology in some important ways. Most important for present purposes is the fact that metallurgical processes are intimately enmeshed with the physical environment, linked up with qualities of the natural resource inputs - qualities which simply were not understood in any serious scientific sense until the latter part of the 19th century. Metallurgy in the period under review was still essentially an activity relying upon crude empiricism. Variations in resource inputs affected the success of the productive process in ways which could be observed experimentally but not understood or predicted. Resource inputs which were best suited could be found by a trial-and-error process much as the 19th century frontier farmers might experiment with a variety of wheat seeds in a new climate until they found the kinds which "worked best." Thus, although a long list of Englishmen had attempted - unsuccessfully - to introduce mineral fuel into the blast furnace, Abraham Darby succeeded where others had failed partly because of the simple geological fact that the coal which was found conveniently near the surface at Coalbrookdale happened to be of a chemical composition peculiarly appropriate for smelting.[30] Similarly, Bessemer's success in the mid-19th century had been conditioned by the fact that he used imported Swedish iron in his experiments - iron which, as it happened, was singularly free of phosphorus, although Bessemer was not aware of this fact. Indeed, the acute distress of British iron makers who attempted, unsuccessfully, to reproduce Bessemer's method with British ores containing substantial amounts of phosphorus was a major event leading to the emergence of the modern science of metallurgy.[31]

It is the combination of (1) this critical role of particular qualities of resource inputs as they determined their economic usefulness with (2) the specific locational matrix in which these resources were embedded, which accounted for the large American lag in the acquisition of new British iron technology. The economic importance of locational factors is seen in the rapid shift of the British iron industry to the local fields of the Midlands, Yorkshire, and

Darbyshire, and South Wales, as coal was substituted for charcoal. Since it required, with the early 19th-century iron making technology, several times as much coal as iron ore to produce a ton of iron, proximity to coal became a dominant economic consideration.[32]

In the first decades of the 19th century, when it still employed the charcoal technology, the blast furnaces of the U.S. were spread out along the eastern seaboard. Pig iron production, requiring large amounts of wood fuel, was widely diffused geographically - as it had to be. The increasing demand for iron was met essentially by increasing the number of charcoal blast furnaces. The new British blast furnace technology was not adopted because bituminous coal was required for coking purposes, and the known deposits of such coal suffered from two deficiencies: (1) they contained substantial amounts of sulfur, which made for the production of poor quality pig iron, or (2) they were located west of the Allegheny Mountains, far from the country's population centers.[33]

America *did,* however, possess some very rich coal deposits in eastern Pennsylvania. But this was anthracite coal, containing neither gas nor sulfur. Although the absence of sulfur meant that it could be used to produce high quality pig iron, the absence of gas made ignition much more difficult and meant that it could not be used employing the blast furnace technology developed by the British in the 18th century.[34]

This situation was altered by the development in England of the hot blast, an innovation which, by preheating the blast before entering the furnace, and later by employing waste gases from the furnace itself, permitted substantial fuel economies.[35] *Here we observe virtually no lag at all in the transfer of the technique from Britain to the U.S.* The method was first employed in England in 1828 and by 1834 it was successfully introduced into a blast furnace in New Jersey.[36] But the important point for present purposes is that during the 1830's a Welshman, David Thomas, discovered that the use of the hot blast made it possible to introduce anthracite as a fuel into the blast furnace. Thomas actually came to the U.S., where he constructed the first successful anthracite blast furnace in 1839. From that point on, the expansion of pig iron output employing anthracite proceeded rapidly.

Thus, when the U.S. finally introduced a mineral fuel into the blast furnace, it was based upon a newly developed British technique. This technique was rapidly adopted because it permitted the exploitation of a resource which was highly abundant and readily accessible to America's main population centers. Anthracite furnaces accounted for one-half of pig iron output in 1856.[37]

The eventual adoption of the earlier coke smelting technology came only in the post-Civil War years. In the late 1850's the proportion of pig iron smelted with coke still did not exceed 10% of total output. By 1870 this figure was 31%, by 1880 45%, 1890 69% and, in the first years of the 20th century the proportion rose over 90%.[38] The major breakthrough was linked, again, to the subtle interplay between the specific resource requirements of a particular technology and the sequence in which the resource base was gradually uncovered.[39] Along with the westward movement of population and the more intensive exploration of the trans-Appalachian west, came the discovery of the high-quality coking coal in the Connellsville region of Pennsylvania. The first blast furnace designed specifically for the exploitation of Connellsville coke was built in Pittsburgh in 1859. The physical structure of the coal and the absence of sulfur made it possible to produce pig iron of high quality. This, together with the development of a low-cost transportation network and further technical developments favorable to coke, assured the eventual domination of this fuel in the blast furnace. The dominance of anthracite in 1860, then, when it accounted for over one half of U.S. pig iron production, proved to be very short-lived.

Conclusion

This paper has stressed the roles of selectivity and modification in the transfer process. It has emphasized how qualitative as well as quantitative differences in resource endowments have affected the rate and the sequence of the transfer of technology. Certain implications of this approach should be emphasized in closing.

First of all, the recipient of these transfers was a nation whose level of mechanical competence and technical skills (aside from its slave population) was already very high. The fact of the matter is that it required considerable technical expertise to borrow and exploit a complex foreign industrial technology. This should hardly be a surprising proposition, but it seems to be worth repeating in view of the vast number of foreign aid and economic assistance programs in the years since the Second World War which have come to grief because of the absence of the appropriate skills in the receiving country. The *selection* of a technology as appropriate in a particular context, and its *adaptation* and *modification* in order to enable it to function efficiently in an environment different from the one in which it originated, are activities which typically require a very high degree of technological sophistication. Although it is difficult to find data which measure directly the technical competence of a population,[40] it is interesting to note that the U.S. compares very favorably with western Europe as early as 1830 in certain crude measures of enrollment for formal education. Although the manner in which a society acquires and diffuses technical skills is still only very imperfectly understood, and although formal education is certainly only a part of that process, it is worth noting that there is a high correlation in the early 19th century between such indexes of educational attainment and the speed with which individual countries were able to adopt and modify Britain's new industrial technology.[41]

Secondly, our whole conception of the way in which the new technology replaces the old requires drastic modification. This is particularly important because our notions concerning the *social consequences* of technical change are often based on an exaggerated sense of the pace with which the new techniques replaced the old. Our understanding of the nature and the pace of historical change can be substantially improved by more detailed, quantitative knowledge of the displacement of one set of techniques by another. We have already seen the stubborn persistence of water power in the face of steam. In *absolute* terms the amount of power generated by water in New England continued to increase right into the 20th century. Similarly, in 1800, by which time Watt's patents had all expired, Newcomen engines were not only still being used but, due to their low construction and maintenance costs and long life expectancy, still continuing to be built.[42] Although the primitive flatboat would seem to have been no match for the Mississippi steamer, flatboat arrivals at New Orleans reached their all-time high in 1846-47, at an absolute level nearly five times as great as that of 1814.[43] For, although the steamboat had decisive economic advantages in the upstream traffic, its advantages in traveling downstream were, not surprisingly, not nearly so great. By 1860 the introduction of the steamboat had resulted in a reduction of downstream transportation costs by a factor of between 3 and 4 but it had reduced upstream transportation costs by a factor of ten. Even the windmill, which might have been expected to be an early casualty after the introduction of the steam engine, experienced a considerable growth in at least one English county.[44]

Finally, the absolute output of pig iron produced by charcoal in the U.S. reached its peak as late as 1890, several decades after the mineral fuel technology had been introduced.

The new technology usually asserted its advantages over the old only slowly. Partly this is because the new technique has many "bugs" at first which need to be eliminated; partly because the capital goods sector takes time to learn to produce the new machine efficiently, and the diffusion of the new technique is closely linked to the gradual decline in price which is associated with this learning process; partly because there is another learning process in the efficient *utilization* of the capital good after it has been produced and installed; partly because improvements continue to be made in the old technique, as was the case with the water turbine long after the development of Watt's steam engine, and the major improvements in sailboat design after the introduction of the ocean steamer;[45] partly because the geographic distribution of resources frequently gives specific localities transport, power or other cost advantages even with the old technology; partly because the process of modification to local conditions is often, as we have seen, a very time-consuming one.

Finally, it seems apparent that we need to descend from the rather Olympian heights from which history is often written and examine the detailed characteristics of individual technologies as they affect the transfer process. Clearly technologies differ very much with respect to the ease of their transfer. A theme which emerges from this paper, for example, is the contrast between the relative ease with which the steam engine was transferred and modified and the difficulties and complexities with respect to metallurgy. Although the pervasive economic forces influencing the selection and the timing of the transfers have been emphasized here, it has not been my intention to suggest that the economic factors constitute an exhaustive explanation. It should be clear, for example, from our earlier discussion, that *underlying* the economic factors are deeply rooted technological considerations influencing the responsiveness of supply side variables which are, as yet, only very imperfectly understood. Thus, the sharp difference in the historical experience between the realms of the mechanical and the chemical suggest that an important variable may be the degree of scientific understanding which is required before some particular realm of the physical environment can be effectively exploited. Even within the iron industry those techniques relying more heavily on mechanical principles were transferred with less difficulty than those depending upon chemistry. Thus, as we saw earlier, the hot blast crossed the Atlantic very quickly. Moreover, innovations in the rolling mill were much less problematical than innovations in smelting and refining. Refining, in turn, was less problematical than smelting because, in the puddling process, the use of the reverberatory furnace eliminated the direct contact between the fuel and the iron. Obviously, this distinction between the mechanical and chemical is exceedingly gross - as gross, perhaps, and as unrevealing as that between "agriculture" and "industry" - and much more subtle distinctions and descriptive categories will have to be introduced before we can satisfactorily analyze and understand the complex historical phenomena which are associated with the processes of technological diffusion and economic growth.

11

Factors affecting
the diffusion of technology*

I

The rate at which new techniques are adopted and incorporated into the productive process is, without doubt, one of the central questions of economic growth. New techniques exert their economic impact as a function of the rate at which they displace older techniques and the extent to which the new techniques are superior to the old ones. Although we are still a very long way from being able to assess the exact role of technological change - as distinct from all other factors - in generating the rise in resource productivity which is at the heart of the growth process, it is, I think, clear that the contribution of technological change itself will have to be established through the study of diffusion. Only in this way can we develop a closer understanding of the rate at which new techniques, once invented, have been translated into events of economic significance.

Although these remarks are, I believe, sufficiently uncontroversial, it is a striking historiographical fact that the serious study of the diffusion of new techniques is an activity no more than fifteen years old.[1] Even today, if we focus upon the most critical events of the industrial revolution, such as the introduction of new techniques of power generation and the climactic events in metallurgy, our ignorance of the rate at which new techniques were adopted, and the factors accounting for these rates is, if not total, certainly no cause for professional self-congratulation. Much of the history of the past two centuries or so has been written by scholars with impressive credentials for technological history and its minutiae, but with a more limited appreciation of the economic consequences of techno-

*The author wishes to acknowledge the helpful comments and criticism received on an earlier version of this paper by the participants in the economic history workshops at the University of Chicago, Indiana University, and the University of Wisconsin. I am also grateful to Peter Lindert, who was particularly helpful at a later stage.

logical events. Thus, H.W. Dickinson, in his classic account of the history of the steam engine states, with respect to the consequences of the improvements which Watt had affected, that by the end of the eighteenth century, "The textile industry had been transformed from handicraft into a machine industry."[2] Although Dickinson wrote with a commanding authority on purely technological matters, it is necessary to record that he was almost a half century off the mark with respect to his economic history. By the end of the eighteenth century a suitable technology simply did not exist for the application of power to fully mechanized spinning or weaving operations.[3]

If we turn to the recent works of economic historians for further assistance on an issue as central to industrialization as the diffusion of steam power in the nineteenth century, we are offered some illumination but not nearly as much, surely, as we are entitled to expect.[4] Habakkuk states that "steam did not begin to play an important role in powering the British economy until the 1830s and 1840s, and was not massively applied until the 1870s and 80s. Even as late as 1870 less than a million horsepower was generated by steam in the factories and workshops of Great Britain."[5] Although Habakkuk does not indicate the source, these assertions are clearly based on Mulhall's estimates which included a figure of 900,000 horsepower for the fixed steam-power of Great Britain in 1870.[6] Mulhall was, of course, a great statistical pioneer, but his estimates were necessarily crude and, in some cases, no more than informed guesses. It is, nevertheless, symptomatic of the limited research conducted on the subject of technological diffusion so far that our knowledge of the diffusion of steam power has not been advanced, in the twentieth century, substantially beyond Mulhall's venerable *Dictionary of Statistics.* This late-nineteenth-century work is still the last word on the subject. Our knowledge of the sequence of events at the purely technical level remains far greater than our knowledge of the *translation* of technical events into events of economic significance.

The present paper attempts to take some steps toward closing the gap between the technical and the economic realms of discourse. In doing so, I am hopeful that we will eventually end up with a better understanding of the nature of the linkages between these two realms. Such an improved understanding should enable us to get a better handhold on the timing of the diffusion process and thereby make it possible to formulate and to test more precise hypotheses concerning the spread of new technologies. This paper does not attempt to examine the whole range of factors influencing diffusion. Rather, it concentrates on certain supply side considerations. Needless to say, alterations in relative factor and commodity prices also affect the rate of diffusion of new technologies - but that is obviously so. We argue here, not so obviously, that, factor and commodity prices aside, the rate at which new technologies replace old ones will depend upon the speed with which it is possible to overcome an array of supply side problems. These problems have not been uniformly resistant to efforts to overcome them, but have proven to be of varying degrees of intractability. This is the basic justification for the focus of this paper.

II. The continuity of inventive activity

That the diffusion of inventions is an essentially economic phenomenon, the timing of which can be largely explained by expected profits, is by now well established. The extended labors of Griliches and Mansfield have clearly demonstrated the power and scope of purely economic explanations in the diffusion of individual inventions. However, if one examines the history of the diffusion of many inventions, one cannot help being struck by two characteristics of the diffusion process: its apparent overall slowness on the one hand and the wide variations in the rates of acceptance of different inventions, on the other.[7] It is argued in this paper that a better understanding of the timing of diffusion is possible by probing more deeply at the technological level itself, where it may be possible to identify factors accounting for both the general slowness as well as wide variations in the rate of diffusion.

How slow is slow? When we speak of diffusion as being relatively slow, we are obviously implying some sort of dating procedure as well as expressing a comparative or absolute judgment. It should be noted at the outset that whether inventions are measured as diffusing rapidly or slowly depends in large part upon the selection of dates. If one dates the steam engine from the achievements of Newcomen around the first decade of the eighteenth century rather than from the work of Watt in the last third of the eighteenth century, as is commonly the case, one gets a much slower rate of diffusion. But on the basis of criteria commonly employed, the case for dating the steam engine from Newcomen is a perfectly compelling one. His atmospheric steam engine was not only technically workable, it was commercially feasible as well. To be sure, it experienced great heat loss in its operation and was a voracious consumer of fuel, but it nevertheless survived the market test and was widely used in the eighteenth century, primarily for pumping water out of mines. One very cautious study has identified 60 Newcomen engines for the period 1712-1733 and no less than 300 for the years 1734-1781. [8] Moreover, even in 1800, by which time Watt's patents had all expired, Newcomen engines not only continued to be used but, due apparently to their low construction and maintenance costs and long life expectancy, still continued to be *built*. One might almost be tempted to say of James Watt that he was "just an improver," although such a statement would be comparable to saying of Napoleon that he was just a soldier or of Bach that he was just a court musician. That is to say, Watt's improvements on the steam engine transformed it from an instrument of limited applicability at locations peculiarly favored by access to cheap fuel, to a generalized power source of much wider significance. Nevertheless, even if we date the steam engine from Watt's accomplishments of the 1760s and 1770s, it still took a full century of improvement and design change before this new power source surpassed water power in manufacturing and displaced the sail on ocean-going vessels.

The essential point to be grasped here is that inventive activity is, itself, best described as a gradual process of accretion, a cumulation of events where, in general, continuities are much more important than discontinuities. Even where it is possible to identify major inventions which seem to represent entirely new concepts and therefore genuine discontinuities, sharp and dramatic departures from the past, there are usually pervasive technological as well as economic forces at work which tend to slow down and to flatten out the impact of such inventions in terms of their contribution to raising resource productivity. Thus, even the big technological breakthroughs which are associated with such names as Darby, Watt, Cort, and Bessemer, usually have much more gently declining slopes of cost reduction flowing from their technical contributions than the historical literature would lead us to expect.

The fact is that the period which is looked at as encompassing the diffusion of an invention is usually much more than that: It is a period when critical inventive activity (what Usher called "secondary inventions") and essential design improvements and modifications are still going on. Although we might be tempted to dismiss this later work as much less important than the initial technological break-through, there is no good *economic* reason for this attitude, for it is precisely this later work which first establishes commercial feasibility and therefore shapes the possibilities for diffusion. We need to approach this whole area of research with a clearer appreciation of the continuum of inventive activity, running from initial conceptual-ization (the "Eureka! I have it!" stage) to establishment of technical feasibility (invention) to commercial feasibility (innovation) to subsequent diffusion. By concentrating our attention upon the sharp discontinuities associated with major inventions, we are misrepresent-ing the manner in which the gradual growth in the stock of useful knowledge is transformed into improvements in resource produc-tivity.

Although we find it a convenient verbal shorthand to speak of the "displacement" of one technique by another, the historical process is often one of a series of smaller and highly tentative steps. Thus, there were several intermediate stages in the displacement of sail by steam: Steamships were at first fully rigged and long continued to carry at least auxiliary sails, particularly as an insurance against breakdowns (which were not infrequent) on long ocean voyages, whereas sailing ships, in their twilight days, were often furnished with auxiliary engines.

The inadequate conception of invention as an intermittent and discontinuous phenomenon has been shared by the historian and the economist. The historian finds that it immensely simplifies the writing of chronological history if particular events - in this case inventions - can be precisely pinpointed in time, just like the Great Fire of London, the accession of the House of Hanover to the British throne, the Treaty of Paris or the abolition of the Corn Laws. The economist, for his part, finds it enormously convenient for analytical purposes to distinguish between movements along existing isoquants in response to alterations in factor prices, and shifts in the isoquants themselves due to the intermittent phenomenon of technological change. Both disciplines, therefore, have in the past found it convenient to focus upon technological change as a discontinuous phenomenon which can be precisely located in historical time. But by breaking into the continuum of inventive activity in this way, we are, subtly but inevitably, distorting our approach to the diffusion process. Once an invention has been "made," after all, the natural expectation is that all that remains in the sequence is for it to be adopted. Any delay becomes a "lag."

However, in viewing the problem in this way, we are very much underestimating the technological and economic importance of the subsequent "improvements." We are engaging in a sort of conceptual foreshortening which distorts our view of later events. For we are led to treat the period after the conventional dating of an invention as one where a fairly well-established technique is awaiting adoption whereas, in fact, highly significant technological and economic adaptations are typically waiting to be made. It is this same foreshortening of perspective which greatly increases our general impression of the slowness of diffusion.

Thus, Enos has studied the length of the lag between invention and innovation for 35 important innovations and has subjected his results to statistical analysis. In his sample the arithmetic mean for the interval between invention and innovation is 13.6 years.[9] Enos defines the date of an invention as "the earliest conception of the product in substantially its commercial form" and dates an innovation from "the first commercial application or sale." [10] It is difficult, however, in view of his definition of invention, to know just what significance to attach to the size of these lags. In some cases the date selected for invention corresponds fairly closely to a time when the technical problems were reasonably well solved. In other cases his choice of dates seems to be based more closely upon "earliest conception," when many important technical problems still remained to be solved. For example, Enos places the invention of the cotton-picker in 1889 and its innovation in 1942 - a lag of 53 years. It was in 1889 that Angus Campbell first made use, on an experimental basis, of a spindle-type picker, but the machine was far from constituting a satisfactory picker. As Jewkes and his co-authors had pointed out, "the machine left cotton on the ground and caused damage to the bolls and blooms."[11] It will be readily conceded that these are serious deficiencies in a device for picking cotton (even though one can conceive of possible price relationships where the adoption of such a machine would have been economically rational). Numerous important technical problems therefore remained to be solved after 1889. Similar complaints can be raised over many of the other dates employed in the Enos study - such as the 79-year interval for the fluorescent lamp (invention dated 1859), the 56-year interval for the gyro-compass (invention dated 1852), or the 22-year interval for television (invention dated 1919).[12]   In these and in many other cases, much critical and indispensable inventive activity remained to be performed. The length of the interval, in other words, was at least partly - and in many cases primarily - due to the time required for carrying out *further* inventive activity. It is therefore extremely difficult to know what significance ought to be attached to Enos's numerical findings concerning these "lags."

III. Improvements in inventions after their first introduction

It follows from the conception of invention adopted in this paper that most inventions are relatively crude and inefficient at the date when they are first recognized as constituting a new invention. They are, of necessity, badly adapted to many of the ultimate uses to which they will eventually be put; therefore, they may offer only very small advantages, or perhaps none at all, over previously existing techniques. Diffusion under these circumstances will necessarily be slow because the clear superiority of the new technique over the old has not yet been established or, perhaps, because the new technique or process alters the quality of the final product in unfortunate or unpredictable ways. Thus, as John Nef has shown, the emergence of a coal-using technology in England was seriously impeded in the seventeenth century by the fact that the use of the mineral fuel damaged the final product - as in the case of glass-making and the drying of malt for the brewing industry.[13]   Indeed, it was out of this attempt to maintain high quality of final product while using coal as a fuel that major advances were made in furnace design and the technique of coking was eventually developed.[14]   More important, the very slow diffusion of the coke-smelting of iron after Abraham Darby's first success in 1709 was due in part to the fact that "coke pig-iron produced a bar inferior in tensility and ductility to that made from charcoal pig: it was 'cold-short' and unsuitable for the production of wares of quality."[15]   Consequently the use of coke pig-iron was confined to the much smaller, cast iron branch of the iron industry. The adoption of a new technique, then, is often limited by imperfections in the product which, in turn, are only gradually overcome or bypassed. Such problems connected with quality control were long a source of persistent difficulties in the use of iron and steel. The inability to achieve a rigid and precise control of the quality and therefore of the performance characteristics of the metal was a major handicap in the application of ferrous materials to a range of industrial uses. Such control came only with the introduction of the Bessemer and open-hearth methods, since these made it possible to control the carbon content within very narrow limits.[16]

If it is true that inventions in their early forms are often highly imperfect and constitute only slight improvements over earlier techniques, it also follows that the pace at which subsequent improvements are made will be a major determinant of the rate of diffusion.[17] Indeed it may very well be the case that such improvements will reduce total costs by an amount greater than the reduction in costs of the initial invention over the older techniques which it eventually replaced. Mak and Walton argue that, on the Louisville-New Orleans route, "The introduction of the steamboat, 1815-20, led to a significant fall in real freight costs, but the absolute as well as the relative decline in real freight rates was greatest during the period of improvement, 1820-60."[18] Not all of the improvement in productivity, of course, was attributable to technological change. In addition to the increase in cargo carrying capacity per measured ton and the extension of the navigation season, which *were* primarily the result of technological changes, there were significant reductions in cargo collection times and passage times, which were not. Nevertheless, it is clear that the overall increase in total factor productivity associated with this major transportation innovation came in the years *following* its initial introduction.[19]

A similar conclusion is reached by Enos in his study of technical progress in the petroleum refining industry in the twentieth century. Enos examined the introduction of four major new processes in petroleum refining: thermal cracking, polymerization, catalytic cracking, and catalytic reforming. In measuring the benefits for each new process he distinguished between the "alpha phase" - or cost reductions which occur when the new process is introduced - and the "beta phase" - or cost reductions flowing from the subsequent improvements in the new process. Enos finds that the average annual cost reductions generated by the beta phase of each of these innovations considerably exceeds the average annual cost reductions generated by the alpha phase (4.5% as compared to 1.5%). On this basis he asserts: "The evidence from the petroleum refining industry indicates that improving a process contributes even more to technological progress than does its initial development."[20]

One final, general point needs to be made in concluding this section. It seems to be extraordinarily difficult to visualize and to anticipate the uses to which an invention will be put. Railroads were originally thought of as essentially feeders to canals and other forms of water transportation. In the early days of radio at the turn of the century, it was regarded primarily as a supplement to wire communication services, to be used only where wire was not practicable - as in certain isolated locations or for ships at sea. [21] Finally, even so versatile an inventor as Thomas Edison is said to have thought that the phonograph would be useful principally to record the wishes of old men on their death beds.

It is easy to sneer at such failures of vision and poverty of imagination - especially since, in history, we always have the immense advantage of knowing how the story ended. Nevertheless, past experience suggests that the prediction of how a given invention will fit into the social system, the uses to which it will be put, and the alterations it will generate, are all extraordinarily difficult intellectual exercises.[22] Such difficulties, in turn, must have played an important role in slowing down the pace of diffusion.

Even when an invention genuinely contains important elements of novelty, there is a strong tendency to conceptualize it in terms of the traditional or familiar. Thus the transition to a new technique is often slowed by the extreme difficulty of breaking away from the old forms and embracing the different logic of a new technique or principle.[23]

IV. The development of technical skills among users

Closely associated with this gradual improvement in the innovation itself is the development of the human skills upon which the use of the new technique depends in order to be effectively exploited. There is, in other words, a learning period the length of which will depend upon many factors, including the complexity of the new techniques, the extent to which they are novel or rely on skills already available or transferable from other industries, etc.[24] There is abundant evidence from a variety of sources showing sustained reductions in real labor costs per unit of output in situations where labor was employed in a plant using unchanged facilities. Indeed, the phenomenon is sufficiently well established that it has come to be known as the "Horndal Effect," after the Swedish steelworks where output per manhour was observed to increase at about 2% per year for fifteen years in spite of the fact that the plant and production techniques remained unchanged. The phenomenon has been further documented in several industries, most notably air-frame production, machine tools, shipbuilding, and textiles.[25]

While the existence of learning curves within the framework of an *established* technology is well recognized, the role of learning experiences in accounting for the gradual improvements of *new* technologies and their slow diffusion has not received much attention. Since it takes time to acquire such skills, it will also take time to establish the superior efficiency of a new technique over existing ones. The point is nicely illustrated with respect to the adoption of Henry Cort's puddling process:

> One of the most important problems associated with puddling in its early years was that of training a labor force that could produce high quality bar iron with the process. The ironmasters who initially adopted puddling had to train workers in the use of a process that was not only new, but was also somewhat of a "mystery" to everyone, including Cort. An efficient puddler was a workman who could not only do the strenuous labor of moving masses of iron in and out of the puddling furnaces, but could also develop a "feel" for the process itself. He had to learn to determine from the color of the flames in the puddling furnace and the texture of the molten metal when the pig iron was fully decarburized or had "come to nature," i.e., when the carbon and other impurities had been sufficiently removed. Puddling was a backbreaking job that also required a great deal of judgment and experience was probably the best teacher. The development of a highly skilled labor force was crucial to the success of the puddling process and the lack of such a labor force was probably the greatest single impediment to its rapid adoption.[26]

The adoption of the steam engine afforded innumerable examples of the importance as well as the slow accumulation of know-how, which is essential to the successful operation of a new technology. The length of life and the frequency of breakdowns of steam engines required that overloading be scrupulously avoided, but this in turn involved the accumulation of experience concerning optimum loads. Again, the life expectancy of engine boilers required careful attention to such matters as pressure levels and appropriate feed-water arrangements - indeed, when these matters were *not* attended to, the life of a boiler was likely to be abruptly, and sometimes disastrously, terminated.[27]

The manner in which these new technical skills are acquired is relevant to the speed of the diffusion process in another way. Many of the technical skills are acquired through direct, on-the-job participation in the work process. Since these include a large component of uncodified skills (or know-how), such skills were not readily transferable through formal education or the printed word, but required the movement of qualified personnel. Where and to the extent that this was so, it placed a serious constraint upon the speed of geographic diffusion.[28]

V. The development of skills in machine-making

The next portion of my argument deals with the development of the skills involved not in *using* the new techniques but in developing the skills and facilities in machine-making itself. This involves the broadest questions of industrial organization and specialization and lies at the very heart of the industrialization process. Successful invention and successful *diffusion* of inventions in industrializing economies has required, above all, a growth in the capacity to devise, to adapt, and of course, to produce at low cost, machinery which has been made suitable to highly specialized end uses. Before a new invention can join the family of technical options genuinely available to the economy, elaborate arrangements must sometimes be made. The mere conceptualization of a solution may be, and often has been, very far removed in calendar time from the availability of a method which is technologically workable, much less commercially feasible.

It is an often-told tale in the history of inventions that they have to sit on the shelves long after their initial conceptualization because of the absence of the appropriate mechanical skills, facilities, and design and engineering capacity required to translate them into a working reality. This is why the emergence of a capital goods industry, with a sophisticated knowledge of metallurgy and the capacity to perform reliable precision work in metals, was so critical to industrialization in its eighteenth- and nineteenth-century form. The desperate and unsuccessful improvisations which otherwise had to be resorted to is perfectly captured in the picture of James Watt stuffing soaked rags in the gaps between his pistons and cylinders in an effort to prevent loss of steam, until Wilkinson's boring mill finally provided him with reasonably accurate cylinders. The commercial practicability of Watt's steam engine with its separate condensing chamber really dates from 1776, the year in which Wilkinson's boring mill, invented in 1774, became available for the preparation of his cylinders.[29]

It might be said of Watt that he was singularly lucky in having a cannon-maker such as Wilkinson nearby, but the essential point, surely, is that his conceptualizations were not so far in advance of the technical capacities of the metal-workers of his day as to render his ideas unfeasible.[30] Such was the lot of da Vinci whose notebooks are crowded with machinery sketches far in advance of the technical skills of early sixteenth-century Florence or Europe. Breech-loading cannon had been made as early as the sixteenth century but could not be fired in reasonable safety (to the user at least!) until precision in metal-working made it possible to produce an airtight breech and properly-fitting case. Christopher Polham, a Swede, devised many techniques for the application of machinery to the quantity production of metal and metal products, but could not successfully implement his conceptions with the power sources and clumsy wooden machinery of the first half of the eighteenth century. Charles Babbage had already conceived of the main features of the modern computer over a hundred years ago and had incorporated these features into his "analytical engine," a project for which he actually received a large subsidy from the British Government. Babbage's failure to complete his ingenious scheme was due to the inability of contemporary British metal-working to deliver the components which were indispensable to the machine's success.[31]

People like da Vinci and Babbage were, to use the popular phrase, "far ahead of their time." But, to give the phrase some operational content, it is the state of development of the capital goods industries, more than any other single factor, which determines whether and to what extent an invention is ahead of its time. Each important invention goes through a gestation period of varying length, while the capital goods industries adapt themselves to the special needs and requirements of the new technique. Therefore the pace of technical advance in the user industry may depend critically upon events in the capital goods sector. This process of problem-solving and accommodation is central to a better understanding of the timing of technical change and the rate of diffusion of new inventions.[32] For it is the speed with which performance characteristics are improved, techniques modified to meet the needs of specialized users, and the price of the invention gradually reduced, which determine its acceptability among an increasingly widening circle of potential users.

VI. Complementarities

A further element significantly affecting the timing of the diffusion process, and one where the capital goods sector also plays an important role, lies in the complementarity in productive activity between different techniques. That is to say, a given invention, however promising, often cannot fulfill anything like its potential unless *other* inventions are made relaxing or bypassing constraints which would otherwise hamper its diffusion and expansion. It is for this reason that a single technological breakthrough hardly ever constitutes a complete innovation. Before the productivity-increasing benefits of any single breakthrough can be realized, many other accommodations need to be made. The expansion of a productive activity runs into a series of new constraints or bottlenecks. As one bottleneck is overcome, others eventually assert themselves and need to be expanded. Although these bottlenecks can often be overcome by committing more resources to a particular activity, frequently inventive activity is called for. In both cases, however, time-consuming procedures are involved which hold back the further expansion and diffusion of the new technique.

The history of American railroads in the second half of the nineteenth and the early twentieth centuries provides compelling evidence that the growth in productivity was the product of many subsequent inventions, none of which was available in the early years of railroad building - say 1840. The growth in productivity was, to begin with, very great. It has been calculated that the incremental expenses required for meeting the railroad demands of 1910 traffic loads with the technology available back in 1870 would have amounted to about $1.3 billion. The cumulation of small innovations and relatively modest individual design changes brought about, between 1870 and 1910 alone, a more than tripling of freight car capacities with only a small increase in dead weight, and a more than doubling of locomotive force with the introduction of more powerful engines. The greater loads and greater speeds made possible by the improved rolling stock could not have been achieved, however, without several other significant inventions: the control of train movements through use of the telegraph, block signalling, air brakes, automatic couplers, and the substitution of steel rails for iron. Not all these inventions were equally significant in reducing costs, nor were they, as a result, adopted with equal speed. Air brakes and automatic couplers (first employed in 1869 and 1873 respectively) were coolly received and were eventually adopted only after the passage of national legislation.[33]   Steel rails, however, in spite of their considerably higher price, were rapidly adopted. Steel rails were first used by the Pennsylvania Railroad in the early 1860s, and by 1890 they accounted for 80% of all track mileage.[34]   The critical importance of steel rails to the growth in railroad productivity was that they were far more durable, lasting more than ten times as long as iron rails, and that they were capable of bearing far heavier loads than iron rails without breaking. Indeed, the old iron rails were simply incapable of supporting the 1910 locomotives, and would have crushed under their average weight of 70 tons.[35]

The argument made here with respect to complementarities in railroad operation could, if space permitted, be expanded to encompass other classes of invention not so far mentioned, such as bridgebuilding. Bridgebuilding in both America and Great Britain underwent drastic changes in structure, design, and materials as the railroad network expanded and confronted engineers with problems and requirements respecting such matters as strength, rigidity, and fire resistance for which there was absolutely no precedent in the pre-railroad age. The need to provide bridges suitable to the requirements of the railroads led, in Great Britain, to a systematic study of iron as a building material. The outcome of this study was a major advance in knowledge concerning the structural properties of iron - resistance of beams and plates, strength of girders, compression and tensile strengths, etc.[36] The knowledge thus acquired soon had a wide range of applications wherever iron was used as a building material - in shipbuilding, multistoried buildings, cranes, steam engines, etc.

The present discussion is merely illustrative of a much larger class of complementarities which exert significant effects upon the pattern of diffusion. The argument could readily be duplicated almost endlessly from different sectors of industry and agriculture.[37]

VII. Improvements in "old" technologies

The discussion so far has examined a variety of factors which have the effect of slowing down the diffusion of new techniques. We have considered several classes of reasons why the full productivity-increasing effects of a new technique may take a great deal of time to assert themselves. There is, however, an additional powerful explanation for such slowness which has received surprisingly little attention. That is, the "old" technology continues to be improved after the introduction of the "new," thus postponing even further the time when the old technology is clearly outmoded. Yet, curiously, it is a very general practice among historians to fix their attention upon the story of the new method as soon as its technical feasibility has been established and to terminate all interest in the old. The result, again, is to sharpen the belief in abrupt and dramatic discontinuities in the historical record.

A closer look at this record discloses not only the persistence of old technologies in places where location or resource availability provided special advantages, but often major improvements in these technologies long after they were supposed to have expired. Thus, the slowness with which the stationary steam engine, as noted earlier, established itself as a new power source in the first half of the nineteenth century, was due in part to important improvements which continued to be made in design and construction of water wheels. Many of these improvements centered around the introduction of iron as a building material, and men like William Fairbairn could achieve international reputations well into the nineteenth century as builders of water wheels.[38] In fact, early steam engines were commonly used to supplement the action of a water mill by pumping the water from the lower mill pond back to the upper pond - thus enabling it to run over the water wheel many times. Needless to say, this was a cumbersome and inefficient arrangement. Nevertheless, until the development of the rotative steam engine which converted the oscillation of the beam into rotary motion, it was an essential expedient since the steam engine prior to that was really no more than a water pump.[39]

An additional fillip was given to the utilization of water after 1840 as a result of the introduction of the water turbine, which further reduced the cost of water power.[40] In America, where early industry had been heavily concentrated in New England, a region with highly favorable water power locations, water power was probably the main power source for manufacturing until well into the second half of the nineteenth century, Even as late as 1869, steam power accounted for barely over one half of primary-power capacity in U.S. manufacturing - 51.8% as compared to 48.2% for water.[41] In general it may be said that steam power tended to be adopted earliest in locations where water power sources were scarce and where fuel was either abundant or easily transportable.

Even so "primitive" a power source as the windmill, which might have been regarded as a certain early casualty to the steam engine, experienced a considerable growth in at least one English county. According to Finch, the number of windmills in Kent employed in the grinding of corn grew from 95 to 239 between 1769 and the 1840s.[42] By the 1860s steam had decisively established its superiority over wind in cornmills, and windmills continued to be operated only so long as they provided quasi-rents to their owners. Jevons, writing in the middle of the 1860s, pointed out: "Wind-cornmills still go on working until they are burnt down, or go out of repair; they are then never rebuilt, but their work is transferred to steam-mills."[43]

In iron and steel, where the location of particular resource inputs and their chemical properties played an extremely important role (especially because of the intimate connection between the presence of such chemicals as sulfur and phosphorus and a poor quality final product), the introduction into the United States of the modern, mineral-based British technology was long delayed. Furthermore, the old and the new technology coexisted for long periods of time after the new technology was finally introduced. As late as 1840 almost 100% of all pig iron in the U.S. was still being produced with charcoal. Although the proportion subsequently declined rapidly as anthracite and, later, bituminous coal were introduced,[44] the tonnage of pig iron produced by charcoal continued to rise through the 1880s and reached its alltime annual peak in 1890.[45] The most interesting point here for present purposes is that during the 1840s and 1850s, when the new mineral fuel technology was being introduced, it was primarily the reduction in the demand for charcoal iron which accounted for the relative decline in the charcoal sector as compared to the mineral fuel sector. Indeed, Fogel and Engerman actually conclude that, between 1842 and 1858, the growth in total factor productivity in the "backward" charcoal sector probably exceeded the growth in total factor productivity in the "modern" anthracite sector.[46]

My point so far has been that one of the reasons new technologies seem to displace old ones slowly is that the old technologies continue to improve. But the point needs to be seen within a larger framework, for there is often an intimate connection between innovations, on the one hand, and improvements in older technologies on the other. That is to say, innovations often appear to *induce* vigorous and imaginative responses on the part of industries for which they are providing close substitutes. What is being suggested here is a possible lack of symmetry in the manner in which business firms respond to alterations in their profit prospects. The imminent threat to a firm's profit margins which are presented by the rise of a new competing technology seems often in history to have served as a more effective agent in generating improvements in efficiency than the more diffuse pressures of intra-industry competition. Indeed, such asymmetries may be an important key to a better understanding of the workings of the competitive process, even though they find no place at present in formal economic theory.

Thus it has often been asserted that, by the 1850s, the iron hull cargo steamship had displaced the sailing ship and that Britain built its worldwide trading empire on the new vessel. In fact, while the complex problems of designing an efficient steamship, with its iron hull, engines, screw propellers, and very high fuel requirements, were being worked out, the wooden sailing ship also underwent a series of drastic changes. Builders of sailing ships responded to the competition of iron and steam by a number of imaginative changes in hull design, including the use of iron itself in a "composite" hull - wood placed on an iron skeleton. They adopted a range of labor-saving machinery to reduce crew requirements. According to Graham:

Although the steam ship had successfully wedged its way into the overseas trade, mainly by carrying passengers and subsidized mails, the evolving sailing ship of the 1860's and 1870's - faster than its predecessors, with double the space for cargo in proportion to tonnage, and manned and navigated by about one-third the number of men - retained on broad oceans a predominance almost as marked as that of the screw steamer in the coastal and neighbouring waters of Europe. Even when the opening of the Suez Canal in 1869 reduced the longest gap between coaling stations from some 5,000 to 2,000 miles, although the China trade was eventually lost, most of the traffic to the Bay of Bengal, the East Indies, South America or Australia, was still conducted by the sailing ship which continued to be the more economical carrier for the greatly expanding trade in bulky commodities, such as iron and coal, the jute and rice of India and Burma, the wool of Australia, the nitrate fertilizer of Chile and the wheat of California.[47]

Even the rapid growth of the steamship fleet in the 1870s offered some consolation to the sailing ship: "it was coal more than any other article that brought a new lease of life to the commercial sailing ship in the latter days of her ascendancy. As the cheapest coal carrier, as well as the cheapest warehouse in the world, the sailing ship became the chief replenisher of overseas coaling bases and depots."[48]   Furthermore, some of the design changes which improved the performance of the sailing ship were also made possible by steam, in this case by the steam tugboat which, "taking them in and out of harbor, relieved the windjammers of need for handiness, enabling greater length and fine lines, and enabling guaranteed sailings out of a harbor."[49]

By the 1880s the sailing ship finally lost its dominance even on long distance hauls to steamships, which were now equipped with high pressure compound engines and a range of superior components provided by the recent breakthroughs in steel-making technology - including the very important boiler plates and boiler tubes, so essential to high pressure and fuel economy.[50]   The sailing ship of the 1880s was far superior to its predecessor of 1850 or so, and it seems plausible to attribute this improvement to the strong competition of steam.[51]   Obviously one cannot assert this with authority, because we do not know what the sailing ship of the 1880s would have been like in the absence of such intertechnological competition. But it seems like a reasonable conjecture for which there is analogous evidence in the experience of other industries. Thus, technological competition recently appears to have been a powerful force among materials producers (where, for example, the increasing competition from aluminum seems to have led to the setting up of product-research and engineering laboratories in the steel industry), among suppliers of transportation services, and among the major kinds of fuel. Not only does this form of competition generate economically beneficial consequences; it also plays a significant role in explaining the rate of diffusion of some new techniques.

VIII. Diffusion and its institutional context

This paper has discussed, at some length, several categories of technological considerations which have influenced the pace of diffusion. Needless to say, the treatment has been suggestive rather than exhaustive; in fact, the number of variables - social, legal, and institutional as well as economic and technological - which might retard the diffusion process is virtually limitless. Nevertheless, it is important that an effort be made to maintain conceptual clarity among these categories because our understanding of the process of long-term economic growth is influenced, in important ways, by this conceptual apparatus. Ever since Abramovitz and Solow opened up the problem of "The Residual," economists have been attempting to sort out the contributions of various factors to economic growth and, particularly, to measure the contribution of technological change as distinguished from all other possible factors. Whereas the entire residual was for some time uncritically attributed to technological change (although not by Abramovitz or Solow) a later, more discriminating approach has attempted to isolate other factors - changes in organization, improvements in the quality of the labor force, etc. - and to measure their separate contributions.[52] In this difficult but essential process of "cutting technological change down to size," however, there is a danger of going too far, by assigning an independent and separate role to factors which really exert their effects upon the growth of productivity by retarding or accelerating the rate of technological diffusion.

A recent example of the position I am criticizing is North's otherwise admirable study of "Sources of Productivity Change in Ocean Shipping, 1600-1850."[53] North states that his objective is to "identify as precisely as possible those sources of productivity usually lumped into the general category of technological change." And, he adds: "The conclusion which emerges from the study is that a decline in piracy and an improvement in economic organization account for most of the productivity change observed."[54]

With a portion of North's argument there is no disagreement whatever. North makes an important contribution to our understanding of productivity growth during the period by demonstrating that organizational and marketing improvements were highly significant. He shows that, in the tobacco trade before the Revolutionary War, the increased number of round trips per year was due, not to increased speed resulting from technological change, but rather to the introduction of a system of factors and an increasing centralization of inventories. These organizational improvements, by making it easier and much quicker for ships to secure cargoes, substantially reduced the ratio of port time to sea time, and thereby sharply increased the quantity of freight which could be carried by a given stock of ships.[55]

The other portion of North's argument, involving an attempt to downgrade the contribution of technological change to the growth of productivity in ocean shipping, is more questionable. North finds that, before 1800, the most important source of the great reduction in crew size requirements was the decline in piracy and privateering. His "downgrading" of the role of technical change proceeds as follows:

One can ask at this point ... the extent to which technical changes in shipping and in ship construction account for the changes in manning requirements and, indeed, in observed ship speed. There is no doubt that the ship of the nineteenth century was in striking contrast to the ship of the early seventeenth century. Except for one crucial point it could be argued that smaller crews were made feasible precisely by technological improvements in sail and rigging. The obstacle to this argument is that by 1600 the Dutch had developed a ship, the flute, which cost less to construct than existing ships, had a tons-per-man ratio similar to that of nineteenth-century ships on the Atlantic route, was at least as fast as existing ships, and could be (and was) constructed of 500-600 tons burden. While the design was copied and modified over the next two centuries, the essential economic characteristics were not basically altered. The enigma to be explained, therefore, is why the flute (or ships of similar design) took so long to spread to all the commodity routes in the world, once it had entered the Baltic route and the English coal trade in the first half of the seventeenth century. The answer lies in the very nature of the flute and its great advantages in that it was lightly built, frequently carried no armament, was easy to sail, and had simple rigging. These characteristics had all come about because the Dutch enjoyed a large-volume bulk trade in the Baltic, where piracy had already been eliminated. Only as privateering was driven from other seas and as improvements took place in market organization was it possible to put into general active service ships designed exclusively for the carrying trade.[56]

The trouble with this paragraph is that the diffusion process has been completely lost from view. A superior technology in the form of the Dutch flute existed by 1600, but security considerations long confined its adoption to a small portion of ocean trade. As piracy and privateering were suppressed in the course of the eighteenth century, the superior vessel was widely adopted on new routes with the expected rise in productivity. But, if all this is so, the elimination of piracy and privateering emerge as factors which influence shipping productivity only as intervening variables: i.e., the threat which they posed to the security of shipping was responsible for *the very slow diffusion of a superior technology* - "the flute (or ships of similar design.)" There seems to be general agreement that, in the *absence* of the security threat, the flute design would have been adopted much earlier.[57]   North, however, in his legitimate concern with deflating the overblown spectre of technological change, gives the impression - doubtless unintended - that it was scarcely of any significance whatever in the period with which he is concerned. It goes unmentioned in his final sentence, which states: "The conclusion one draws is that the decline of piracy and privateering and the development of markets and international trade shared honors as primary factors in the growth of shipping efficiency over this two-and-a-half century period."[58]

The interpretation which North has placed upon his historical account has been restated by Fogel and Engerman in their volume, *The Reinterpretation of American Economic History,* where North's article has been reprinted. Fogel and Engerman regularly refer to North's article as showing that factor productivity did not rise as a result of *new* inventions.

In the case of ocean shipping, Douglass North . . . found that a rapid and protracted increase in total factor productivity took place despite the absence of a single major *new* invention. According to North the rise of efficiency was due largely to the change in the proportion of large ships in the Atlantic fleet. This diffusion of large ships was set off, not by *new* technological knowledge, but by a change in institutional conditions.[59]

And, earlier: "Thus, *new* equipment plays virtually no role in Douglass North's explanation of the 50 per cent fall in the cost of ocean transportation that he finds for the 250-year period between 1600 and the middle of the nineteenth century."[60]

But if a superior ship designed specifically for improved cargo-carrying capacity had been developed by 1600, it is no verbal quibble to say that the improvements in ocean shipping productivity due to the eventual adoption of this design should correctly be regarded as belonging to the category of technological change. The portion of North's paper dealing with piracy is not an explanation of productivity growth which is *independent* of technological change, although it is frequently made to sound that way. Rather, it is a cogent and forceful explanation for the very slow *diffusion* of a major technological innovation. What seems to be at issue here - and this emerges with particular clarity in the writing of Fogel and Engerman - is that benefits attributable to technological change are being arbitrarily confined to recent or "new" developments. But there is no obvious reason why the productivity-increasing effects of technological change should be confined to changes of recent vintage. Surely the essential point, on which all would agree, is that the productivity of any technology is never independent of its institutional context and therefore needs to be studied within that context. North's paper should be interpreted as a striking demonstration of this point, for he shows how this institutional context can account for the very slow diffusion of a superior shipping technology.

IX

The several arguments of this paper all add up to a perspective and a program for research rather than a sharply defined set of conclusions. A variety of reasons have been advanced for believing that a new technique establishes its advantages over old ones only slowly, and it has been argued that the apparent slowness of the *diffusion* of technologies is linked to this process and needs to be studied in relation to it. In spite of the occasional appearance of inventions which seem to be spectacular for their *technological* novelty, the *economic* impact of such inventions is much more diffuse and gradual. Their introduction into the texture of the economy is more accurately - if less dramatically - viewed as occurring along a gradual downward slope of real costs rather than as a Schumpeterian gale of creative destruction. At the same time, it is perfectly apparent that the question posed earlier has not been answered: How slow is slow? (How fast is fast?). But it should be clear by now that I would not worry excessively about that failure so long as we can advance our understanding of the reasons for the *actual* historical pace of technological diffusion. Once that pace has been established, we can each go our separate ways in deciding what we choose to regard as fast or slow.

Part 4

Natural resources, environment and the growth of knowledge

12

Technology and the environment:
an economic exploration

Technology is and always has been a two-sided phenomenon. On the one side it encompasses all those forms of knowledge and technique which account for man's growing mastery over his physical environment and for his increasing ability to achieve specific human goals - whether these goals be improved diet, entertainment, shelter, more rapid transportation or communication, the prolongation of human life via an increasingly sophisticated technology of death control, or the prevention of the formation of new lives via the technology of contraception. On the other side, all technologies are accompanied by various side effects, some of which may be unpleasant or dangerous - these the economists call "negative externalities" (there are also positive externalities, but, perhaps characteristically, economists give them much less attention). In the extreme case these negative externalities may be so obnoxious that, by common consent, the activity giving rise to the externality should be terminated. I recall a cartoon strip some time ago in which an allergist was holding a consultation with his patient - Jones - and, at first, congratulating himself on having brought Jones's allergy symptoms under control. Eventually, in the final strip, the allergist concedes, reluctantly, that the treatment produced some rather nasty side effects which would have to be suppressed. Only then do we actually see Jones for the first time: He is approximately three inches tall! The cartoon is funny, obviously, precisely because the costs of the treatment in this case so absurdly outweigh its putative benefits. We instantly - and correctly - reject the employment of a technology to which very costly side effects are attached, as was the case in the thalidomide tragedy of a few years ago. Much more difficult questions arise where the costs of employing a technology are lower than this but still very substantial: where the costs, although high, make possible the attainment of highly valued goals not otherwise attainable: or where all the alternative technologies for the achievement of a particular goal contain substantial side effects; or where a high degree of uncertainty surrounds the employment of a beneficial new technology.

My purpose in this paper, therefore, will be to explore the nature of the technology-environment problem from an economic perspective, with a view to clarifying some of the major issues and alternatives. I shall focus first on these problems as they confront members of an affluent society such as the United States and turn later and more briefly to the special perspective of poor countries.

The urban context

One of the distinctive characteristics of modern technology is that it has been associated with new kinds of human interactions. The nature of these interactions is highly significant for both economic and social policies because they restrict the area in which the market mechanism, even under the most favorable competitive conditions, may be said to bring about an optimal allocation of resources. In some cases these new interactions flow directly from certain peculiar characteristics of modern technology; in others they flow from forms of social organization, particularly urbanization, closely related to the requirements of modern technology; in yet others the interactions are of a kind which have long existed but which have acquired much greater significance in recent years.

The market mechanism is an efficient allocator of resources, and the private pursuit of wealth maximizes the economic welfare of all members of society (for a given distribution of income) when there are no significant external effects. If, however, (1) my actions impose costs on other persons for which I am not required to provide compensation, or if (2) my actions confer benefits upon others for which I cannot receive remuneration, purely self-seeking behavior as mediated through the signals of the marketplace may lead to inefficient patterns of resource use. In the first case, too much of society's resources will be employed, and in the second case an insufficient amount. Modern technology-*cum*-urbanization has multiplied many times the frequency and the significance of such interactions. Furthermore, there is a large class of goods and services in which consumption is in some sense necessarily joint, where more of something for X does not mean that there is less available for Y or Z: flood control projects, public health measures, additions to the stock of human knowledge. Here, too, a reliance upon purely private cost-benefit calculations will lead to an insufficient volume of society's resources being devoted to the activity.

Consider the case of the supersonic jet as an example of a negative externality. If the jet were built and placed in operation, I could not choose not to listen to the sonic boom. It would be a cost imposed on me regardless of whether I made use of air travel. Were some people to travel at supersonic speed they would produce intensely unpleasant experiences for people who were *not* traveling. I should not have a choice in the matter, short of withdrawing from civilization - and note that, so long as Boston's Logan Airport continues to operate, Walden Pond will by no means serve as a sufficiently remote retreat.[1] Similarly, when large numbers of people opt to travel by car they produce smog and traffic congestion even for those who choose to walk. Or, when the teen-ager in the next apartment decides to play her latest rock records, I find myself a distraught member of a captive audience.

It is a distinctive feature of a technologically dynamic society that more and more of the choices and actions of individual consumers directly affect the welfare of others. This they do by altering the nature of the environment. The automobile and the supersonic jet plane are only two of the more notorious examples. We should have to include all cases involving the joint use of certain facilities from, quite literally, the air we breathe in varying degrees of impurity and the rivers which we pollute, to the common use of public highways where A's reckless driving or negligence poses a threat not only to his own life but to that of others driving on the same road. How my neighbor disposes of his leaves in the fall or clears his sidewalk of snow in the winter is of significance to me. Whether or not he smokes in bed is more than a matter of casual interest. Similarly, whether in a supermarket, subway, movie house, library, or laundromat, our behavior as consumers impinges in important ways upon the environment or "living space" of others and thereby affects their welfare. This is in the very nature of cities, which may be regarded as man-made environments specifically designed for the large-scale sharing of common facilities.

The beer bottle which bears the ostensibly reassuring message "No deposit, no return" neatly symbolizes many of a growing number of negative externalities. The cost of disposing of the bottle is not borne directly by either the brewery or the consumer, but the replacement of bottles (on which a deposit was paid) by cans or by nonreturnable bottles which can be conveniently discarded has significantly added to the volume of trash which has to be collected. The disposal of solid waste in fact poses a problem of increasingly serious proportions in affluent societies. I do not know the cost of disposing of the Sunday edition of the *New York Times,* but I should be most surprised if it is not a substantial percentage of the original purchase price. Further, in New York City alone, more than 57,000 automobiles were simply abandoned by their owners in 1969,[2] with the huge disposal problem left to the municipal authorities. The underlying economic forces, of course, are the high price of labor and low prices for scrap materials which are fast rendering uneconomic the traditional activities of the junkman. The end result, however, is that new and costly burdens are added to the services performed by the public sector. To consider another example, the use of tungsten carbide studs in automobile tires has been shown to provide significantly improved traction on ice and hard-packed snow. As a result, the sale of such studs has been booming in the past few years. While this undoubtedly represents a desirable innovation from a safety point of view, the properties of tungsten carbide which result in increased traction also result in considerable destruction to road surfaces. Such additional cost will be borne by the public sector, and this has caused some states to prohibit the use of such studs by law.

It seems to be a technological condition of modern life, therefore, that we engage in more numerous activities which generate costs we do not bear directly. The converse, however, is equally true. Members of urban communities not only generate costs for which they are not directly responsible but also generate benefits to other members of the community for which they receive no compensation. My neighbor's rose garden and azalea bushes cost me nothing yet constitute a considerable source of pleasure to me. The fire extinguisher he stores in his closet may vitally affect my family's welfare if it enables him to bring under control quickly a fire which might otherwise spread to my house. More generally, all expenditures made by my neighbors on the education of their children accrue as a benefit to me insofar as such expenditures improve my environment by raising the level of cultural and political sophistication in the community in which my family lives.

My welfare, in other words, is at least partly a function of other people's activities, tastes, and consumption expenditures.[3] While this has always been true in some respects, the thrust of modern technology and its implications for social organization is to increase sharply the number and importance of such interdependencies. Whereas they might have warranted an occasional footnote in a Jeffersonian society of small holders, they constitute some of the most vital issues of the densely populated, late 20th-century world. One may even conjecture that our growing national concern with poverty reflects more than just a sense of equity or human compassion. It reflects also the discomfort and even danger which the existence of poor people poses to those more comfortably situated. My welfare may be affected not only by the well-known disutility of envy but also by other people's poverty, whether the effect takes the form of social disturbances generated by urban ghettos, driving hazards compounded by motorists who neglect to replace worn-out tires or brake linings, or the hazards of living in a society where emotionally disturbed people are too poor to afford the costs of psychiatric care.

The plausibility of this argument is supported by the fact that the recent increasing concern over the plight of the poor has coincided with a general improvement in their material well-being. This is not really surprising. Social goals and individual aspirations are both closely related to our capacity to realize them. In this sense, one of the most important consequences of the rising productivity which flows from technological change is our growing realization that poverty and its attendant suffering and indignities can be eliminated in the American economy. The elimination of poverty has, in fact, only recently become an integral part of our public dialogue. No less interesting, from a longer-term perspective, is the fact that the very definition of poverty has, itself, undergone drastic upward revision in the past several decades.[4]

In a sense, our growing technological capacity is creating a powerful challenge by producing what amounts to a national "moment of truth." Although the abolition of poverty has long been part of our national rhetoric, it is only recently that a vague, utopian possibility has been transformed into something within the reach of our present productive capacities. In the past, poverty and its various manifestations have been tolerated on the grounds of unavoidability. Today it is becoming clear that the abolition of poverty is something concerning which we now have a genuine choice. It seems apparent that a growing awareness of these new possibilities has a good deal to do with the rising tensions of our society. The real challenge which technology currently presents to our egalitarian values is that it has deprived us of any technological excuse for failing to fulfill long-professed ideals. Technological change and rising productivity have been responsible for placing the elimination of poverty high on our national agenda. By thus confronting us with social options which were once inconceivable, it has forced certain questions to the forefront of our attention. The apparent malaise of a large proportion of American youth today, I suggest, is due in part to our collective failure to exploit the possibility presented by modern technology for the fulfillment of long-expressed ideals.

The ecological context

So far my discussion of the technology-environment problem has concentrated primarily on the man-made environment - the city. I should like to turn more explicitly now to the natural environment. I do not think it is necessary to rehearse the numerous ways in which the functioning of modern technology is inflicting damage - in some cases irreversible - upon our natural environment. This has been extensively catalogued and graphically depicted elsewhere. Indeed, every ecologist seems to have his own special chamber of horrors on industrial man's (or would-be industrial man's) apparently unlimited capacity for fouling his own nest. The evidence is indeed overwhelming that we have, in the past few decades, introduced drastic new agents into our environment - insecticides, herbicides, fertilizers, radiation, industrial waste - with an astonishing indifference to the possible ecological consequences. All sensible men must agree on the need to award a much higher priority to the improvement of our understanding of these interactions and to the monitoring and anticipating of future deteriorations. There is no excuse for allowing actions which may have significant ecological consequences to be undertaken without the best possible information or estimate of the likely nature and scope of these consequences.

Such assessments, however, should not be expected to make the decision-making procedure easy, but merely better informed. That is, they will clarify and perhaps in some cases quantify the alternatives with which society is confronted. For the simplistic, ultraconservationist view, which assumes that the mere identification of an activity as a source of pollution is a sufficient reason for terminating the activity, is not and never has been viable. We must face the fact that it is quite impossible to utilize any sophisticated technology - or, for that matter, many primitive ones - without engaging in numerous actions which alter some important dimension of the environment.[5] Indeed, in view of the tendency to wax nostalgic over a rural and bucolic past, it is worth remembering that one of the most drastic of all ecological disturbances, with respect to its impact on wildlife, was the conversion of forest and plain to cropland.[6]

The important point to be made here, however, is not only that ecological disturbances are unavoidable but that the objectionable forms of these disturbances are usually the outcome of a decision to adopt the cheapest technique available for a given productive process. This is most obviously the case where the offending substances are waste materials which are indiscriminately discharged into the atmosphere or the nearest body of water. That is, pollution is in large measure the product of deliberate decisions to minimize the direct costs of production on the part of a business firm, public utility, or municipal government. We could eliminate much of it if we decided - or allowed ourselves to be persuaded - that we were prepared to give up some portion of our material output in exchange for a more attractive and livable natural environment. Everyone wants unpolluted rivers and streams, just as everyone (at least everyone *I* have spoken to) laments the death of Lake Erie. But not everyone (certainly not everyone with whom I have discussed the matter) is prepared to pay his share of the huge costs of alternative methods for the disposal of industrial and municipal waste. Such unwillingness, surely, is the critical element. Merely to deplore the deterioration of our natural environment is to be on the side of the angels - that is, it does not genuinely confront the real issues. After all, if such destruction were merely wanton, if it could be terminated at no cost whatever, it would presumably be done immediately. The relevant question is how much we are prepared to pay to purify the Potomac or the Hudson or the Charles. How much urban renewal, medical research, foreign aid, and vocational rehabilitation are we prepared to forego so that our rivers may run pure once again? To such questions, such scrutiny of our individual and collective values, we should, I submit, now be addressing ourselves.

The fact is that techniques are at present available for achieving a high degree of purity of our water systems. One rough estimate places the annual cost of returning effluents to water courses in a state of "pristine purity" at approximately $20 billion.[7] When confronted with an annual bill of this magnitude, I confess that I find myself prepared to tolerate a certain amount of pollution. Twenty billion dollars could purchase a very considerable quantity of food or provide a very large amount of medical care or vocational training. Some portion of it, if intelligently spent, might even bring our high infant mortality rate (a national scandal if ever there was one) down to something like western European levels.[8] Happily, the decisions confronting us here are not of the all-or-none variety. There exists a gradation of degrees of water purity, each stage more pure than the very low stage we at present tolerate, which is available at progressively higher cost, all the way to the upper-bound estimate for pristine purity. I strongly support the expenditure of more public funds (for which I will happily pay increased taxes) to reduce environmental pollution substantially below its present levels. I must also add, however, that even if we were somehow to experience a sudden windfall of $20 billion I am quite certain that I should not opt for spending it all on water purification. Obviously, in saying this I am not delivering a professional pronouncement but rather expressing something of my own sense of social priorities.

We can, moreover, with our *present* technology, substantially reduce the amount of air pollution emitted by the electric utilities in big cities, either by removing the offending substances prior to combustion or by burning fuels which are more expensive but contain smaller amounts of these substances. In this respect fuel oil is preferable to bituminous coal, and natural gas is preferable to both. A large part of our present air pollution, in other words, is not an inevitable product of our present technology but rather the product of decisions to utilize low-cost fuel supplies.[9]

A related complication in addition to the high absolute cost of reducing pollution, to which we need to give a great deal of further thought, is the likely distributional consequences of such measures. A rise in electricity bills due to the employment of higher-cost fuels is likely to weigh most heavily upon low-income receivers; and the compulsory installation of an expensive automobile exhaust filter will push very few Cadillacs off our highways but will almost certainly result in the disappearance of numerous battered old jalopies which often constitute the only practical form of transport available to the rural and urban poor.[10] Specific measures to reduce pollution may turn out to be in direct conflict with the goal of reducing poverty.

The painfulness of the choices with which we are confronted may be forcefully seen in a different context. The public has become increasingly familiar in the past few years with some of the deplorable side effects of insecticides, herbicides, and commercial fertilizers. Yet these inputs have played a crucial role in the growth of farm output in the United States over the past thirty years. The use of such inputs was instrumental in the production of the large wheat surpluses which have recently been shipped to India during years of crop failure in that country. Without question, our willingness to employ these new chemical inputs in agriculture has had as a direct consequence, on at least one occasion, the avoidance of mass starvation in India. When this evidence is placed in the balance, the moral self-righteousness of some of the unqualified opponents of the use of such inputs may be seen in a distinctly different light. What, after all, is the final value judgment placed upon the use of a chemical agent which damages birds and wildlife at home and at the same time provides lifesaving food for the Indian subcontinent?

Obviously, the relevant issues may be more complicated than I have suggested in this last example, but this is a matter for careful investigation and demonstration and not bald assertions. If, for example, the use of such chemicals were to have a long-run cumulative effect in reducing soil fertility, through, say, ecological or bacteriological disturbances, the situation might then become un-stable. That is, higher productivity today would be purchased at the price of lower productivity at some future date. Although the use of these agents might of course still be economically justified, the circumstances providing such justification would clearly have to be more compelling if the process involved were unstable or irreversible. It is urgent that we vastly improve our understanding of the impact of such actions upon the long-term productivity of our resource environment. Clearly there is sufficient work in hand here for whole generations of ecologically trained specialists. Does a particular practice on some particular scale constitute a nuisance to the present generation - an offense to the olfactory senses perhaps - but nothing more? Does it, by its impact upon the food chain, threaten the prospects for a particular species we are anxious to preserve - say, hawks or eagles? Or does it threaten some wholly intolerable transformation of the environment? There are mechanisms in Nature through which she occasionally wreaks a fearful revenge (Nature's backlash) upon the human enterprise, and we have, without question, enormously increased our capacity to evoke this revenge. Unwise irrigation practices in ancient Mesopotamia eventually ruined her topsoil (and thereby destroyed her civilization) through the resulting excess salinity. Overgrazing of arid grasslands in the Middle East and elsewhere has converted untold millions of acres of productive soil into useless desert. The upsetting of the delicate ecological balance in swidden farming has resulted in apparently irreversible transformations in Southeast Asia - transformations leading to a replacement of forest by *Imperata* savannah grass which has turned parts of that region into an ecological system sometimes described as a green desert.[11]   Finally, closer to home, agricultural malpractices created for the United States the awesome dust bowl of the 1930s. Clearly there are ecological disturbances of varying degrees of severity, and they should not be treated as a homogeneous collection of offensive phenomena. Some of them we may decide we can - and should - safely ignore. Others we can ignore only at our great peril. It seems to me, therefore, to be extremely important that we learn to discriminate among the many ecological interactions going on around us. They are not equally urgent, although it is of course true that we often simply do not know how urgent some of them are or may become.[12] To secure a public hearing for the truly urgent problems, it may be wise to recognize - insofar as possible - the lower social priority of the less urgent interactions. The American public, I am afraid, possesses a very limited capacity for indignation.

In some measure, technology itself may be invoked to reduce the painfulness of the choices confronting us. It is possible to put our sophisticated technology to work at the task of reducing the destructiveness of some of the obnoxious side effects of modern technology or even of providing new uses for waste materials. It is really an old story that new technologies create problems which require offsetting or remedial actions within the technological sphere, just as, throughout history, new offensive weapons have generated responses in the form of improvements in the technology of defense. Similarly, I am sure that many of the gloomy, ecologically based forecasts will prove to have been excessively pessimistic because they underestimated our capacity for undertaking corrective action by using the tools of modern science and technology.[13] It seems to be unfortunately true, however, that governments are crisis oriented and fail to initiate remedial action until a problem appears to assume crisis proportions.[14] Nevertheless, in some limited areas progress is already being made. Many agricultural chemicals of recent development decompose into harmless substances more rapidly than did their predecessors. Modern detergents are incapable of rising up into billowing mountains of foam as detergents made by an earlier formula did. But even the tamer detergent formula took some fifteen years of diligent and expensive chemical research to develop. Furthermore, although detergents are now "biodegradable," their high phosphate content continues to act as a fertilizing agent encouraging the growth of algae. In this respect detergents remain an important contributor to algae pollution in rivers and lakes.

It seems fair to say that one of our greatest collective failures concerning technology has been our failure to direct research into the development of new technologies which would contribute to the solution of pressing urban and ecological problems. This, it seems to me, is a major factor in the recent disillusionment with both technology and economic growth. To a substantial degree, it is obvious that basic scientific research has become socialized, that both the size and the direction of research efforts are increasingly the product of allocative decisions within the public sphere. Yet, to an extraordinary degree, we have allowed the expansion of basic knowledge and technology to take place as an incidental by-product of the needs of our military establishment. Although 60 percent or so of all research and development activity in recent years has been financed through the public sector, almost 90 percent of this federal spending has been accounted for by three agencies: the Department of Defense, the National Aeronautics and Space Administration, and the Atomic Energy Commission. Surely our growing sense of failure in some areas of our national life reflects, not an intrinsic failure of some vast, impersonal, transcendent force called Modern Technology, but much more the failure to allocate research resources to areas of urgent national concern - education, health, low-income housing, public transportation and a whole range of urban services, and, not least, environmental pollution. Furthermore, it is obviously becoming a matter of growing importance that we develop new institutional mechanisms for formulating and for achieving goals in areas of our national life (other than in the military area) where the market mechanism has proved to be an unsatisfactory instrument. Among such mechanisms in particular is the not-for-profit organization, which performs much of the research financed by government agencies. Such organizations are playing, and will doubtless continue to play, an increasingly important role in the production and distribution of knowledge. But, more generally, there is an urgent need for developing new forms of collaboration between the public and private sectors which will direct our technological potential much more effectively toward the solution of nonmilitary problems. The size and the complexity of this challenge should not be underestimated. We have, as a nation, shown much less ingenuity and innovative capacity where the required innovations have been social and institutional than where they were merely technological.[15]

It simply will not do, then, to state, as do some of the more apocalyptic writers on the subject, that many of our modern ills are readily traceable to "modern technology" and leave the matter at that. This is, at best, simply a half-truth. The other half of the truth is that these ills are a product of human choices. We have *chosen* not to reduce the unpleasant side effects of our technology in at least two important ways: First, we have failed to devote more of our resources to seeking out new techniques for reducing these side effects. Second, within the spectrum of alternatives offered by the present state of our technology, we often select alternatives which are less costly in terms of money outlays but which generate higher levels of pollution - as our choice of fuels illustrates. From this perspective, environmental pollution is simply the result of a decision to adopt a less costly method of production.

The poor-country context

It is no accident that the growth of concern over problems of environmental pollution has been almost entirely restricted to the wealthy countries of the world, and that it has been a very recent development. A demand for an unpolluted environment is likely to emerge only when high levels of income have been reached and basic needs fully gratified. Even within the wealthier countries of the world, concern over ecology is largely felt by the more affluent members of those societies, and it is not a conspicuous concern of the poor. There have been no protest movements on this issue, that I am aware of, in Harlem, Watts, or Roxbury. For people still concerned with finding regular jobs, filling empty stomachs, or providing minimum standards of shelter for their families, an unpolluted environment will be regarded as a luxury good - like a station wagon or winter vacation in the Caribbean - which they simply cannot afford.

For the poor countries of the world, their present poverty is compounded by rapid rates of population growth (itself a product of modern technology) which compel them to seek increments in output wherever they may be available. Countries whose present income levels are very low and whose rates of population growth are 2-3 percent per year simply cannot afford to be very fastidious about exploiting any sources of productivity increase available to them. Such a fastidiousness at the present juncture would be tantamount to population control by a deliberate increase in the death rate. Although we in the United States can - and I think should - ban the use of DDT because of growing evidence of its noxious long-term effects on all forms of life, a similar self-denying ordinance might well have catastrophic effects elsewhere. One of the undeniable benefits of high income is that it makes it possible to treat longer-term problems with greater solicitude. Poor countries must contend with a cruel variant of Lord Keynes's well-known dictum that in the long run we are all dead. It states that many more people may be dead or malnourished in the short run if we preoccupy ourselves excessively with long-run considerations. In poor countries where DDT makes an important contribution to the food supply and where no other agent of comparable effectiveness - and lacking the side effects - is available, the case for continued use of DDT seems to me to be compelling. For, although the long-term hazards of its continued application may well be significant, the short-term consequences of an abrupt reduction in food supply will *certainly* be lethal.[16] In an ideal world, of course, we should not be dependent upon the use of such hazardous chemical agents, but we do not - yet - inhabit such a world (in any case, in an ideal world there would be no hungry people to begin with). Counsels of perfection are of doubtful value in an imperfect world where real choices are not between good and evil but among evils of differing sizes and shapes. Estimates have been made which suggest that perhaps as much as a quarter of the food grown in India is destroyed or damaged by pests - rats and insects. An Indian minister of health presented with a pesticide which would save this food for human consumption would, in my judgment, be guilty of gross irresponsibility if he rejected its use on the same grounds on which we in America are currently rejecting the use of DDT. Similarly, fertilizer plants are major producers of atmospheric pollution. However, given the unsatisfied food requirements of the Indian subcontinent and the major contribution to food production which remains to be made in that region by the use of fertilizers, I should advise an Indian minister of agriculture essentially to ignore this problem for the immediate future. I should certainly advise him against introducing expensive devices into factories which would reduce the pollution by significantly raising the cost of fertilizer to Indian peasants.

At the present stage of their development most of the really poor countries have suffered not so much from the deleterious side effects of modern technology as from the fact that a good deal of the technology of the Western developed world is inappropriate or irrelevant for their purposes. It has taken us a long time to realize this. For years we assumed that we had the technological knowledge and skills to bring about vast improvements in living standards throughout the world if only we could overcome the social and cultural resistances involved in getting our "know-how" accepted. But we are sadder and wiser now than in the halcyon "technical assistance" days following President Truman's Point Four program of 1949. Much of our technology was uneconomic when transplanted - that is, it was designed in countries with abundant supplies of capital and relatively limited supplies of labor, although the available labor was highly skilled. In poor countries, by contrast, capital is scarce and unskilled labor highly abundant. Because of these differences in factor proportions, Western technology has generally been of only limited applicability. In agriculture, by far the world's most important industry, an additional complication limited the usefulness of Western technology even further - the natural environment participates in a very direct way in the productive process. As a result, agriculture is always immersed, as manufacturing is not, in a unique ecological context. The success of an individual crop will depend on a delicate combination of qualities supplied by the environment - topography, rainfall, sunlight, temperature variations, chemical composition of the soil, etc. Cereal strains which are successful in one latitude will frequently fail when transferred even a relatively short distance to a different latitude because of a high degree of sensitivity to changes in diurnal rhythm. Therefore, each region and often each subregion has to develop, through on-the-spot research, optimal adaptations to local ecological characteristics. Happily, such research has been under way for some time and is now producing dramatic results. The application of scientific knowledge in disciplines such as genetics, biochemistry, and botany to the breeding of new seed varieties, especially of the basic cereal crops, is currently producing results which promise to revolutionize the whole process of food production. The work of the Rockefeller Foundation in Mexico, where wheat output per acre almost doubled in just over a decade and far more than doubled over a somewhat longer period, provides a model for the application of scientific techniques to the improvement of agricultural technology. This experience is now being duplicated elsewhere - for example, by the work of the International Rice Research Institute in the Philippines (where again it is significant that private foundations - the Rockefeller and Ford Foundations - took the initiative). The new high-yield rice varieties are full of spectacular possibilities, amounting to nothing less than the prospect of eliminating hunger as a major Asian problem.[17] This experience suggests that the most effective way of using our technological expertise for the benefit of poor countries may be by assisting them in setting up first-rate research organizations which, once established, can apply themselves to developing technologies relevant to local problems. For the fundamental technological problem of poor countries has been that, while Western technology has been of only limited usefulness, they have lacked the knowledge and technical skills for developing their own.

The momentous breakthroughs in agriculture virtually assure that poor countries will possess the necessary technological inputs (chemical, biological, and mechanical) for increasing food production to feed their rapidly growing populations over the next couple of decades. While the numerous Cassandras in our midst have been making dire (and often quite irresponsible) predictions about mass starvation just a few years in the offing, a sophisticated application of modern scientific knowledge has been quietly bringing about a revolution in man's food-producing capabilities. Even if population continues to grow at its present rate or slightly higher, there is no longer any serious question of our possessing the technological capability to feed such numbers for the foreseeable future. This is not to deny that a reduction in rates of population growth is eminently desirable but merely to assert that mass starvation is no longer one of the imminent dangers. For what is at issue is, not merely the capacity to feed a growing population, but rather the quality of life of such an expanded human population and the capacity of this population to enjoy something *beyond* its basic subsistence needs. In this respect I am happy to find myself in full agreement with my ecologist friends. Perhaps I can reinforce this point by quoting, in closing, the words of a distinguished economist, writing during the 1840s, who certainly would not have needed reminding (as noneconomists quite unaccountably seem to think *all* economists need reminding) that Gross National Product does not measure every significant aspect of human welfare.

A population may be too crowded, though all be amply supplied with food and raiment. It is not good for man to be kept perforce at all times in the presence of his species. A world from which solitude is extirpated is a very poor ideal. Solitude, in the sense of being often alone, is essential to any depth of meditation or of character; and solitude in the presence of natural beauty and grandeur, is the cradle of thoughts and aspirations which are not only good for the individual, but which society could ill do without. Nor is there much satisfaction in contemplating the world with nothing left to the spontaneous activity of nature; with every rood of land brought into cultivation, which is capable of growing food for human beings; every flowery waste or natural pasture ploughed up, all quadrupeds or birds which are not domesticated for man's use exterminated as his rivals for food, every hedgerow or superfluous tree rooted out, and scarcely a place left where a wild shrub or flower could grow without being eradicated as a weed in the name of improved agriculture.[18]

13

Technological innovation and natural resources: The niggardliness of nature reconsidered

I

Concern over the adequacy of natural resources has always been a central preoccupation of economists. The management of the human enterprise at both the household and larger societal levels has always been perceived to be dependent upon the intelligent management of the limited resources made available by the natural environment. Indeed, this is evident in the very etymology of the word "economy," which derives from the Greek words *oikos* (house) and *nomos* (manage), or the management of household affairs. When economics as a discipline is conceived as dealing with the problems of the household writ large the obvious and central question is how the resource endowment constrains the production of goods and services.

Some of the basic insights of classical economics emerged out of this preoccupation with natural resource constraints. Malthus and Ricardo were both pessimistic about the long-term prospects for economic growth in an economy experiencing substantial population growth and with only a fixed supply of land available for food production. Indeed, as Malthus argued, in an "old" country population growth was always pushing society against the limits imposed by the fixed supply of land and subsistence incomes were, as a consequence, inevitable - at least for the working class. Indeed, it was in speculating about the consequences of Great Britain attempting to grow its own food supply as population continued to grow that classical economics formulated one of its central relationships - the law of diminishing returns. It is worth noting that the formulation of this law clearly distinguished both a quantitative and a qualitative dimension.[1] For not only did continual population growth lead - necessarily - to a decline in the land/man ratio; it also, in its Ricardian variant, created the necessity for resorting to qualitatively inferior soils - that is, soils where an equal dosage of capital-and-labor led to a smaller increment to output than was the case with land already under cultivation. There was, in other words, both an extensive and an intensive margin of cultivation. The classical economists' concern with the benefits of free trade and the theory of comparative advantage followed directly from this vision of the deteriorating resource position of a growing economy in possession of a fixed supply of land.

The impact of the Malthusian conception is difficult to exaggerate. It not only forcefully focused upon the implications of limited resources for the prospects for economic growth, but it specifically linked the problem, as it must be linked, to the rate of population growth. Malthus, it turned out, was not much of a demographer, but his formulation nevertheless has had an overwhelming influence upon the framework within which these matters have been discussed right up to the present day.[2] If the Malthusian view seemed to play a subordinate role in the writings of professional economists after the "Marginal Revolution" of the 1870's, it was not so much because the Malthusian spectre had been exorcised as because economists began to address themselves to a very different range of problems - problems dealing with the optimal allocation of a fixed amount of resources. John Bates Clark summarized this new focus very well in his *Distribution of Wealth:*

In any given society, five generic changes are going on, every one of which reacts on the structure of society by changing the arrangements of that group system which it is the work of catallactics to study.
1. Population is increasing.
2. Capital is increasing.
3. Methods of production are improving.
4. The forms of industrial establishments are changing.
5. The wants of consumers are multiplying.[3]

Clark regards all of these forces as belonging to the area of economic dynamics. Economic statics, on the other hand, is the study of an economic system where none of these forces is present. Therefore in the study of economic statics, Clark argues, we must rigorously abstract from the operation of these forces.

One function of economic society is that of growth. It is becoming larger and richer, and its structure is changing. As time passes, it uses more and better appliances for production. The individual members of it develop new wants, and the society uses its enlarging process to gratify them. The organism is perpetually gaining in efficiency, and this is promoting the individual members of it to higher planes of life.[4]

But all of these "generic" changes connected with economic growth, Clark argues, must be firmly placed aside in order to permit the study of economic statics. Clearly this is a level of abstraction within which there is no room, by assumption, for the operation of Malthusian forces.

Although Malthus and Ricardo had been primarily concerned with the adequacy of the supply of arable land to provide food for a growing population, the voracious appetite for natural resources of an industrializing economy shifted the locus of concern to other resources in the second half of the nineteenth century. As early as 1865 the distinguished English economist W.S. Jevons published his book, *The Coal Question,* in which he warned that the inevitably rising costs of coal extraction posed an ominous and urgent threat to Britain's industrial establishment. Having demonstrated the British economy's extreme dependence upon coal - especially in the form of its reliance upon steam power - Jevons went on to argue explicitly that there were no prospective reasonable substitutes for coal as a fuel.[5] After examining recent trends in coal consumption he demonstrated the inevitability of rising coal costs associated with the necessity of conducting mining operations at progressively greater depths. As a result,

I draw the conclusion that I think any one would draw, that *we cannot long maintain our present rate of increase of consumption; that we can never advance to the higher amounts of consumption supposed. But this only means that the check to our progress must become perceptible considerably within a century from the present time;* that the cost of fuel must rise, perhaps within a lifetime, to a rate threatening our commercial and manufacturing supremacy; and the conclusion is inevitable, that our present happy progressive condition is a thing of limited duration.[6]

II

Even in the United States, a country of continental proportions that possesses a resources/man ratio far more favorable than that of Great Britain, expressions of concern over the adequacy of natural resources became widespread before the end of the nineteenth century. The Conservation Movement emerged into the consciousness of American political life at just about the time that Frederick Jackson Turner was announcing *The End of the Frontier* (1893). The movement was a significant force in American political life from 1890 to 1920.[7] Although the spokesmen of the Movement did not attack their concerns in an analytical or rigorous way, they did forcefully present a conceptualization of the problem which was to exercise a profound influence upon later thinking. In the words of Clifford Pinchot, Chief Forester of the United States, and an articulate leader of the Movement,

We have a limited supply of coal, and only a limited supply. Whether it is to last for a hundred or a hundred and fifty or a thousand years, the coal is limited in amount, unless through geological changes which we shall not live to see, there will never be any more of it than there is now. But coal is in a sense the vital essence of our civilization. If it can be preserved, if the life of the mines can be extended, if by preventing waste there can be more coal left in this country after we of this generation have made every needed use of this source of power, then we shall have deserved well of our descendants.

And,

The five indispensably essential materials in our civilization are wood, water, coal, iron, and agricultural products. . . . We have timber for less than thirty years at the present rate of cutting. The figures indicate that our demands upon the forest have increased twice as fast as our population. We have anthracite coal for but fifty years, and bituminous coal for less than two hundred. Our supplies of iron ore, mineral oil, and natural gas are being rapidly depleted, and many of the great fields are already exhausted. Mineral resources such as these when once gone are gone forever.[8]

The Conservationist view, therefore, was that nature contained fixed stocks of useful inputs for man's productive activities, which could be readily identified and measured in terms of physical units. These resources need to be husbanded carefully. Preference should be given to exploitation of renewable resources - agricultural crops, forests, fisheries - which should be operated on a sustained yield basis. Waste should be avoided in the exploitation of nonrenewable resources so as to pass on the largest possible inventory to future generations.

When, in the years immediately after the Second World War, economists turned their attention once again to the problems and prospects for long-term economic development, the Malthusian pressures once again figured prominently. Undeveloped countries all appeared to confront some serious demographic obstacle - either that of already high population densities, or rates of population growth which had sharply accelerated in the twentieth century, or both. Not surprisingly, Malthusian-type models made their appearance. In these models, that is to say, there existed a "quasi-equilibrium" such that temporary improvements which raised per capita income triggered off demographic changes which eventually restored the low level of per capita income at a higher absolute population size.[9] Concern over the implications of population growth subsided in the United States and most high income countries by the late 1950's when the postwar "baby boom" appeared to have exhausted itself. However, it reappeared forcefully in the 1960's as a consequence of the growing concern over pollution and environmental deterioration; and its possible consequences have been more recently dramatized in the apocalyptic Meadows Report, which attempts to focus attention on the long-term implications of continued growth of population and industrial production.[10] In this model, exponential growth of population and industrial output confront a world of finite re-sources - mineral deposits, arable land, etc. - and limited capacity of the natural environment to absorb the growing pollution "fallout." The finite limits appear to guarantee, not merely a ceiling to future growth, but a precipitous and disastrous decline in approximately a century.

III

In assessing the Malthusian view and its implications over the past century or so, a convenient starting point is the recognition that Malthus and Malthus-like models have led to predictions which have been demonstrably and emphatically wrong. Indeed, it is difficult to understand the persistence of widespread attachment to a hypothesis which has been so decisively refuted by the facts of history. As Stigler has said:

What evidence could have been used to test the theory? If the subsistence level has any stability, and hence any significance, Malthus' theory was wrong if the standard of living of the masses rose for any considerable period of time. He did not investigate this possibility . . . and ignored the opinions of such authorities as Sir Frederick Eden that it had been rising for a century. His theory was also contradicted if population grew at a constant geometrical rate in an "old" country, for then the means of subsistence were also growing at this rate, since population never precedes food. Despite the rapid increase of population in almost all western European nations at the time, which he duly noted, he persisted in considering this as only a confirmation of his fecundity hypothesis. . . .

The "principle of population" had the dubious honor of receiving from history one of the most emphatic refutations any prominent economic theory has ever received. It is now fashionable to defend Malthus by saying that his theory applies to other places and times than those to which he and his readers applied it. This may be true, but it is tantamount to scientific nihilism to deduce from it any defense of Malthus. It is an odd theory that may not some day and somewhere find a role; for every answer one can find a correct question.[11]

Not only have per capita incomes in the western world experienced a sustained increase over long periods of time while population has grown at exponential rates.[12] The long-term historical record also fails to reveal any convincing evidence that the limited supply of natural resources and associated diminishing returns have significantly hampered long-term growth of these countries in the past. Clearly, no amount of evidence from past history will enable us to predict future relationships, for there is certainly no necessary reason to believe that the future will resemble the past. However, it is not our purpose to attempt anything so foolhardy as prediction, but rather to grasp more firmly and precisely the nature of the interrelationships between shifting patterns of resource scarcity and the innovative process.

Classical models have failed to make relevant predictions primarily because they adopted an excessively static notion of the economic meaning of natural resources and because they drastically underestimated the extent to which technological change could offset, bypass or provide substitutes for increasingly scarce natural resources. A large part of what economists have had to say in recent years about the innovative process in relation to natural resources needs to be seen as a prolonged effort to break out of the restrictive conceptualizations inherited from Malthus and Ricardo.

The need to break out of this framework has been made increasingly apparent by numerous studies, each of which in its different way, has shown that natural resources have played a role of declining importance, at least within the favored circle of industrializing countries. Whereas classical economic models based upon fixed resources, population growth and diminishing returns, lead us to expect rising relative prices for extractive products and an increasing share of GNP consisting of the output of resources, such features obviously have not dominated the growth experiences of industrial economies. The agricultural sectors, to begin with, have declined in their relative importance. A declining agricultural sector, as Kuznets has shown, has characterized all economies which have experienced long-term economic growth.[13] Indeed, perhaps the most distinctive characteristic of growing economies has been the complex of structural changes associated with the declining importance of the agricultural sector. Moreover, within agriculture itself, the implicit assumption that there were no good substitutes for land in food production has been belied by a broad range of innovations which have sharply raised the productivity of agricultural resources and have at the same time made possible the widespread substitution of industrial inputs for the more traditional agricultural labor and land - machinery, commercial fertilizer, insecticides, irrigation water, etc. As Schultz observed in 1951,

It is my belief that the following two propositions are historically valid in representing the economic development that has characterized Western communities:
1. A declining proportion of the aggregate inputs of the community is required to produce (or to acquire) farm products;
2. Of the inputs employed to produce farm products, the proportion represented by land is not an increasing one, despite the recombination of inputs in farming to use less human effort relative to other inputs, including land.[14]

IV

The growing appreciation of the importance of technological change in raising resource productivity has led to several notable studies of the working of this process in agriculture. Attempts have been made to identify and measure the contribution of separate variables to the growth in agricultural productivity (including the impact of individual innovations), as well as to studying the factors accounting for the rate of diffusion of new innovations. The most notable attempt to measure the contribution of a single innovation was Griliches's pathbreaking study of hybrid corn, in which he estimated that the social rate of return over the period 1910-1955 to the private and public resources committed to research on this innovation was at least of the order of 700%.[15] Parker, in his study of cereal production (wheat, corn, and oats) found that output per worker more than tripled in the United States between 1840 and 1911. Parker estimated that 60% of the increase was attributable to mechanization, which raised the acreage/worker ratio, and that practically all of the growth in productivity can be explained by the combination of mechanization with the westward expansion of agriculture. The most important improvements came in those activities which had previously been highly labor-intensive - especially the harvesting and post-harvesting operations. In fact, two innovations alone - the reaper and the thresher - accounted for 70% of the total gain from mechanization.[16]

It is worth emphasizing at this point that major improvements in productivity in agriculture were the result of innovations in other sectors of the economy. For example, the growth in agricultural productivity due to westward expansion and an increasing degree of regional product specialization was, in turn, dependent upon improvements in transport facilities. Indeed, by the end of the nineteenth century the railroad, the iron steamship, and refrigeration had created, for the first time in human history, a high degree of regional agricultural specialization on a worldwide scale.[17]

The impact of an innovation upon productivity growth in agriculture, as is the case elsewhere, is not only a function of its potential for reducing costs, but of the speed with which it is adopted. As a result, increasing attention has been devoted to the diffusion process in agriculture. Paul David has employed a threshold model of farm size to explain the very slow adoption of the reaper for the twenty-year period before 1853 and the sudden rapid rate of adoption beginning in that year.[18] Griliches has shown how the spatial and chronological diffusion of hybrid corn can be explained in terms of economic factors shaping the profit expectations of farmers and seed suppliers. His model closely accounts for the early and rapid adoption in Iowa as compared to the later and slower adoption on the western fringes of the Corn Belt.[19] A significant element of Griliches's diffusion model is its explicit recognition of the dependence of the adoption process upon the need to undertake local adaptations.

An important attempt to provide a synthesis of a wide body of literature on the interaction between resource endowment and agricultural technology is represented in the recent book, *Agricultural Development,* by Y. Hayami and V. Ruttan (The Johns Hopkins Press, 1971). In this book the authors present a theoretical framework for examining the patterns of agricultural development in individual countries within which endogenous technological change plays a critical role. A crucial feature of their approach is a theory of induced innovation which incorporates a unique, dynamic response of each country to its agricultural resource endowment and relative input prices. Hayami and Ruttan argue that

... there are multiple paths of technological change in agriculture available to a society. The constraints imposed on agricultural development by an inelastic supply of land may be offset by advances in biological technology. The constraints imposed by an inelastic supply of labor may be offset by advances in mechanical technology. The ability of a country to achieve rapid growth in agricultural productivity and output seems to hinge on its ability to make an efficient choice among the alternative paths. Failure to choose a path which effectively loosens the constraints imposed by resource endowments can depress the whole process of agricultural and economic development.[20]

In developing their induced innovation model, Hayami and Ruttan postulate the existence of a "metaproduction function." This is an envelope curve which goes beyond the production possibilities attainable with existing knowledge and described in a neoclassical long-run envelope curve. It describes, rather, a locus of production possibility points which it is possible to discover within the existing state of scientific knowledge. Points on this surface are attainable, but only at a cost in time and resources. They are not presently available in blueprint form.

Within this framework Hayami and Ruttan study the growth of agricultural productivity as an adaptive response to altering factor and product prices. The adaptation process is conceived as the ability to move to more efficient points on the metaproduction function, especially in response to the opportunities being generated by the industrial sector, which offers a potential flow of new inputs.

From the intercountry cross-section and time-series observations, the relative endowments of land and labor at the time a nation enters into the development process apparently have a significant influence upon the optimum path to be followed in moving along a metaproduction function. Where labor is the limiting factor, the optimum for new opportunities in the form of lower prices of modern inputs is likely to be along a path characterized by a higher land-labor ratio. Movement toward an optimum position on the metaproduction function would involve development and adoption of new mechanical inputs. On the other hand, where land is the limiting factor, the new optimum is likely to be the point at which the yield per hectare is higher for the higher level of fertilizer input. Movement to this point would involve development and adoption of new biological and chemical inputs.

The partial productivity and factor input ratios presented earlier in this chapter suggest that those nations which have achieved relatively high levels of either land or labor productivity have been relatively successful in substituting industrial inputs for the constraints imposed by a relatively scarce factor, either land or labor. It seems possible to explain many of the vast differences in productivity levels and factor input ratios in agriculture among countries by hypothesizing that technical advance in agriculture occurs primarily as a result of new economic opportunities created by developments in the nonagricultural sector. The advances in mechanical and biological technology do not, however, occur without cost. Development of a more fertilizer-responsive crop variety, in response to declining prices of fertilizer, typically requires substantial expenditures on research, development, and dissemination, before it actually becomes available to farmers. Public investment in water control, land development, and other environmental modifications may also be required before it becomes profitable for farmers to adopt the newly developed varieties.[21]

The Hayami and Ruttan approach constitutes a very significant development because it is the most detailed attempt to date to specify the nature of the mechanisms through which technical change and adaptation take place in response to shifting patterns of relative resource scarcities.[22]

V

Barnett and Morse (*Scarcity and Growth*) have attempted a quantitative test of the implications of the classical model that increasing natural resource scarcity should make itself apparent in the secularly rising unit cost of extractive products generally - agriculture, minerals, forestry and fishing. To do this they examined data for the American economy over the period from about 1870 to the late 1950's. Their findings are that unit costs of extractive products, as measured by labor plus capital inputs required to produce a unit of net extractive output, declined through the period 1870-1957. For agriculture, minerals and fishing the trends are persistently downward. Indeed, the time sequence of the trends is particularly damaging to the increasing scarcity hypothesis because in the cases of agriculture and minerals (which together account for about 90% of the value of extractive output) the rate of decline in unit costs was *greater* in the later portion of the period than the earlier. Forestry, in fact, is the only extractive industry in which the long-term trend in unit costs and relative prices has been upward since 1870. Since forestry has accounted for less than 10%, by value, of extractive output in the twentieth century, the influence of this sector has been swamped by the downward trends elsewhere. But, moreover, the rising cost of forest products seems to have induced a large-scale substitution of more abundant non-forest-based products for forest-based ones - for example, metals, masonry and plastics - which may account for the rough constancy since 1920 in the unit cost of forest products.[23]

The declining importance of raw materials is further confirmed in the findings of input-output analysis. Anne Carter in her recent book *Structural Change in the American Economy* (Harvard University Press, 1970) provides important insights into the process of technological change between 1939 and 1961 by studying changing intermediate input requirements. In particular, she neatly documents, for this period, the decline in the quantities of materials required in order to make possible the delivery of the same bill of goods. The great virtue of input-output analysis is that it enables us to understand the structural interdependence of the economic system by providing quantitative measures (input-output coefficients) of the interindustry flow of goods and services. Changes in such coefficients need to be interpreted with care, since they may alter not only as a result of technological change, but also as a result of changes in product mix over time and because of substitution resulting from changes in the relative prices of inputs. However, it can also be counterargued that much of what is called "substitution" is, itself, made possible by *earlier* technological change.[24] Indeed, one of the most significant findings of Carter's book is precisely her careful demonstration that "there now seems to be a technological basis for greater substitutability as relative price conditions change."[25] Carter shows that economic change has been associated with an increasing reliance upon general sectors - producers of services, communications, energy, transportation and trade. What is most interesting for our present purposes is that these increases have been offset by decreases in other sets of coefficients, most conspicuously in the general, across-the-board declines in the contributions of producers of materials. Technological change has been forcefully associated with a significant expansion in the kinds and qualities of materials and in improvements of design generally. Moreover, Carter demonstrates how technological change has been expanding the range of substitutability among materials.[26] The traditional dominance of steel in many uses, for example, has been successfully challenged by improved aluminum, plywood, and prestressed concrete. The growing importance of plastics and chemicals, and the changes in product design associated with such new and versatile materials, is thoroughly documented.

An attempt to provide an overall estimate of the role of resources in the growth of the American economy finds quite conclusively that "Since about the beginning of this century, the resource base has been playing a noticeably smaller, and in many respects different, role in growth than it did before then. General economic growth has been less closely and clearly tied to abundant resources than it was previously."[27] In quantitative terms Fisher and Boorstein find that the output of resources, expressed as a percentage of United States GNP, has declined from 36 in 1870 to 27 in 1900 and to 12 in 1954.[28] Furthermore, it is apparent that a decline of this magnitude can in no way be accounted for by an increasing importation of primary products and resource-intensive goods,[29] although it is of course true that our reliance upon imports of certain resources - for example, metallic ores - has vastly increased in recent decades. In fact, after a comprehensive study of a wide range of prices and other data, Potter and Christy found that the available evidence did not support the apprehensions of the Paley Commission a decade earlier, that the long-term decline in natural resource prices was finally and permanently coming to an end for the United States.[30]

In a suggestive article, "The Development of the Extractive Industries,"[31] Anthony Scott constructs a three-stage model for dealing with the sequence through which man has gone in the exploitation of his natural environment. In the third stage, where the application of science makes possible a sophisticated degree of control and manipulation over the environment, older scarcities are deliberately bypassed in the systematic search for substitutes from abundant and convenient sources - the classic early-twentieth-century example would be the Haber process of extracting nitrogen from the atmosphere. Scott makes the important point that there is no economic demand for specific minerals as such, but rather for certain *properties,* and that an advanced technology makes it possible to obtain these properties from materials available in great abundance.

Demand for minerals is *derived* from demand for certain final goods and services. Therefore, certain properties must be obtainable from the raw materials from which such services and types of final goods are produced. Man's hunt for minerals must properly be viewed as a hunt for economical sources of these properties (strength, colour, porosity, conductivity, magnetism, texture, size, durability, elasticity, flavour, and so on). For example, there is no demand for "tin," but for something to make copper harder or iron corrosion-free. No one substitute for tin has been found, but each of the functions performed by tin can now be performed in other ways. Tin's hardening of copper (as in bronze) has been supplanted by the use of other metals. Food need no longer be packed in tin cans. Hence the immense capital investment that society might have been forced to undertake to satisfy its former needs for tin from the minute, low-grade quantities to be found in many parts of the world have been replaced by simpler investments in obtaining other materials. Chief of the replacements for tin is glass, made from apparently unlimited quantities of sand and with little more energy than is needed to bring metallic tin to the user. Lead and mercury are being bypassed in similar fashion; zinc and copper may be next.[32]

Thus, although there may be no close substitutes for tin in nature, there may be excellent substitutes for each of the properties for which tin is valued. The implications of this position are, of course, far-reaching.

VI

These findings have had a great deal to do with what may be called a steady downgrading of the importance of natural resources. Schultz called attention, in his 1951 article, to the fact that R.F. Harrod, in his notable book, *Toward a Dynamic Economics,* published in 1948, ". . . saw fit to leave land out altogether."[33] Although it is doubtful that many economists concerned with economic growth (as opposed to Harrod's primary concern with cylical phenomena) would wish to go that far, there has certainly been increasing agreement on some weaker propositions. This was made strikingly apparent in 1961 with the publication of a collection of conference papers in a volume titled *Natural Resources and Economic Growth,* edited by J.J. Spengler. Although the participants approach the subject from a wide variety of different perspectives, there are two propositions which recur so frequently as to amount almost to a consensus: (1) the pervasive influence of classical economics in its Malthusian-Ricardian variant had resulted in a vast exaggeration of the importance of natural resources and an overstatement of the constraints which they imposed upon an economy's development possibilities; (2) the relative importance of natural resources is a declining function of development itself. Kindleberger, for example, adopted what almost has to be described as a patronizing tone in conceding the economic significance of natural resources: "It may be taken for granted that some minimum of resources is necessary for economic growth, that, other things being equal, more resources are better than fewer, and that the more a country grows the less it needs resources, since it gains capacity to substitute labor and especially capital for them."[34]

In a similar spirit, Barnett and Morse conclude that

... the Conservationists' premise that the economic heritage will shrink in value unless natural resources are "conserved," is wrong for a progressive world. The opposite is true. In the United States, for example, the economic magnitude of the estate each generation passes on - the income per capita the next generation enjoys - has been approximately double that which it received, over the period for which data exist. Resource reservation to protect the interest of future generations is therefore unnecessary. There is no need for a future-oriented ethical principle to replace or supplement the economic calculations that lead modern man to accumulate primarily for the benefit of those now living. The reason, of course, is that the legacy of economically valuable assets which each generation passes on consists only in part of the natural environment. The more important components of the inheritance are knowledge, technology, capital instruments, economic institutions. These, far more than natural resources, are the determinants of real income per capita.[35]

VII

It should by now be obvious that economics has in the past been burdened with an excessively restrictive definition of natural resources. More precisely, economists (and conservationists) worked with a conception of natural resources which encompassed the prevailing state of technology but failed to recognize how profoundly technological changes required a redefinition of the *economic meaning* of the natural environment. Purely physical or geological definitions of resources, even if they are exhaustive,[36] are not very interesting.

The Plains Indian did not cultivate the soil; neither coal, oil nor bauxite constituted a resource to the Indian population or, for that matter, to the earliest European settlers in North America. It was only when technological knowledge had advanced to a certain point that such mineral deposits became potentially usable for human purposes. Even then the further economic question turns, in part, upon accessibility and cost of extraction. Improvements in oil drilling technology (as well as changing demand conditions) make it feasible to extract oil today from depths which would have been technically impossible fifty years ago and prohibitively expensive twenty years ago. Similarly, low grade taconite iron ores are being routinely exploited today which would have been ignored earlier in the century when the higher-quality ores of the Mesabi range were available in abundance. Oil shale, known to exist in vast quantities - for example, in the Green River formation in Colorado, Utah, and Wyoming - is not yet worth exploiting but might well be brought into production if petroleum product prices rise very much above their present levels. The rich and abundant agricultural resources of the Midwest were of limited economic importance until the development of a canal network beginning in the 1820s with the completion of the Erie Canal, and later a railroad system which made possible the transportation of bulky farm products to eastern urban centers at low cost. Natural resources, in other words, cannot be catalogued in geographic or geological terms alone. The economic usefulness of such resources is subject to continual redefinition as a result of both economic changes and alterations in the stock of technological knowledge. Whether a particular mineral deposit is worth exploiting will depend upon all of the forces influencing the demand for the mineral, on the one hand, and the cost of extracting it, on the other.[37]

The extreme sensitivity of any definition of natural resources to changing economic conditions as well as to changing technology needs to be emphasized. The President's Materials Policy Committee reported an estimate of 30 billion tons of coal were recoverable at 1951 cost levels. However, as Dewhurst has pointed out, "A rise of 50 per cent in prices and costs . . . would mean a 20-fold increase in estimated recoverable reserves."[38] Not only is there a widening range of substitution among resources with respect to end uses - new materials, such as plastics and aluminum, and older ones, such as glass, have been replacing wood as a packaging material - but there is often also a range of *sources* from which a given input may be extracted. Oil, for example, is found in nature not only in its crude form. Enormous reserves of shale oil are known to exist and "The technical feasibility of recovering oil from shale has been clearly demonstrated."[39] Furthermore, oil can be produced from coal, a much more abundant resource. Finally, the extraction of oil from tar sands is a clear possibility. A significant increase in crude oil costs may be expected to activate such possibilities.[40]

The manner in which sharply rising prices of timber products (which quadrupled between 1870 and 1954) triggered off technological and other adaptations is illuminating. Conventional materials, such as iron and steel, were substituted in many uses. In construction, which is by far the largest single user of lumber, traditional masonry and other mineral building materials as well as aluminum were substituted. More recently, technologically new products such as plastics and fiberglass have served as substitutes. Moreover, new techniques for economizing upon wood requirements have been developed which do not involve the substitution of competitive materials, or which, in effect, substitute cheaper woods for more expensive woods - as in plywoods and wood veneers. Other technological changes, such as the self-powered chain saw, the tractor, and the truck have made previously inaccessible forest stands available for exploitation by reducing the cost of extraction and transport. Yet other technological changes have significantly increased the size of our forest resource base by making possible the exploitation of low-grade materials which had previously gone unused.

Although wood pulp has been manufactured for about 100 years in this country, it was the rapid advances in sulphate pulping technology in the 1920's that released the industry from its dependence on spruce and fir of the Northeast and made it possible to utilize southern pine for pulp. This led to the "sulphate revolution" of the South and was largely instrumental in increasing national consumption from 6.8 million cords in 1926 to 33.4 million in 1955. The South's share of production rose from 1.1 million to 19.2 million cords in this period while pulpwood consumption in the West was increasing from 0.5 to 6.4 million cords. By the mid-1950's the South's share of woodpulping capacity had risen to over 55 percent of the national total.[41]

VIII

The current round of intense concern over natural resources, which began in the late 1960's, contains an important new element. Not only does it assert the inadequacy of the resource base to sustain continued growth of population (because of the limited supply of arable land) and industrial production (because of the limited supply of mineral resources). Insofar as models such as *The Limits to Growth* and its companion, Jay Forrester's *World Dynamics* (Cambridge, 1971) do this, they are essentially, for all their elaborate systems dynamics and computer methodology, Malthus in modern dress. That is, they deduce the conclusion, from certain restrictive behavioral assumptions, that continued exponential growth is impossible in a world of finite resources. The new element in the present debate is the prominence, not to be found in Malthus, of concern over the pollution problem and the assertion of additional limits to growth imposed by the increasing incidence of environmental pollution.[42]

Although an extended appraisal of the unique aspects of the new concerns is impossible here, it is important to stress that they are subject to the same criticisms which have been made of earlier static approaches to the resource problem. That is to say, all such extrapolations of recent trends which point to some collapse of the social system t years hence, fail fundamentally to take account of the human capacity to adapt and to modify technology in response to changing social and economic needs. Indeed, a central feature of such models (like the dog that did *not* bark in the night) is the total *absence* of any social mechanism for signalling shifting patterns of resource scarcity and for reallocating human skill and facilities in response to rising costs, either the rising direct costs of raw materials or the rising indirect costs associated with environmental pollution.[43] One need not be a capitalist apologist to argue that the price mechanism does perform such a function, albeit imperfectly, or to believe that it might perform these functions even better under altered arrangements which no longer allowed private firms to treat our water courses and atmosphere as if they were free goods.[44]

These models constitute a reversion to Malthusian Fundamentalism in another, closely related way. They allow no possibility that mankind's taste for offspring and therefore his reproductive practices will be altered *as a result of the growth experience itself.* In both the demographic and technological realms, therefore, such purely deductive models ignore the mass of dramatic evidence of the past 150 years or so that human behavior undergoes continuous modification and adjustment as a response to the changing patterns of opportunities and constraints thrown up by an industrializing society.[45]

IX

The thrust of these critical remarks is *not* that population growth, pollution, and increasing scarcity of key natural resources are unimportant problems, or that technology may be confidently relied upon to provide cheap and painless solutions whenever these key variables, and their interactions, begin to behave in problematic ways. Quite the contrary. The main complaints against such apocalyptic, neo-Malthusian models, built upon a global scale and incorporating naive behavioral assumptions, are first, that they define the problems incorrectly, and second, that they deflect attention from more "modest" but genuine questions which ought to be placed more conspicuously upon the agenda of social science research. Consider the following sample.

What are the determinants of human fertility? How is it likely to respond to future changes in mortality, income, urban densities, pollution, education levels, female employment opportunities, new goods (including new contraceptive techniques)? The very recent reductions in American fertility levels, which may indeed turn out to be a short-run phenomenon, indicate forcefully that demography is a subject about which our present state of ignorance is truly momentous. Malthusian-type models which simply accept certain population growth rates as somehow exogenously determined and then go on to demonstrate the awesome power of exponential growth over time periods of a century or more, should surely be losing their capacity to fascinate all but the most intellectually immature.

What sort of alterations in our system of property rights and our tax structure hold the greatest promise for the control of pollution? What are the technological possibilities that an altered incentive system might lead to the development of "cleaner" technologies, and at what cost?

What kinds of technological and social adaptations can we visualize if the cost of key resources, such as fuels, should increase substantially in the future?

What improvements can be made in our present mix of private and public institutions devoted to the "production" of useful knowledge?

Instead of the almost total preoccupation with supply considerations, is it possible to anticipate future shifts in income and social organization which may reduce the need for resources by shifting demand away from resource-intensive goods?

What can we find by opening up the "black box" of technological change? *How* responsive are human agents with the requisite talents, to the market forces which continually shift the prospective payoff to inventive and innovative activities? And - a very different question - what can be said in a systematic way about the responsiveness of nature? That is, given the incentives, what factors shape the prospects for success in overcoming different kinds of scarcities? Is there any meaningful sense in which it may be harder or easier to make resource-saving innovations than capital-saving innovations, or either of these innovations than labor-saving innovations? Is a dollar spent on agricultural invention likely to yield the same eventual social return as a dollar spent on manufacturing or on service? What of petrochemicals vs. textiles, or building materials vs. transportation, or pharmaceuticals vs. machine tools? Instead of computing the number of years before we "run out" of specific deposits, can we project reasonable estimates for the changing resource costs of alternative technologies?

Finally, of what relevance is the historical experience of the presently rich countries to the prospect for the presently poor countries? For one thing, the industrializing countries of the nineteenth and twentieth centuries were often able to overcome their own resource deficiencies by imports of primary products from the less developed world. When - or if - these regions are able to enter successfully into the stream of industrialization, what alternatives will they be confronted with in overcoming their own resource deficiencies? But, perhaps most fundamentally, what factors have determined the differential effectiveness with which societies have responded to the problems and constraints posed by their unique resource endowments? For, as Simon Kuznets has observed:

It need not be denied that in the distribution of natural resources some small nations may be the lucky winners at a given time (and others at other times). But I would still argue that the capacity to take advantage of these hazards of fortune and to make them a basis for sustained economic development is not often given. In the 19th century, Brazil was commonly regarded as an Eldorado - and indeed enjoyed several times the position of a supplier of a natural resource in world-wide demand; yet the record of this country's economic growth has not been impressive, and it is not as yet among the economically advanced nations. The existence of a valuable natural resource represents a permissive condition, facilitates - if properly exploited - the transition from the pre-industrial to industrial phases of growth. But unless the nation shows a capacity for modifying its social institutions in time to take advantage of the opportunity, it will have only a transient effect. Advantages in natural resources never last for too long - given continuous changes in technology and its extension to other parts of the world.

To put it differently: *every* small nation has some advantage in natural resources - whether it be location, coastline, minerals, forests, etc. But some show a capacity to build on it, if only as a starting point, toward a process of sustained growth and others do not. The crucial variables are elsewhere, and they must be sought in the nation's social and economic institutions.[46]

14

Innovative responses
to materials shortages*

The central concern of this paper is with the adequacy of natural resource supplies to support an indefinite continuation of high rates of economic growth in advanced industrial economies. It is inspired - if that is the right word - by a recent spate of apocalyptic literature purporting to show that natural resource constraints impose an insuperable obstacle to such growth. I will suggest that this extreme pessimism is unwarranted because it attaches insufficient weight to an impressive array of adaptive mechanisms through which a market economy responds to shifting patterns of resource scarcity. Most important, I will argue that technological change is, in the long run, the most powerful mechanism of response.

Some disclaimers are in order. It should be obvious that no amount of historical analysis can provide an adequate basis for optimism concerning our *future* prospects. Evidence of past successes in dealing with natural resource constraints cannot, by itself, disprove the possibility that we may now have arrived at some crucial turning point or reached the end of an historical epoch. Equally obviously, arguments drawn from the past history of countries which have demonstrated a high degree of technological innovativeness are not likely to be directly applicable to countries where such skills and talents have been notably lacking. The purpose of my recourse to history is, therefore, relatively modest. It is simply to provide some insight into the manner in which economies possessed of a high degree of technological versatility have adapted to changing patterns of resource scarcities. History can, I believe, enlarge our awareness of the nature and possible range of these mechanisms. In so doing, it can also suggest the reasonableness of a number of possible alternative scenarios to the essentially Malthusian one which is currently receiving so much uncritical acclaim.

*I would like to express my indebtedness to Stanley Engerman for his customarily astute comments on an earlier draft. The paper draws in a few passages upon my book, *Technology and American Economic Growth,* New York 1972.

249

Let me first sketch out some basic historical trends. A broad view of the long-terms trends in American natural resource use (agriculture, timber products, minerals) reveals a substantial upward movement in resource consumption. An index of consumption of resources (1947-49 = 100) rose from 17 in 1870 to 41 in 1900 to 110 in 1954.[1] When consumption is expressed in per capita terms, the rise is far more modest, growing from $174 in 1870 to $221 in 1900 and to $279 in 1954 (all in 1954 dollars). If, over this period, the growing scarcity of resources had been acting as a serious constraint upon economic growth, this role would presumably have become apparent in the form of an increase in the relative prices of extractive products and a rising share of *GNP* consisting of the output of resources. The latter, based upon a production function with an implicit inelasticity in the demand for resources, would surely have been Ricardo's expectation. In fact, however, this has not been the case. Since 1870 there have been extreme short-run fluctuations in the prices of extractive products relative to the general wholesale price index. But, at most, one can argue that there is some evidence of a slight upward drift in the long-term trend. There is, however, one important exception to this statement: timber product prices have moved sharply and unmistakably upward, roughly quadrupling between 1870 and the 1950's. Timber products, therefore, offer an excellent opportunity for studying the nature of the economy's response to a rapid increase in the relative price of a major class of raw material inputs, both because the price increase has been very substantial and because the materials involved are of a class upon which the economy was once heavily dependent. With respect to *all* resource inputs, it should be noted, the evidence seems reassuring. Output of resources as a percentage of *GNP* has been declining consistently, from 36 in 1870 to 27 in 1900 to 12 in 1954. The magnitude of this decline cannot nearly be accounted for by increasing imports of resource-intensive goods.[2]

Although it is not usually thought of in such terms, concern with problems of raw material supply was at the heart of what has come to be called the Industrial Revolution. The Industrial Revolution in England may properly be said to have begun when techniques were successfully developed for overcoming - or, more properly, bypassing - the resource constraints which restricted the growth in industrial output of an earlier age. Preindustrial societies were heavily dependent upon wind, water power, and animal power. Above all, however, they were dependent upon wood, which served several purposes: it was the major source of fuel, it was a building material, it was an industrial raw material of unsurpassed versatility, and it was a critical source of chemical inputs, as in the use of potash in the production of alkalis. The urgent need to find a substitute for increasingly scarce wood fuel was being signaled in Elizabethan England, when wood fuel prices were rising some three times as fast as prices generally. Laws were passed during the reign of Elizabeth which limited forging and furnace operations in districts where timber had become sufficiently scarce. Indeed, British industrial expansion by the early seventeenth century was confronted with nothing less than a "national crisis," as John Nef has called it, as a result of the severe depletion of her timber supplies. The substitution of coal for wood was long delayed, partly because the chemical interchange between the mineral fuel and the final product frequently damaged or produced a highly inferior final product - as in the cases of glassmaking, baking, the drying of malt for the brewing industry, and, most important of all, the production of iron. By the end of the eighteenth century these problems were satisfactorily solved in metallurgical uses, and the massive utilization of the mineral fuel made possible the cheap supplies of fuel, power, and iron products upon which early industrial societies were built. The Industrial Revolution in Britain essentially substituted cheap coal for wood as a source of fuel and power, and cheap and abundant iron for vanishing timber resources.

In the United States where resource supplies were drastically different, the historical pattern of resource use also developed differently. Americans, for whom wood constituted an abundant and cheap fuel, continued to use wood long after the British had adopted coal - as in metallurgy. Moreover, Americans early began to develop a new technology specifically geared to the intensive use of this abundant resource. In order to exploit the vast forest resources of the country, the United States in the first half of the nineteenth century brought to an advanced stage of perfection a whole range of woodworking machines for sawing, planing, mortising, tenoning, shaping, and boring, in addition to innumerable other specialized machines - and in addition to important improvements in the design of that much more venerable instrument, the axe (both head and handle). If manufacturing sectors are ranked by value added by manufacture, the lumber industry in 1860 was, according to the U.S.

Census of that year, the second largest industry in the United States, after cotton goods. It was the largest manufacturing industry in the West and South. During this same period, per capita lumber consumption in the United States may have been as much as five times as high as in England and Wales. Wood was a by-product - indeed in some cases a *waste* product - as agricultural expansion pushed its way into the great forest regions east of the Mississippi River.

American inventive effort in the early decades of the nineteenth century focused heavily upon developing techniques for exploiting her abundant forest resources. By the 1850's American woodworking machinery was generally acknowledged by Europeans to be the most sophisticated and advanced in the world. The relatively limited degree to which these machines were adopted in Europe, however, seems to have reflected the fact that they were, in many ways, wasteful of wood - a consideration much less important in the United States than in Great Britain in the first half of the nineteenth century. American circular saws, for example, while very fast, had thicker blades, with their teeth spaced widely apart, and they converted a distressingly large portion of the log into sawdust instead of lumber. Indeed, an observer writing in the early 1870's, who was intimately familiar with British and American woodworking methods, stated categorically that "Lumber manufacture, from the log to the finished state, is, in America, characterized by a waste that can truly be called criminal . . ." (J. Richards, p. 141). This characterization might have been reasonable had American techniques been employed in Britain. Given the relative factor scarcities in the United States, however, these techniques, by substituting abundant, cheap wood for scarce and expensive labor, may well have been optimal. In England, by contrast, handicraft technology, which amounted to the substitution of relatively cheap labor for relatively expensive wood, continued to prevail.

A similar profligacy in wood consumption persisted within the household so long as wood supplies were locally abundant. Under these circumstances, fireplaces were designed to accommodate large logs, an arrangement which was wasteful of fuel wood but economized upon the labor-intensive activities of cutting or chopping wood. (Stoves, which utilized wood supplies more efficiently, but were more expensive and raised the labor cost of preparing the wood, became increasingly popular whenever and wherever wood prices began to rise substantially.) America's abundance of forest resources, in fact, led to innumerable adaptations which involved substituting natural resource inputs for other, scarcer factors of production. The American builder relied on wood, a highly tractable material, in uses where his European counterpart would have employed stone, iron, or other materials. In the construction of houses, this led to the development in the 1830's of a distinctively American technique of housebuilding: the balloon-frame design, which was not only highly utilitarian, but the method of construction of which was uniquely suited to a labor-scarce, resource-abundant society. Similar adaptations took place elsewhere. Americans employed wood in uses which astonished European visitors - not only in building bridges and aqueducts, but in more improbable uses such as the framing of steam engines, canal locks, and pavements. They even - *mirabile dictu!* - built roads (the famous plank roads) out of wood.

In looking at these adaptations, by the way, the analytical distinction between technological change and mere factor substitution becomes extremely difficult to maintain. *Historically,* establishing new possibilities for factor substitution has typically been the outcome of a search process involving substantial financial costs and the use of specialized knowledge and creative skills. The kinds of new knowledge underlying both substitution and innovation possibilities are, in other words, the historical *outcome* of research activities. The range of substitution possibilities conveniently summarized on a single isoquant are the product of such past research efforts and their resulting technological changes. *Today's* factor substitution possibilities are made possible by *yesterday's* technological innovations.

The increase in the relative prices of forest products and the cheapening of coal and iron in mid-nineteenth century America signaled a shift away from the use of wood both as a fuel and as a building material. In 1850 mineral fuels still supplied less than 10 percent whereas wood supplied more than 90 percent of all fuel-based energy. In the second half of the nineteenth century, however, the changes in the relative costs of fuel sources as well as technological changes favoring the use of mineral resources both in the manufacture of iron and steel and the production of steam power brought about a rapid shift to coal. Some of these changes occurred very rapidly. For example, although almost all the energy needs of the railroads were supplied by cordwood at the outbreak of the Civil War, twenty years later the railroads were using twenty times as much coal as wood, and the railroads were consuming more than 25 percent of the country's bituminous coal output. In the last decades of the nineteenth century, steam power relying upon coal grew to the peak of its influence and in 1899 steam power accounted for over 80 percent of primary power capacity in manufacturing, whereas waterpower accounted for less than 15 percent. Moreover, among material sources of energy, coal had largely displaced wood by the early years of the twentieth century - from its position of overwhelming dominance in 1850, wood declined to less than 10 percent in 1915, whereas coal sources accounted for three-quarters. Subsequent decades are largely the story of the declining importance of coal - a decline in which dieselization in transportation and the loss of household markets played major roles - and the rise of liquid and gaseous fuels. In contrast to the prolonged dominance of coal in other industrial countries, the supremacy of coal as an energy source in the United States was relatively short. Coal accounted for over one-half of energy sources only for the period between 1885 and 1940 (S.H. Schurr and B.C. Netschert, p. 36). Just as fuelwood was rapidly displaced by coal as an energy source in the second half of the nineteenth century, so coal was displaced by oil and natural gas in the half century after the First World War.

This drastic reduction in the reliance upon increasingly expensive wood as a fuel source had a direct counterpart with respect to the use of wood as an industrial raw material. The price of timber products, unlike the prices of agricultural products and minerals, has risen dramatically, quadrupling between 1870 and 1950. This increase has triggered off substitution of other inputs, including that of minerals, and appears to have induced significant technological changes which have limited the utilization of timber. Thus, whereas consumption per capita of mineral products increased almost ten times between 1870 and 1954, consumption of timber products per capita rose to a peak in the first decade of the twentieth century and then declined to almost one-half of its 1900 level by 1954. Even in absolute terms the consumption of timber products was no greater in the mid-1950's than it had been in 1900. Timber products, which accounted for fully 4 percent of *GNP* in 1870, accounted for a mere .69 percent in 1954 (J. Fisher and E. Boorstein, p. 43). This was brought about through a broad spectrum of responses. Iron and steel were substituted for wood across a whole range of investment goods in the nineteenth century, going back to the pre-Civil War period. Machinery, ships, and bridges which were made of wood in 1800 were made of iron or steel in 1900. In construction, by far the largest consumer of lumber, there has been an increasing reliance upon traditional masonry and other mineral building materials and upon aluminum. More recently, technological change has produced plastics and fiber glass materials which have served as substitutes for wood. New materials, such as plastics and aluminum foil, and older ones, such as glass, have replaced forest products as a packaging material. Further technological changes have also generated methods which economize upon wood requirements without the substitution of competitive materials, or which substitute cheaper woods for more expensive woods - for example, in the cases of plywood and wood veneers. (In addition, wood waste is now utilized in the manufacture of fiber board and synthetics.) Other technological changes, such as the self-powered chain saw, the tractor, and the truck have reduced the cost of extracting and transporting the timber from its forest stands. Finally, other technological changes have, in effect, significantly increased the size of our forest resource base by making possible the utilization of low-grade materials which previously had gone unused. Until the 1920's the woodpulp industry utilized only the spruce and fir trees of the northern portions of the country. Improvements in sulphate pulping technology during the 1920's made possible the exploitation of faster growing southern pine which was previously unusable, as a result of which the South accounted for over half of the country's woodpulping capacity by the mid-1950's.

The discussion up to this point has obviously been illustrative rather than exhaustive. As forest resources became increasingly costly in Elizabethan England and later, a technology was gradually devised in the seventeenth and eighteenth centuries which effectively exploited a new set of abundant mineral resources and drastically curtailed dependence upon the traditional scarcer ones. When British industrial technology was transplanted to the New World, the abundance of forest resources induced a large number of inventions and substitutions which made possible a more intensive exploitation of these resources. In turn, the eventually rising cost of these inputs led to an increasing and successful concentration of inventive effort to providing substitute inputs for forest products with the result that they now account for much less than 1 percent of *GNP*. Let me now advance the generalization that this sort of experience has, in fact, been quite representative of the experience of industrial economies in dealing with shifting patterns of resource availability. Modern industrial economies possess a remarkably wide range of options with respect to the exploitation of the natural resource environment. At any one time the range of substitution possibilities among material resource inputs is far higher than is generally recognized. From a historical point of view, these possibilities are, in large measure, the product of past technological change which has produced new substitute inputs or raised the productivity of old ones. The ways in which it has done this defy simple categorization, but they have included the following:

1. Raising output per unit of resource input - as, for example, the decline in the amount of coal required to generate a kilowatt-hour of electricity, which fell from almost seven pounds in 1900 to less than nine-tenths of a pound in the 1960's.

2. Development of totally new materials - synthetic fibres, plastics, etc.

3. Raising the productivity of the extractive process.

4. Raising the productivity of the process of exploration and resource discovery.

5. Development of techniques for the reuse of scrap or waste materials.[3]

6. Development of techniques for the exploitation of lower-grade, or other more abundant, resources.

One of the main effects of these technological developments is to reduce the economy's dependence upon any specific resource input and to widen progressively the possibilities of materials substitution.[4] As a result, although particular resources of specified quality do inevitably become increasingly scarce, the threat of a *generalized* natural resource constraint upon economic growth by no means follows from this. It now seems clear that the discussion of the role of natural resource constraints upon economic growth cannot be usefully pursued within the framework of asking how long it will be before we "run out" of specific resource inputs defined and estimated in physical units. That is simply not an interesting question, partly because there are seldom sharp discontinuities in nature, and partly because, by making it possible to exploit resources which could not be exploited before, technological change is - in economic terms if not in geological terms - making continuous *additions* to the resource base of the economy. What we are more normally confronted with are limited deposits of high quality resources and then a gradually declining slope toward lower-grade resources, which typically exist in abundance. The much greater profusion in the earth's crust of low-grade resources than high-grade resources is one of those geological facts of life which we can - and, if the past is any guide at all, will - learn to live with. Our technological adjustment to this fact has been apparent in recent decades. As the high quality Mesabi iron ores approached exhaustion, the major steel companies not only turned increasingly to foreign sources, but directed their research toward the exploitation of the enormous deposits of hard, low-grade taconites. Techniques such as beneficiation have proven to be highly successful in resisting the pressures toward higher cost, even though the iron ore which we exploit today is inferior in quality to the iron ore which we exploited fifty or sixty years ago. Similarly, although we import large quantities of bauxite because our domestic reserves of high-grade bauxite are now inadequate, alternative sources of aluminum such as clays are, quite simply, immense. As a matter of fact, the earth's crust contains far more aluminum than it does iron. The economic exploitation of these and many other low-grade resources turns primarily upon the question of fuel and power costs, since the technologies of low-grade resource utilization are highly fuel-intensive in nature.

Although there is cause for pessimism over the long-term prospects for crude oil, in spite of great improvements in techniques of oil exploration and extraction, oil may also be recovered from other sources, such as shale. Indeed, the technical feasibility of such a process has already been established and these deposits are estimated to be several times as great as the combined total of petroleum and natural gas reserves. Beyond this alternative loom the possibilities of producing oil from coal - a much more abundant resource - or tar sands. And beyond all fossil fuels, of course, looms the eventual prospect of widespread reliance upon atomic energy.[5] These alternatives may, in spite of technological change, turn out to be more costly than our present arrangements, but it should be apparent that the *cost* of alternative sources of fuel is the question that needs to be addressed and not the prospect of resource exhaustion. Elementary though it may seem, it would constitute a major step forward in the public discussion of long-term growth prospects if the focus could be shifted from the prospects (imminent or remote) for resource *exhaustion* to the prospects for significant alterations in the resource costs of alternative technologies. Even in the absence of further technological change, it must be noted, the question of what constitutes a recoverable natural resource may be extremely sensitive to price changes. It was estimated for the Paley Commission, for example, that, at cost levels prevailing in 1951, our recoverable coal reserves amounted to some 30 billion tons. However, at prices 50 percent above those prevailing at the time, it was estimated that reserves which were worth recovering were no less than twenty times greater than the 30 billion tons estimate.

There is no obvious reason why the further growth of technological skills should not make it possible to continue the shift from dependence upon scarce sources of materials to dependence upon more abundant sources - a shift already dramatized in the present century by the extraction of nitrogen from the air and magnesium from seawater. Indeed, the ability to manipulate the raw material inputs of nature has been multiplied several times in the past century by fundamental advances in our knowledge of the physical world. These have been based upon a continuous deepening of the understanding of the general rules which determine how atoms and molecules combine together into progressively larger and more complex groups. A mastery of these rules makes it possible to alter the characteristics of traditional materials and even to create entirely new and synthetic materials with desired combinations of properties. Moreover, this knowledge promises to be of enormous use in making it possible to substitute abundant material inputs for scarcer ones. This recently acquired knowledge of "molecular architecture" is greatly expanding the already large arsenal of technological responses which have long been available in progressive economies. Indeed, one of the main economic consequences of scientific progress is to enlarge continually the range of substitution possibilities which confront advanced industrial economies. Thus, one may readily agree with the statement that an increasing scarcity of natural resources is an inevitable correlate of economic development and population growth - indeed, as the problem is often defined, with a known, fixed quantity of geologically given resources and population growth, it is tautological to say that resources are becoming increasingly scarce. If my argument is reasonable, however, such statements, while they may be correct, are not economically-interesting propositions.

References

J. Fisher and E. Boorstein, *The Adequacy of Resources for Economic Growth in the United States,* Study Paper No. 13, prepared in connection with the Study of Employment, Growth, and Price Levels, 86th Congress, Dec. 16, 1959.

J. Richards, *A Treatise on the Construction and Operation of Woodworking Machines,* London 1872.

S.H. Schurr and B.C. Netschert, *Energy in the American Economy, 1850-1955,* Baltimore 1960.

J. Vanek, *The Natural Resource Content of United States Foreign Trade, 1870-1955,* Cambridge, Mass. 1963.

U.S. Dept. of Commerce, *Historical Statistics of the United States: Colonial Times to 1957,* Washington 1960.

15

Science, invention,
and economic growth*

I

Not too many years ago most economists were content to treat the process of technological change as an exogenous variable. Technological change - and the underlying body of growing scientific knowledge upon which it drew - was regarded as moving along according to certain internal processes or laws of its own, in any case independently of economic forces. Intermittently, technological changes were introduced and adopted in economic activity, at which point the economic *consequences* of inventive activity were regarded as interesting and important - both for the contribution to long-term economic growth and to short-term cyclical instability. Schumpeter, for example, saw the engine of capitalist development as residing in this innovative process in the long run, and at the same time he developed a business cycle theory which centred upon the manner in which the capitalist economy absorbs and digests its innovations. In Schumpeter's model, exogenous technological changes stimulated investment expenditures, the variations of which, in turn, generated cyclical instability.

In the years after the Second World War the economist's attitude gradually changed. The vast expenditures on Research and Development made it increasingly obvious that inventive activity was - or could be made to be - responsive to economic needs (or even to noneconomic needs if such needs received sufficient financial support). Clearly much of the search activity of R and D was highly purposive: business firms were looking for new techniques in specific categories of products, they spent much money upon this search, and they were sometimes highly successful. Similarly, government agencies had long directed research into specific problem areas and in some cases had achieved conspicuous successes - as in agriculture.

*The author is grateful to Professors S. Engerman, W. B. Reddaway and E. Smolensky, and to an anonymous referee for their helpful comments on earlier drafts of this paper. They are, however, accorded the usual absolution for all remaining deficiencies.

In addition, the growth of interest in technological change after the Second World War was closely connected with the increasing concern over the prospects for economic growth in underdeveloped countries. When economists turned their attention to this range of problems, they brought with them an intellectual apparatus which placed overwhelming emphasis upon the role of saving and the growth in the stock of capital goods as the engine of economic growth. But it soon became clear that long-term economic growth had taken place at rates far beyond what could plausibly be accounted for by mere growth in the supply of conventionally measured inputs. It became increasingly obvious that economic growth could not be adequately understood in terms of the use of more and more physical inputs, but rather that it had to be understood in terms of learning to use inputs more productively. With this realization came, of course, a renewed interest in technological change as the source of rising resource productivity.

The growing interest in the role of technological change as a contributor to economic growth led to a considerable amount of empirical research on technological change, particularly in two areas: (1) attempting to quantify the contribution of technological change to the growth in long-term resource productivity; and (2) attempting to study the rate at which new inventions, once made, were diffused throughout the economy, since clearly inventions exert an impact upon resource productivity only to the extent that they are actually adopted in the productive process. The work of Griliches was particularly important in showing that one could explain the diffusion process in considerable detail as a response to economic forces - *i.e.,* on the basis of profit expectations as shaped by market size.[1]

Increasingly, therefore, economists have become more and more confident of their ability to deal with technological events in economic terms. This growing confidence was capped by the publication of a major book by Jacob Schmookler in 1966, called *Invention and Economic Growth* (Cambridge: Harvard University Press). Schmookler argued, quite persuasively, not only that one could explain the *diffusion* of existing inventions in economic terms - *à la* Griliches - but that one could even explain the pattern of inventive activity itself.

As a result of these developments, the attitude of the economics profession toward technological change seems to be coming full circle. Whereas technological change was once regarded as an exogenous phenomenon moving along without any direct influence by economic forces, it is now coming to be regarded as something which can be *entirely* explained by economic forces. Indeed, factors on the technological and scientific levels are increasingly coming to be regarded as not constituting very interesting problems, because we already "know" that we can explain their particular timing in economic terms.[2]

Schmookler's book is obviously very appealing to the economist because it argues that inventive activity is an essentially economic phenomenon, and that it can be adequately understood in terms of the familiar analytical apparatus of the economist. Perhaps I should anticipate my conclusions by saying that I propose to start off from Schmookler's analysis, not because I am in search of a convenient straw man, but rather because I am in substantial agreement with much that he has to say. Moreover, Schmookler's analysis is so rich and so suggestive that it has to be the starting point for all future attempts to deal with the economics of inventive activity and its relationship to economic growth.

II

Schmookler's ultimate interest is, to quote the opening sentence of his book: "What laws govern the growth of man's mastery over nature?" His book represents an attempt to supply building blocks for the answer to that very big question by systematically studying two smaller questions: (1) how to explain the variations in inventive activity in any particular industry over time; and (2) how to explain different rates of inventive activity between industries at a given moment of time. Schmookler's fundamental answer to these questions involves the attempt to link up inventive activity with the structure of human wants and therefore with changes in the composition of demand which are associated with rising *per capita* incomes and other related aspects of economic growth.

The empirical core of Schmookler's book is an attempt to demonstrate, through the study of several American industries, that demand-side considerations are the major determinant of variations in the allocation of inventive effort to specific industries. In examining the railroad industry, for which comprehensive data are available for over a century, Schmookler found a close correspondence between increases in the purchase of railroad equipment and components, and slightly lagged increases in inventive activity as measured by new patents on such items. The lag is highly significant because, Schmookler argues, it indicates that it is variations in the sale of equipment which induce the variations in inventive effort. Schmookler finds similar relationships in building and petroleum refining, although the long-term data on these industries are less satisfactory.

Furthermore, and no less important, in examining cross-sectional data for a large number of industries in the years before and after the Second World War, Schmookler finds a very high correlation between capital goods inventions for an industry and the volume of sales of capital goods to that industry. These data support the view that inventors perceive the growth in the purchase of equipment by an industry as signalling the increased profitability of inventions in that industry, and direct their resources and talents accordingly.[3] Thus, Schmookler concludes that demand considerations, through their influence upon the size of the market for particular classes of inventions, are the decisive determinant of the allocation of inventive effort.

Far from being an exogenous variable as most economists had earlier believed - an activity which, although it had important economic *consequences* was not *controlled* by economic forces - Schmookler concludes that we can treat invention just like any other economic activity. Just as we can analyze production and consumption in terms of revenues and costs and the desire to maximize some relevant magnitude, so we can analyze inventive activity in precisely the same terms.

Schmookler not only attempts to incorporate inventive activity into an economic framework. *Within* that framework he attaches overwhelming importance, as already indicated, to demand forces, and regards supply side considerations as relatively subordinate and passive. Thus, in discussing consumer goods inventions, Schmookler argues that it is the changes in consumer demand over time which are the primary determinant of shifts in the direction of inventive effort.

. . . (I)f we start out at a given point of time with relative outlays on the different classes of goods given, and allow capital accumulation, technical progress, education, and so on, to occur, then per capita income will gradually rise. In consequence the proportion of income spent on different classes of goods will also gradually change. As different classes of goods become relatively more important than before, the yield to inventive effort in different fields will tend to change correspondingly. And if we further grant that inventive effort is influenced by prospective yield, the direction of inventive activity will shift. Thus, even under the extreme assumption that the *structure* of generic wants is permanently fixed, economic progress will bring successive sections of that structure into play over time, thereby altering the reward structure confronting inventors and rechannelling their efforts accordingly. This is why, for example, American inventors concentrated on food production in the first part of the nineteenth century but gave much more attention in the twentieth century to the requirements of leisure, by creating motion pictures, radio, television, and so on.[4]

Schmookler's argument, as presented so far, would seem to be subject to the fatal objection that its overwhelming emphasis upon demand simply ignores the whole thrust of modern science and the manner in which the growth of specialized knowledge has shaped and enlarged man's technological capacities. Such growing technological sophistication, surely, suggests that at least some of the initiative in the changing patterns of inventive activity lies on the supply side and not on the demand side where Schmookler has placed it.

Schmookler has anticipated this objection, and his answer is in fact an ingenious one. He argues that the *commodity classes* towards which inventors direct their efforts are determined by expectations concerning financial payoffs which, in turn, are shaped by the familiar considerations of demand and market size. Developments on the side of science and technology are highly relevant to the inventive process, but only in determining the technical realms - mechanical, electrical, chemical, biological - upon which the inventor will *draw*. While the growth in knowledge at the scientific and technological levels will thus influence the specific *characteristics* of inventions, the *purposes* for which inventions are undertaken will depend upon the state of the market for classes of final commodities.

The point is that, while a marketable improvement in envelope-making equipment is probably about as easy to make as one in glass making, it may be easier today to make an improvement in either field via electronic means than through some mechanical change. . . . If differences exist in the richness of the different inventive potentials of the product technologies of different supplying industries, the pressure to improve an industry's production technology tends to be met by the creation of relatively more new products in supplying industries with richer product inventive potentials. For example, if new electrical machines are easier to invent than are non electrical machines, then the aggregate demand for new machinery tends to induce relatively more electrical than non electrical machinery inventions. In brief, inventors tend to select the most efficient means for achieving their ends, and at any given moment, some means are more efficient than others.[5]

Schmookler thus argues for the primacy of demand-size considerations, not by suggesting that shifts on the supply side have been unimportant. Quite the contrary. Science and technology have brought about a great transformation in man's capacity to pursue his material ends. But it is precisely because of the *versatility* of man's enlarged inventory of scientific and technical skills that demand-side forces retain their primacy.

Oddly enough then, science and technology play a subordinate role in influencing the *direction* of inventive activity within Schmookler's analysis, not because his analysis downgrades their historical significance, but rather because he regards science and technology in the modern age as being, in a significant sense, omnicompetent. Schmookler looks upon the body of modern science and technology as constituting a kind of "putty clay" out of which almost anything can be shaped. As he states, "... mankind today possesses, and for some time has possessed, a *multi-purpose knowledge base*. We are, and evidently for some time have been, able to extend the technological frontier perceptibly at virtually all points."[6]

Now this is precisely the aspect of Schmookler's argument which seems to be most inadequate. If Schmookler is right, then economists need not pay too much attention to the internal histories and structures of the sciences and technologies in order to understand the direction of inventive activity. If he is right, then science and technology have not functioned as major independent forces in shaping the timing and the direction of the inventive process. If economic forces can so powerfully shape, not only technology, but science as well, in the achievement of its own ends, then these subjects retain little interest for the economist or economic historian.[7] On the other hand, if Schmookler is wrong in this respect, then his analysis needs to be supplemented by a more careful examination of the manner in which the state of knowledge at any time shapes and structures the possibilities for inventive activity.

III

To establish the independent importance of supply-side considerations, it is necessary to demonstrate several things: (1) That science and technology progress, in some measure, along lines determined either by internal logic, degree of complexity or at least in response to forces independent of economic need; (2) that this sequence in turn imposes constraints or presents opportunities which materially shape the direction and the timing of the inventive process; and (3) that, as a result, the costs of invention differ in different industries.

As soon as one speaks of the "costs of invention" it is necessary to recognize that the economic analysis of inventive activity is seriously handicapped by our present inability to specify the production function for inventive activity with any pretence of precision. Inventions, unfortunately, do not come in units of equal size, whether considered from the point of view of their usefulness or their costs of production. Both the inputs and the outputs in the production of invention are appallingly difficult to measure. Schmookler's basic unit of measurement is, in fact, not an "invention" but a "patent" which serves as a surrogate for an invention. Schmookler's primary interest is in illuminating the process through which society allocates resources to inventive activity. The extreme heterogeneity which is the essence of inventive output is, Schmookler believes, less serious a problem for his interests than it would be in an attempt to link up the number of inventions with the larger phenomena of technological progress and economic growth.[8] Schmookler appears content to regard inventive output as adequately measured by the mere number of inventions since, it is important to note, he is not attempting a direct link-up between the inventive process and the larger question of the historical growth in resource productivity. His results, he is careful to point out, ". . . apply only to the number of inventions made, not to their importance. . . . One of the problems of research now is to establish the nature of the connection between the number of inventions in a field and the rate of technological progress."[9] Within this framework the attempt to compare a unit of invention in one industry with a unit of invention in another industry (or even two inventions in the *same* industry) is obviously fraught with difficulty. Schmookler is content to observe that the prospective *value* of inventive output is likely to be greater in industries undertaking large amounts of investment than in industries where such investment is smaller. An industry's volume of investment activity, in other words, is the primary determinant of the profitability of a unit of invention.

This leaves us very much in the dark in attempting to attach a larger significance to a unit of invention. It would be most convenient, for analytical purposes, if there were an identifiable unit of invention which lowered the cost of production in a plant by, say, 1%. This would enable us to assess the importance of a unit of invention by relating it to the size or to the rate of growth of the adopting industry. Unfortunately, the extreme heterogeneity of inventive output simply does not allow us to assume any simple relationship between the number of inventions and the number of such units of invention or productivity growth.[10] Schmookler does, however, hold the view that the *cost* of invention is likely to be the same in all industries. He points out that "... the very high correlations obtained ... between capital goods invention and investment levels in different industries, and the substantial similarity in the patent-worker ratio of durable and nondurable goods industries indicate that *a million dollars spent on one kind of good is likely to induce about as much invention as the same sum spent on any other good. Hence, doubling the amount spent on one kind of good is likely to induce about as much invention as the same sum spent on any other good.*"[11] This position raises serious difficulties to which we will shortly return.

Although Schmookler's treatment of the relationship between demand forces and invention is, in general, highly illuminating, his conceptual apparatus even here contains some disturbing gaps. This is apparent when he states that, "From a broader point of view, demand induces the inventions that satisfy it."[12] One wishes to rush in at once with qualifications: *some* demand induces the inventions that satisfy it. But which, and when? As soon as these questions are raised we are compelled to consider the different rates at which separate branches of science have progressed. Many important categories of human wants have long gone either unsatisfied or very badly catered for in spite of a well-established demand. It is certainly true that the progress made in techniques of navigation in the sixteenth and seventeenth centuries owed much to the great demand for such techniques in those centuries, as many authors have pointed out. But it is also true that a great potential demand existed in the same period for improvements in the healing arts generally, but that no such improvements were forthcoming. The essential explanation is that the state of mathematics and astronomy afforded a useful and reliable knowledge base for navigational improvements, whereas medicine at that time had no such base. Progress in medicine had to await the development of the science of bacteriology in the second half of the nineteenth century. Although the field of medicine was one which attracted great interest, considerable sums of money, and large numbers of scientifically trained people, medical progress was very small until the great breakthroughs of Pasteur and Lister. Improvements in the treatment of infectious diseases absolutely required progress in a highly specific discipline - bacteriology - and the main thrust of medical "inventions" in the past one hundred years would be difficult to conceive without it. Indeed, it is highly doubtful that, with the single exception of vaccination against smallpox, medical progress was responsible for any significant contribution to the decline in human mortality before the twentieth century.[13]

The point at issue here is one of general importance to Schmookler's argument. The role of demand side forces is of limited explanatory value unless one is capable of defining and identifying them *independently* of the evidence that the demand was satisfied. It would not require a very lively imagination, as the references to medical progress suggest, to compile an extensive list of "high priority" human needs which existed for many centuries, which would have constituted highly profitable commercial activities, but which yet remained unsatisfied. Schmookler's formulation is such that it is capable of being fitted to almost any conceivable set of historical observations. For his argument to be nontautological, however, it would have to be formulated in such a way that the component elements of demand could be identified *independently* of our observations concerning inventive activity. Until this is done it is difficult to conceive of any set of observations which could directly refute Schmookler's hypothesis. In the absence of a reasonably clear, independent specification of the composition of demand, one can never demonstrate either that important components of demand have gone unsatisfied or that supply side factors played an important role in laying down the time pattern of inventive activity.

In fact, the argument of this paper is that, if we want to explain the historical sequence in which different categories of wants have been satisfied *via* the inventive process, we must pay close attention to a special supply-side variable: the growing stock of useful knowledge. Historical evidence confirms that inventions are rarely equally possible in all commodity classes. The state of the various sciences simply makes some inventions easier (*i.e.,* cheaper) and others harder (*i.e.,* more costly). In considering the manner in which the stock of scientific knowledge has grown, and the manner in which this growth has, in turn, shaped the possibilities for inventive activity, one basic fact stands out: the world of nature contains many subrealms, which vary enormously in their relative complexity. If one considers the broad sweep of scientific progress over the past 300 or 400 years, the timing and sequence of the growth of knowledge in these separate disciplines is closely related to the complexity of each - as well as to the complexity of the technology upon which scientific research in the discipline depends. For example, the microbial world and to a great extent the biological world could not be examined without the assistance of the microscope, and the contemporary study of the atomic structure of giant molecules awaited the technique of X-ray crystallography. On the other hand, it is not surprising that the disciplines which were carried to the most advanced state in antiquity were astronomy, mathematics, mechanics and optics. These were each disciplines which could be carried far on the evidence of unassisted human observations, with little or no reliance upon complex instruments or experimental apparatus.[14] Thus, a mastery of the principles under-lying the mechanical world was attained long before a similar mastery was achieved over the principles of chemistry - almost 200 years, if we use as our benchmark dates the publication of Newton's *Principia* on the one hand and Mendelejeff's periodic table of the elements on the other. Similarly, within the discipline of chemistry itself, progress was more rapid in inorganic than in organic chemistry. Even though it had long been apparent that there were huge economic benefits to be reaped throughout the vegetable and animal worlds from a greater knowledge of organic chemistry, such knowledge persistently lagged behind the growing knowledge of inorganic chemistry. Organic chemistry long remained intractable and unresponsive to an obvious and compelling demand. Even after it had become apparent that all organic substances are composed of small numbers of elements - mainly carbon, hydrogen, oxygen and nitrogen - science quite simply remained baffled at the mysteries of the organic world. Progress in organic chemistry, we now know, lagged far behind inorganic chemistry because of a basic and unyielding datum of the natural world: the far greater size and structural complexity of organic molecules.[15] Similar considerations underlie a broad range of research activities and go far towards explaining the timing with which commercially marketable results are extracted from such activities. Thus, the molecular structure of vitamin B_{12}, essential in the treatment of pernicious anemia, is much more complex than vitamin B_1 or C and, as a result, it took far longer to isolate, synthesise and place in commercial production. Similarly, the comparative lateness of the organic chemist's successful assault upon the structure of protein molecules is largely attributable, we now know, to their great complexity. Amorphous materials, as a group, are much more complicated in their atomic structure than crystalline solids and have therefore required a much greater research effort to understand. Progress in the treatment of diabetes has long been held up by the inability to decipher the insulin molecule. Recent research utilising X-ray crystallography has finally revealed a remarkably complex three-dimensional structure consisting of no less than 777 atoms. This finding goes a long way toward explaining why a more effective medical program has taken so much longer to launch in the case of diabetes than in the relatively "simple" diseases such as malaria, syphilis or cholera. Much scientific research at the microbiological level is, in fact, preoccupied with mapping out the highly complex structural arrangement of the component atoms of organic molecules.[16]

Thus, while I believe that Schmookler has supplied an essential corrective to an earlier, widely held view which looked upon the scientific enterprise as not only totally exogenous to the economic sphere but even as a completely autonomous force, propelled by a purely internal logic, I also believe that he has overstated his case in some important aspects. Although economic forces and motives have inevitably played a major role in shaping the direction of scientific progress, they have not acted within a vacuum, but within the changing limits and constraints of a body of scientific knowledge growing at uneven rates among its component subdisciplines. The shifting emphasis of inventive activity over the past two centuries - mechanical, chemical, electrical, biological - is deeply rooted in the history of science, and it is difficult in the extreme to visualize how any plausible set of social and economic forces could have brought about a total reversal of that order.[17] *Given* that sequence in the development of science, inventive activity in some commodity classes was much easier than in others. Furthermore, although

Schmookler is doubtless correct that we have an *increasingly* multipurpose knowledge at our disposal, it is easy to exaggerate the extent to which separate subrealms of knowledge offer genuine options in the satisfaction of given categories of human wants, in the sense of presenting methods which are *substitutes* for one another. Such substitution is frequently nonexistent and usually highly imperfect. Moreover, in many cases the inventive process confronts relationships of complementarity rather than substitution. Thus, the great twentieth century transformation in world agriculture is largely a product of biological knowledge - the mastery of the principles of heredity which have made it possible to develop entirely new, highly productive strains such as hybrid corn in the 1930s and 1940s and, more recently, new wheat and rice varieties. But a fundamental characteristic of these life-science "inventions" is their high degree of complementarity with chemical inputs. Indeed, the new high-yielding rice varieties recently introduced into southeast Asia are often no more productive than the traditional varieties if they are grown under the old techniques of crop and soil management. Their unique feature is a high degree of fertilizer-responsiveness brought about by genetic manipulation. A much better name than "miracle" rice would be "fertilizer-responsive." There are no miracles. In fact, the sharp increases in output per acre, which superficially suggest massive improvements in resource *productivity,* are really the result of large increases in fertilizer and other chemical inputs combined with rigorous attention to techniques of water management.[18]   Thus, these biological inventions require, for their success, large doses of chemical inputs: fertilizer on the one hand, and pesticides to protect them from the many pests to which they are peculiarly vulnerable, on the other.[19] In this critical area of agricultural technology, then, and in other areas as well, the dominant relationships are those of complementarity and not substitution. In this respect, therefore, our freedom of choice in drawing upon different realms of science and technology for ways of increasing food production is largely illusory. The range within which we can exercise genuine *options* in the achievement of specific goals is in fact, severely attenuated.

IV

When we move from the realm of science to that of technology, we enter a world where economic motives are much more direct, immediate and pervasive. Since technological concerns are dealt with primarily within a matrix of profit-seeking business firms, one would expect to find, as one does, a high degree of responsiveness to conditions of market demand and profit expectations generally. But here too it is abundantly clear that an understanding of demand forces alone provides only very limited insight into the direction and the timing of inventive activity. Here, too, differences in the inherent complexity at the technological level shed a flood of light on the inventive process as it has occurred in historical time. If this is correct, then the Schmookler position that technological problems will be solved (one way or another) when the demand for such a solution is sufficiently pressing (*i.e.,* profitable) is seriously incomplete, and needs to be supplemented by a careful scrutiny of supply side variables.

Consider one of the central events of the industrial revolution: the substitution of a mineral fuel for wood in industrial activities. The growing scarcity of wood and the desirability of substituting coal became increasingly clear in Great Britain as early as the second half of the sixteenth century, during which time the price of firewood rose far more rapidly than prices generally. By 1600 the growing pressure upon the limited supplies of firewood and timber had already produced numerous attempts to introduce coal into individual industries. And yet, in spite of strong and pervasive economic inducements, it took over 200 years before this substitution was reasonably complete. But what is particularly interesting from our present vantage point is that, in *some* industries, the transition to the new fuel was effected very rapidly, whereas in others, including some of the most important such as metallurgy, a span of 200 years was required.

Why? A complete answer would be long and complex, but a major part of the answer is that the substitution presented no technical problems at all in some industries, while it created very serious problems in others. No major problems arose in using coal in the evaporation of salt water in salt production, or in lime-making or in brick baking. But in other industries the use of the new fuel seriously reduced the quality of the final product - as in glass-making, the drying of malt for breweries and most importantly, in the smelting of metallic ores. Throughout the seventeenth century considerable effort and experimentation were devoted to these problems. The problems of glass production were solved relatively early by the use of closed crucibles which protected the glass from the destructive effects of the mineral fuel (although, significantly, the method could be used only to produce a coarse cheap glass). In malt production a more palatable beer was being produced by mid-century by first reducing coal to coke and thus eliminating some of the offending elements. Later in the century a reverberatory furnace was introduced which was eventually successfully employed in the smelting of lead, tin and copper. The coke-smelting of iron was first achieved by Abraham Darby in 1709, but the method produced only a very inferior quality of iron. As a result the use of coke pig iron was restricted to the small, cast-iron branch of the iron industry, and charcoal pig iron continued to be used for almost another century for all high-quality purposes. It was only after Henry Cort's introduction of the puddling process in the 1780s for the refining of pig iron that the transition to mineral fuel was finally completed. [20]

Thus the timing of a whole series of inventions connected with the introduction of coal can be understood only in terms of a protracted effort at maintaining quality control while introducing coal into industrial uses. The use of coal created a series of new problems, of varying degrees of complexity, in different industries. Moreover, the fuel itself varied considerably in its chemical composition from one region to another. Since the nature of the chemical interchanges between the new fuel and the various raw materials with which it was employed were not understood, a great deal of time was required (in some cases hundreds of years) before crudely empirical methods finally sorted out the economic opportunities presented by the new fuel. Moreover, the sequence in which solutions were found to the problems of different industries varied considerably, depending upon the technical difficulties involved. Indeed, it may be confidently asserted that the solution came *last* in precisely that industry where the economic payoff was greatest: the iron industry.[21]

The burden of my argument here is that the allocation of inventive resources has in the past been determined jointly by demand forces which have broadly shaped the shifting payoffs to successful invention, together with supply-side forces which have determined both the probability of success within any particular time frame as well as the prospective cost of producing a successful invention. But even if one were to accept the proposition, which I do not, that demand-side forces alone determine the allocation of inventive resources, it would still remain true that supply-side forces exercise a pervasive influence over the actual *consequences* of such resource use: *i.e.*, the *output* of successful inventions, and the timing of these inventions. The explanation of the nature and composition of inventive *output* necessarily requires an understanding of the operation of supply-side forces. These supply-side forces determine whether the output is of the kind associated with the medieval alchemist or the modern scientific metallurgist, the medical quack and patent medicines or broad spectrum antibiotics. Even if knowledge of demand forces alone yielded sensible predictions about the direction of inventive effort, such knowledge, in the absence of further information about supply side forces (the state of scientific knowledge, the prevailing levels of technological skills, the specific characteristics of raw material inputs, etc.) is likely to provide only limited insight into the flow of inventive output.

If we turn to the sequence of invention in textiles, the first major industry to experience full mechanization, one overriding fact stands out: mechanization at all stages in the productive process came much earlier to the new cotton branch of the industry than to the older woolen branch. There were several economic reasons for this, which were rooted in the underlying conditions determining the supply of the basic raw materials on the one hand, and the nature of the demand for each of the final products on the other. But, in addition, there was again a fundamental technological fact: cotton production lent itself to mechanization far more easily than did wool production for reasons intrinsic to the nature of the two materials. As Landes has aptly pointed out:

... (C)otton lent itself technologically to mechanization far more readily than wool. It is a plant fibre, tough and relatively homogeneous in its characteristics, where wool is organic, fickle, and subtly varied in its behaviour. In the early years of rudimentary machines, awkward and jerky in their movements, the resistance of cotton was a decisive advantage. Well into the nineteenth century, long after the techniques of mechanical engineering had much improved, there continued to be a substantial lag between the introduction of innovations into the cotton industry and their adaptation to wool. And even so, there has remained an element of art - of touch - in wool manufacture that the cleverest and most automatic contrivances have not been able to eliminate.[22]

If we consider the sequence in which machine technology was introduced into separate operations in American agriculture, the relative difficulty of applying machine methods to different operations again looms up as a critical variable. Why did the reaping and threshing of wheat come so much earlier than mechanization in cotton picking, corn picking and husking, and milking? Here again, conditions affecting the demand for such individual inventions spring readily to mind. The harvesting of wheat was especially constrained by weather conditions in a way that the other crops were not. The peculiar history of the cotton-growing South provided that region with more abundant labor than other parts of the country and thus considerably weakened the incentive to introduce labor-saving machinery. Yet, as Parker has pointed out, milking operations were also subject to a very strong time constraint and were concentrated in labor-scarce regions of the country where the incentive to invent labor-saving machinery should have been correspondingly strong. Moreover, there is abundant evidence - *e.g.,* from the Patent Office - that considerable, if unsuccessful, inventive effort had been directed toward these operations in the nineteenth century. "Surely the most plausible single answer," Parker suggests,

is that these operations were all inherently difficult to mechanize without radical alteration and improvement of basic elements in the prevailing technology. In the case of the corn harvester, the problem of harvesting the ear separately from the stalk, while preserving the stalk for forage, was hard to solve. In cotton picking, the need to make several passes over the field as the bolls ripened prevented a crude solution. The possibility of mechanical milking was hardly dreamed of, except by cranks, before the gasoline engine and electric power. It is no accident that in all three cases, the mechanical problem was to imitate complex motions of the human hand rather than the simple sweeping actions of the arm required in reaping and threshing.[23]

A large part of the economic history of the past 200 years is, in fact, the story of an enormous outward shift in industrial man's capacity to solve certain kinds of production problems. This growing capacity has been fitful and highly selective. For most of the nineteenth century it involved the exploitation of new power sources and an increasing mastery over the use of large masses of cheap metal (iron and, later, steel). These techniques became available with no fundamental accretions to basic knowledge. They nevertheless were developed slowly because it took time to develop and then to diffuse new techniques in the precision working of metals and to devise the innumerable small improvements and adaptations which were often required to enable them to operate successfully. There is always a gap, moreover, between the ability to conceptualize a mechanism or technique and the capacity to bring it into effect. Thus, da Vinci's notebooks are full of sketches for novel machinery which could not be realized with the primitive metal-working techniques at his disposal. Breech-loading cannon had been made as early as the sixteenth century, but could not be used until precision in metal working in the nineteenth century made it possible to produce an air-tight breech and properly fitting case. (Without the air-tight breech, a breech-loading cannon was likely to present far greater danger to the persons engaged in firing it than it did to those at whom the first was being directed.) Christopher Polhem, a Swede, devised many techniques for the application of machinery to the quantity production of metal and metal products, but could not successfully implement his conceptions with the power sources and clumsy wooden machinery of the first half of the eighteenth century. Although the principle of compounding was embodied in a patent in 1781, compound steam engines were not introduced into ocean-going vessels until the 1880s, a full century later, in spite of strong economic incentives. Not until major breakthroughs in steel-making technology was it possible to provide high quality components such as boiler plates and boiler tubes upon which the operating efficiency of the compound engine depended. Charles Babbage had conceived of the main features of the modern calculator over a century ago, and had incorporated these features in his "analytical engine," a project which was even favoured with a large subsidy from the British Exchequer. Babbage's failure to complete this ingenious scheme was due to the inability of the technology of his day to deliver the components which were essential to the machine's success.

The purpose of this recitation of frustrations and failures is simply to argue that, given the state of purely scientific knowledge, society's technical competence at any point in time constitutes a basic determinant of the kinds of inventions which can be successfully undertaken. Of course it is possible to argue, as it has been with respect to the long delay in the introduction of a mechanical cotton picker, that if factor prices and/or cotton prices had been significantly different, a practical machine would have been introduced much earlier. If, for example, the available labor supply had been much more expensive, more inventive effort would presumably have been devoted to solving the complex technical problems of a cotton-picking machine much sooner. While this is probably true, it is also incomplete. Because it is also true that, *given the set of factor and commodity prices which actually prevailed,* the cotton-picking machine would also have been developed more quickly if the technical problems which had to be overcome were less serious. These technical problems and their relative complexity stand independently of demand considerations as an explanation of the timing and direction of inventive activity. Therefore any analytical or empirical study which does not explicitly focus upon both demand and supply side variables is seriously deficient.

V

Where has this analysis taken us? I have argued that the central weakness of Schmookler's approach is his treatment - or, rather, his neglect - of the supply-responsiveness of technology and invention.

Essentially, Schmookler is saying that, given the state of science (and regardless of "how we got here") the supply of inventions is, in effect, perfectly elastic, and at the same price, in all industries. At any moment in time it is possible to get as many inventions as wanted in any industry at a constant price. Therefore the observed *composition* of inventions is entirely a demand side phenomenon, reflecting the manner in which inventive resources have been allocated between industries (or, better, commodity classes) in response to the structure of (demand-induced) profit expectations.

The main objection which I have raised is that inventions are *not* equally possible in all industries. This is because there is a crucial intervening variable: the differential development of the state of subdisciplines of science and bodies of useful knowledge generally at any moment in time. Indeed, I think it is very important that we cease talking about "the state of science" and begin thinking in terms of "sciences." A central problem is to trace out carefully the manner in which *differences* in the state of development of individual sciences and technologies have influenced the composition of inventive activities. Let me suggest further that one way of getting at this is to pay more attention to historical failures.

Our understanding of inventive activity (and perhaps of social change generally) is excessively rooted in success stories. We study the history of successful inventions but devote little attention to inventions which were not made. Yet it is highly relevant to ask why it took so long to do certain things, and why inventors failed for so long at some inventive efforts while they succeeded quickly at others. It is certainly possible to study past patterns of research expenditure and inventive effort, and to seek the reasons for unsuccessful as well as successful outcomes, for very long gestation periods in the development of new inventions as well as for shorter periods.[24] In short, if we want to probe the relations between science, technology and inventive activity more deeply, we must learn much more about what was *not* possible as well as what *was* possible. We need to understand what scientific and technological discoveries were needed for key breakthroughs in invention. For knowledge not only permits - it also constrains. For this reason we can learn much from the study of unsuccessful attempts to invent something for which the market was perceived to be ready. In this respect, the study of failure is essential to a determination of the precise role of supply-side variables in the inventive process. After all, the demand for higher levels of food consumption, greater life expectancy, the elimination of infectious disease, and the reduction of pain and discomfort, have presumably existed indefinitely in the past, but they have been abundantly satisfied only in comparatively recent times. It seems reasonable to suppose that the explanation is to be found in terms of supply-side considerations. It is unlikely that any amount of money devoted to inventive activity in 1800 could have produced modern, wide-spectrum antibiotics, any more than vast sums of money at that time could have produced a satellite capable of orbiting the moon. The supply of certain classes of inventions is, at some times, completely inelastic - zero output at all levels of prices. Admittedly, extreme cases readily suggest arguments of a *reductio ad absurdum* sort. On the other hand, the purely demand-oriented approach virtually assumes the problem away. The interesting economic situations surely lie in that vast intermediate region of possibilities where supply elasticities are greater than zero but less than infinity!

The perspective which I am suggesting, therefore, states that, as scientific knowledge grows, the cost of successfully undertaking any given, science-based invention declines - from infinitely high, in the case of an invention which is totally unattainable within the present state of knowledge, down to progressively lower and lower levels. Perfectly inelastic supply curves of invention gradually unbend and flatten out. (To what *extent* they flatten out is, of course, an empirical question, on which Schmookler has adopted the arbitrary and implausible extreme assumption of perfect elasticity.) Thus, the growth of scientific knowledge means a gradual reduction in the cost of specific categories of science-based inventions. The timing of inventions therefore needs to be understood in terms of such shifting supply curves which gradually reduce the cost of achieving certain classes of inventions. More precisely, we need to think in terms of a number of supply curves for individual industries, depending upon the knowledge bases upon which inventive activity in that industry can draw, and we need to understand more clearly the extent to which different "pools" of knowledge are potential substitutes in the inventive process. Schmookler's hypothesis states, in effect, that there is one supply curve for all industries and that the extent of substitution renders it unnecessary to look at supply conditions in individual industries. It seems to me that a clear articulation of the relations between science, invention and economic growth requires a critical examination of this assertion. The basic economic question, of course, is not an "either or" proposition telling us whether a particular technological achievement is or is not possible at a particular point in time. The economic question is: given the state of the sciences, *at what cost* can a technological end be attained? How does the state of individual sciences differentially structure the cost of society's technological options?[25] Answers to these questions will carry us a long way toward a deeper understanding of both the nature of inventive activity and the process of economic growth by providing further insight into the economy's changing capacity to respond to economic needs.

Epilogue

The preceding chapters may strike the reader as reflecting an underlying mood of unwarranted technological optimism. Certainly they lack the apocalyptic vision (or, at the very least, the obligatory lugubrious tone) which has come to dominate the recent discussion of man's present estate and his future economic prospects. This seems an appropriate opportunity, therefore, for linking some of the themes of this volume to present-day concerns, especially our preoccupation with the supposed inadequacy of our natural-resource base for sustaining continued economic growth and the prospective contribution of science to such growth.

My starting point is that the potential flexibility and adaptability of advanced industrial economies is vastly underestimated. This is particularly the case with respect to dependence upon specific natural-resource inputs. Most thinking about the role of natural resources in economic growth continues to be excessively static. It ignores not only the dynamic interactions between technological change and natural resources, but also the whole range of additional adaptations which are a mixture of pure technological change, redesigning, and substitution. Thus rising fossil-fuel prices may eventually lead to the more rapid development of techniques for the exploitation of atomic or solar energy, but they will doubtless also lead to the utilization of lower-quality oil sources, to the redesign and modification of automobile engines, and to the use of more blankets, insulating material, and thermal underwear. It is even conceivable that the influx of people into Florida and southern California will grow. The point is not that, therefore, the increasing scarcity of fossil fuels is of no consequence. That would be patently absurd. Rising fuel costs have very widespread ramifications. The whole technology of low-grade resource utilization, for example, is highly fuel-intensive in nature and would be seriously affected. Rather, the point is that an advanced industrial society, given sufficient time, has available a wide range of mechanisms by which this rising cost and its consequences may be offset, and the historical evidence at least suggests overwhelmingly that such rising costs have not, *in the past,* constituted an insurmountable obstacle to sustained increases in both total population and per capita incomes. It seems reasonable, therefore, that growth models dealing with the distant future should allow for the possibility of human response to increasing scarcities - responses in which adaptation and modification of behavior, substitution of abundant for scarce materials, and even learning, play a prominent role.

Obviously the adaptive responses do not occur instantaneously. Developing new technologies and installing them is both costly and time-consuming, often involving a time frame, not of months or even years, but decades, and is perfectly consistent with severe temporary shortages, especially when an accustomed source of supply - such as Mideast oil - is suddenly and unexpectedly turned off for political purposes.

Equally obviously, what is being suggested is not a rather mindless complacency based upon the confidence that technology will somehow, surely, find a comfortable and neatly packaged solution to all our problems. Technology, to begin with, seldom offers self-contained solutions. That is to say, it typically also requires some sort of modification in human behavior, often painful modifications. Moreover, technological solutions frequently bring with them new and unanticipated problems - for example, of an ecological sort, to which we have been remarkably indifferent in the past, and which we are only just beginning to examine in systematic ways. Technological solutions, we are discovering, have a notorious capacity for replacing one set of problems with another. This does not render the solutions valueless, but it *does* typically mean that we find ourselves engaged in a continuous problem-solving process. Finally, to assert that a technological solution is available to us is, by itself, extremely uninteresting. For the basic economic question remains: how costly is the solution? Although an obsession with the increasing niggardliness of nature is, I believe, unjustified, there is certainly no warrant for going to the opposite extreme and regarding nature as if it were some sort of Lady Bountiful. And although I am optimistic about the eventual availability of technological solutions to a wide range of pressing problems, I see nothing in the cards which guarantees that these solutions will be cheap. It is one thing to be confident that we will not run out of fuel sources, and quite another to believe that fuel costs in, say, twenty or fifty years will be cheap by present-day standards.

Most of this volume has been concerned with the recent history of the most highly successful portions of the world economy. I have suggested that industrial economies - and particularly the United States - have been highly successful so far in adapting to the shifting patterns of resource scarcity which economic growth has generated, and I have examined some of the most significant dimensions of that adaptation process. But from a worldwide perspective the picture has not been nearly so bright. Specifically, the capacity to generate new, sophisticated, and appropriate technologies to overcome or to bypass resource scarcities is a capacity which has been acquired by only a small portion of the world's population. Or, to put the point somewhat differently, all societies which survive have managed to *adapt* to the resource possibilities of their environment, but they have adapted at very different levels of success and technological sophistication. They show enormous variation in the knowledge and technical skills which they bring to bear upon their environment. The Kalahari Bushman has survived for thousands of years with no more than his essentially unchanged neolithic technology. Some societies in the humid tropics maintain soil fertility by a primitive system of "slash-and-burn" cultivation which supports a precarious existence and requires long fallow periods to permit restoration of natural fertility. In other societies a knowledge of soil chemistry makes it possible to pinpoint the soil nutrients which allow continuous cultivation and far higher levels of output per acre. Nomadic herdsmen represent one level of adaptation to an arid climate, and irrigation agriculture represents another. Let me suggest that a direction in which social science research urgently needs to move is the attempt to understand what factors determine the degree of technological responsiveness which a society is capable of generating. The capacity for such response, clearly, constitutes a large part of what successful economic growth is all about. The point deserves further development. In particular, it needs to be linked to the evidence of the growing role of the scientific establishment as an essential ingredient of flourishing industrial societies.

Before the twentieth century there was no very close correspondence between scientific leadership and industrial leadership with respect to individual countries. English science went into a state of decline in the post-Newtonian eighteenth century, at precisely the time that she began to assert her decisive leadership in the technologies of metallurgy, mining, steam power, and textiles. On the other hand, the flowering of French science in the eighteenth and early nineteenth centuries, when Paris was the acknowledged scientific center of the world and the *Academie des sciences* the world's most important scientific organization, was not accompanied by a corresponding thrust of industrial advance. Even though French inventive talent was remarkably productive, it was notorious that French inventions had to be taken elsewhere in order to be profitably employed.[1] Similarly, Russia produced numerous brilliant scientists and inventors in the nineteenth century, but they seem to have exerted a negligible impact upon the country's economic development. Finally, the rise of the United States to a position of burgeoning economic growth and technological leadership occurred during the nineteenth century, a period in which American achievements in basic science were minimal and, with few exceptions, of no great international consequence. Indeed, as late as 1869, when Mendelejeff published his periodic table of the elements, there was not a single journal devoted to the subject of chemistry in the United States. To be sure, America after World War I represents an instance in which priority in basic scientific research and technological leadership did coincide, and Germany's emerging leadership in the chemistry-based industries in the second half of the nineteenth century coincided closely with the decisive superiority of her scientific work in chemistry in that period. But obviously I am arguing, not that these two never coincide, but merely that they *need not* do so. What seems clear is that, on a country by country basis, a top-quality scientific establishment and a high degree of scientific originality have been neither a necessary nor a sufficient condition for technological or industrial leadership. The proposition is further confirmed by the extraordinary performance of the Japanese economy in the twentieth century and, more particularly, in the years since World War II. There is much evidence suggesting that Japan is now developing a top-quality scientific community, particularly in certain selected areas,[2] but its absence in the past does not seem to have been a serious handicap.

Further corroborative evidence is provided by a study published in 1971 by the Organization for European Cooperation and Development. The study attempted to establish, by an examination of member-country performances, the relationship between R&D expenditures and measured rates of productivity growth for the economy as a whole. Its conclusion may be simply stated: "there is no observed correlation between the proportion of national resources devoted to R&D and rates of growth of productivity."[3]

Some of the reasons for the lack of a closer correlation between scientific and industrial attainments are fairly apparent. On the one hand, until comparatively recently many of the sciences had nothing particularly useful to offer industry. The iron industry stumbled along for centuries on a crude, trial-and-error basis, because it was not until the very end of the nineteenth century that the chemical transformations involved in smelting and refining began to be understood. Science provided only very limited assistance to agriculture until major breakthroughs occurred in biology, organic chemistry, and genetics which created new possibilities for systematic regulation and manipulation of plant and animal growth in the twentieth century. On the other hand, many industries have been able to make great technological strides without science for the perfectly sufficient reason that technological progress did not *require* new scientific knowledge. This was true of the central thrust of nineteenth-century American industrialization, which involved primarily the development of a machine technology. The invention of new machines or machine-made products - cotton gin, reaper, thresher, cultivator, typewriter, revolver, barbed wire, sewing machine, and bicycle - involved the solution of problems which required mechanical skill, ingenuity, and resourcefulness, but not, typically, a resource to scientific knowledge or elaborate experimental methods.

It should be clear that economic growth, when it occurs, is the outcome of a complex social process. Science, even when it constitutes an essential ingredient, as it has increasingly in the past century or so, is merely one component of this larger process. Whether an active and creative scientific establishment in a country will make an important contribution to that country's economic growth depends upon a whole network of institutions and motivations outside of the scientific community. Translating new scientific knowledge into more productive techniques and into new final products raises questions of inventive abilities and commercial talent which are far removed from "mere" scientific originality. Why do some societies have a much greater apparent capacity to generate inventions appropriate to their economic needs than do others? Why are some societies much more receptive than others to the introduction and adoption of inventions made elsewhere? These are very different questions, for the requirements for successful inventive activity seem to be very different from the requirements for the rapid *adoption* of an invention which has already been made. The first question involves the supply of inventive talent, engineering skills, and the ability to bring specialized knowledge to bear upon the solution of technical problems. The second question is likely to turn upon the supply of managerial abilities, highly motivated entrepreneurs, business acumen in perceiving market opportunities, and organizational flexibility and effectiveness. But, of course, a society which is well endowed with technological and engineering expertise and an astute and aggressive business community will readily perceive ways of exploiting the fruits of scientific research in *other* countries, and will not be hopelessly constrained by a modest scientific effort. This seems to constitute a reasonable description of both nineteenth-century America and twentieth-century Japan. Both countries drew freely and energetically, when the economic need arose, upon scientific research abroad, and both were quick to import and to utilize technologies developed elsewhere even when doing so may have involved very substantial adaptation and redesign.

Thus, the appropriate conclusion is not that science has been unimportant to economic growth. Although it was of only limited significance 200 years ago, or even as recently as 100 years ago, it is without question becoming of increasing significance in terms of its actual and potential contribution to economic growth. The appropriate conclusion is that science is not *uniquely* important and *by itself* can contribute relatively little. In the absence of other complementary inputs - high-level technical skills, strong incentive systems, flexible and responsive organizational structures for mobilizing resources - even a highly creative scientific community is of little economic consequence. On the other hand, the evidence suggests that when a country *has* possessed these other inputs in the past, it was usually able to surmount the difficulties imposed by the undeveloped state of its own scientific establishment.

In the larger panorama of world history, a striking feature of the history of European culture was not only the rise of modern science, but the combination of cultural values and institutions which made possible the practical and effective *application* of new knowledge, wherever it may have originated. Certainly inventiveness has never been a European monopoly. As the extraordinary scholarship of Joseph Needham and his associates has so lavishly documented, Chinese civilization was for many centuries remarkably inventive by any standards.[4] Francis Bacon long ago observed in his *New Organon* that three great mechanical inventions - printing, gun-powder, and the compass - had, as he put it, "changed the whole face and state of things throughout the world; the first in literature, the second in warfare, the third in navigation."[5] But what also needs to be observed about these three inventions is that the origin of each is usually attributed to China, and yet the full exploitation of these inventions (for destructive as well as productive purposes) was brought about by Europeans. What western European civilization seems to have developed was a powerful combination of cultural values, incentive systems, and organizational capabilities for transforming inventions into highly utilitarian forms. Marx of course saw all this and attributed it to the emergence of modern capitalism. But from my present perspective that simply reformulates the very big question: why did modern capitalism develop in western Europe and not elsewhere? Interesting answers to this question are unlikely to come from any single social science discipline.

Because so much of my concern has been with the capacity to find technological solutions to economic problems, I would like finally to emphasize how such solutions are shaped and constrained by larger social forces. Part - but only a part - of the forces which I have in mind fall into the category of what the economist calls "tastes and preferences." I referred earlier to the vast differences among societies in their capacity to generate new technologies. It is now necessary to insist on a complementary point: that the role which such a technology can play - and indeed whether it can play any role at all - will depend upon the cultural values and social structure of the society. For the *productivity* of any technology is always dependent upon its institutional and cultural context, and its eventual impact must therefore always be examined· within that context. For example, it is a striking historical fact that, although the United States was at least half a century behind Great Britain in its "takeoff" into an industrializing economy, some of the most distinctive technological features of a modern industrial economy first emerged in the United States. A mass-production technology dependent upon the progressive assembly of interchangeable parts which, in turn, were produced in large quantity by a sequence of highly specialized machines, was a distinctively American achieve-ment. A major reason for this, I would suggest, was differences in taste and demand, which in turn were deeply rooted in the social structure and distribution of income in each country. A result of these differences was that Americans readily accepted highly standardized products so long as they served a utilitarian purpose, whereas the British producer was under much greater pressure to customize his products to the tastes - often the highly idiosyncratic upper-class tastes - of individual consumers in a society where class distinctions were far more conspicuous. This factor played a significant role in America's early assertion of leadership in the machine production of such articles as firearms, ready-to-wear clothing, boots and shoes, and, of course, the automobile. It is difficult in the extreme to imagine a British car manufacturer - say Lord Nuffield - uttering Henry Ford's famous epigram that his customers could have their cars in any color they wanted, so long as it was black.

Tastes and preferences intrude themselves in innumerable ways in dealing with the possibility of providing technological solutions to problems of growing scarcity and also in providing *alternatives to the need for* technological solutions. That is to say, in my earlier discussion I concentrated on substitution possibilities for *inputs* which became increasingly scarce. But, clearly, it is possible to achieve substitutions on the *output* side as well as the input side. One can redesign final products and alter the *mix* of products in order to consume smaller amounts of resources - we can produce smaller cars (or if we cannot, the Japanese can do it for us) or substitute buses for cars, substitute long underwear and blankets for fuel oil, move from separate residential dwellings to apartment houses, and sub-stitute returnable bottles for disposable cans. We can even, if I may be allowed the temerity of a truly radical suggestion, deliberately redesign consumer products so that they will last longer. All of these substitution possibilities are now readily available and require no commitment of resources to the development of new technologies. They *do* require a willingness on the part of the consumer to alter his habits and life-style.[6]

Without question the most important taste of all, in the long run, is the "taste" for children. Here it is clear that the possibility of a technological solution is severely circumscribed by tastes, since even a perfectly reliable, perfectly safe, cheap, and convenient contraceptive technology cannot reduce birth rates by more than the amount by which total births exceed the number of *wanted* births. For much of the world the persistence of high fertility levels is a reflection of the elemental fact that couples continue, for a variety of well-known reasons, to *want* large families. So long as such high fertility values and mores persist, no amount of contraceptive technology is going to have a substantial effect upon high birth rates. But, if the persistence of the desire for large families in most poor countries constitutes "the bad news" (because it suggests the futility of purely technological solutions), "the good news," surely, is the mass of demographic evidence that the taste for offspring and therefore reproductive practices have undergone massive and quite voluntary changes in countries, such as the United States, which have experienced successful long-term economic growth. It is unmistakably clear that in the past, the process of economic growth has produced changes in values and preferences which have led to sharp reductions in human fertility. I certainly do not want to suggest that there has been anything which we could characterize as "socially optimal" about this demographic adjustment process, but merely to insist that it does exist.[7] Back in the days of Malthus and Jefferson, American crude birth rates are estimated to have been over 50/1000 whereas today they are around 15-16/1000. It is apparent that we have gone through changes in our pattern of reproductive behavior which are nothing less than staggering and which, *if continued* (the qualification is all-important) will bring us into the realm of a constant population size (ZPG) within the lifetimes of people who are already alive. Obviously, there exist demographic adjustment mechanisms of a subtle and complex sort which parallel the technological ones with which I have been primarily concerned.[8] It would be remarkable if perceptions of optimum family size did *not* alter as economic growth generated such things as reduced mortality, higher incomes, higher educational levels, new goods and services, increased urban densities along with higher levels of congestion and pollution, and so on. (It would be even more remarkable, to let one final bit of optimism creep in, if, at some threshold level of income, people did not begin to discover a large and unsatisfied demand for a cleaner, less polluted, and more livable environment. Our increasing public concern with problems of pollution reflects, *in part,* the fact that a substantial proportion of our population may now have crossed that threshold income level.) Yet, even at this late date, we know very little about these relationships between rising income levels and changing tastes. This is perhaps the reason why it is still possible for social scientists and systems analysts to develop increasingly elaborate models of social systems which take no account of the capacity of the human animal to *respond* to environmental changes and to make appropriate modifications in his behavior - in brief, to take account of the social learning process. Models such as *The Limits to Growth* do serve a useful purpose, not because they provide a likely or plausible scenario, but because they illustrate the long-term implications of certain highly implausible assumptions. *Of course* it is true that, if population were to grow at a rapid rate over a long period of time, *and* if we lacked the ingenuity to increase the productivity of some resources and learn how to make use of more abundant ones, the human race will be in for a very bad time. Malthus reached that conclusion, by purely intuitive reasoning, and with no access to a computer, almost 200 years ago. But the history of the Western world since Malthus makes it clear that mankind does have the ingenuity to develop a very different scenario, and has in fact been doing so. I cannot help believing that we will improve the allocation of scarce intellectual resources if we devote much more attention to trying to comprehend the social and economic mechanisms underlying these adjustment processes, both in their technological and demographic dimensions, and less to neo-Malthusian models which simply project current or recent trends indefinitely into the future.

Notes

1. Technological change in the machine tool industry, 1840-1910

1. Robert Solow, "Technical Change and the Aggregate Production Function," *Review of Economics and Statistics,* XXXIX, No. 3 (Aug. 1957), 312-20; Moses Abramovitz, "Resource and Output Trends in the U.S. Since 1870," *American Economic Review Papers and Proceedings,* XL, No. 2 (May 1956), 1-23; Benton Massell, "Capital Formation and Technological Change in U.S. Manufacturing," *Review of Economics and Statistics,* XLII, No. 2 (May 1960), 182-88.
2. The two most important works which should be consulted by anyone interested in this area of research are H.J. Habakkuk, *American and British Technology in the 19th Century* (Cambridge [England]: The University Press, 1962), and W. Paul Strassmann, *Risk and Technological Innovation: American Manufacturing Methods During the Nineteenth Century* (Ithaca: Cornell University Press, 1959).
3. W.W. Rostow, "The Take-Off into Self-Sustained Growth," *Economic Journal,* LXVI, No. 261 (Mar. 1956), 25-48, and W.A. Lewis, "Economic Development with Unlimited Supplies of Labour," *The Manchester School,* XXII, No. 2 (May 1954), 139-91.
4. Simon Kuznets, "Quantitative Aspects of the Economic Growth of Nations: VI. Long-Term Trends in Capital Formation Proportions," *Economic Development and Cultural Change,* Vol. IX, No. 4, Part 2 (July 1961).
5. Robert Gallman, "Commodity Output, 1839-1899," *Trends in the American Economy in the Nineteenth Century* (Studies in Income and Wealth, 23 [Princeton, N.J.: Princeton University Press for NBER, 1960]) 13-67; Simon Kuznets, *Capital in the American Economy* (Princeton for NBER, 1961), ch. iv.
6. "There were mules and steam-engines before there were any labourers whose exclusive occupation it was to make mules and steam-engines; just as men wore clothes before there were such people as tailors." Karl Marx, *Capital,* I (Modern Library Edition [New York: Random House, 1936]), 417.
7. The experiences of Eli Whitney with his government musket contract are by now legendary. See Jeanette Mirsky and Allan Nevins, *The World of Eli Whitney* (New York: Macmillan Company, 1952). Compare H.J. Habakkuk, "The Historical Experience on the Basic Conditions of Economic Progress," *International Social Science Bulletin,* VI, No. 2 (1954), 189-96.
   A skeptical note on Whitney's role in developing the "American system" has recently been struck in Robert S. Woodbury, "The Legend of Eli Whitney and Interchangeable Parts," *Technology and Culture,* II, No. 2 (Summer 1960), 235-53.
8. It is not an historical coincidence that specialized machinery producers began to emerge on the national scene during precisely the same period as the development of our national railway network. Until roughly 1840, machinery production was not only relatively unspecialized - each producer typically undertaking a wide range of output - but it was also, because of the high cost of transporting machinery, a highly localized operation - each producer typically producing for a very limited geographic radius. The growing specialization in machine production after 1840, characterized by the emergence of large numbers of producers each of whom typically concentrated on a very narrow range of machines, was closely linked with the transportation improvements and consequent reduction in freight costs during the period.

290

9. *A Chronicle of Textile Machinery 1824-1924* (Boston: Saco-Lowell Shops 1924), pp. 16-20. The number of locomotives built by the Locks and Canals Company and by the Lowell Machine Shop between 1835 and 1861 is given, together with the purchasing railroad companies, in George S. Gibb, *The Saco-Lowell Shops* (Cambridge: Harvard University Press, 1950), appendix 6, p. 641.

10. John L. Hayes, *American Textile Machinery* (Cambridge, 1879), pp. 57-58; Jonathan T. Lincoln, "Machine Tool Beginnings," *American Machinist*, LXXVI (Aug. 3, 1932), 902-3. Although Mattias Baldwin in 1852 described himself as a "Manufacturer of Locomotive, Marine, and Stationary STEAM ENGINES," his advertisements also added: "All kinds of Machinery furnished to order." *Journal of the Franklin Institute*, L (June 1852), opposite p. 444. A report of British observers for the same period states, "The practice which prevails of combining various branches of manufacture in the same establishment, would . . . render separate descriptions of each somewhat complicated. In some cases the manufacture of locomotives is combined with that of mill-gearing, engine-tools, spinning, and other machinery. In others, marine engines, hydraulic presses, forge-hammers, and large cannon are all made in the same establishments. The policy of thus mixing together the various branches arises, in addition to other causes, from the fact that the demand is not always sufficient to occupy large works in a single manufacture." *The Industry of the United States in Machinery, Manufactures and Useful and Ornamental Arts,* (compiled from the official reports of Messrs. Joseph Whitworth and George Wallis [London, 1854]), p. 3.

11. Joseph W. Roe, "Machine Tools in America," *Journal of the Franklin Institute* (May 1938), 499-511.

12. As early as 1855, a team of British engineers which had traveled to the United States for the purpose of inspecting American methods of arms manufacture felt compelled to make the following observations on the extent to which specialized machinery had been developed in American industry: "As regards the class of machinery usually employed by engineers and machine makers, they are upon the whole behind those of England, but in the adaptation of special apparatus to a single operation in almost all branches of industry, the Americans display an amount of ingenuity, combined with undaunted energy, which as a nation we would do well to imitate, if we mean to hold our present position in the great market of the world." *Report of the Committee on Machinery of U.S.* (H.C., 1855), p. 32. Similar observations and admonitions appear through a succession of reports by British observers to international exhibitions through the subsequent decades of the nineteenth century. See, for instance, the report on machine tools in *Reports on the Paris Universal Exhibition, 1867, Presented to both Houses of Parliament* (1868), IV, especially 370-73, and "Machines and Tools for working Metals, Wood and Stone," in *Reports on the Philadelphia International Exhibition of 1876, Presented to both Houses of Parliament* (1877), I, especially 228-35. In the latter report, Mr. John Anderson, the author, states: "To realize the nature of the competition that awaits us, their [American] factories and workshops have to be inspected, in order to see the variety of special tools that are being introduced, both to insure precision and to economize labour; this system of special tools is extending into almost every branch of industry where articles have to be repeated. This applies to furniture, hardware, clocks, watches, small arms ammunition, and to an endless variety of other things. The articles so made are not only good in quality, but the cost of production is extremely low, notwithstanding that those employed earn high pay." (p. 235). Mr. Anderson closes with a rhapsody on the tool-using and tool-making abilities of the Americans, and on the urgency of Britain's girding her loins for the coming industrial and engineering competition.

13. *Twelfth Census, of the United States* (1900), X, Part 4, "Manufactures," 385. This report also refers to the existence of a total of ninety metal-working machinery firms in the five leading centers of Cincinnati, Philadelphia, Providence, Hartford, and Worcester. For a breakdown by value of output of major categories of metal-working machinery in 1900 and 1905, see *Special Reports of the Census Office* (1905), "Metal-working Machinery," p. 227.

14. *Census of Manufactures* (1914), II, "Reports for Selected Industries," 269. For the same year there were 277 metal-working machinery plants, other than those producing machine tools, with an output valued at $17,419,526.
15. *American Machinist,* XL (Jan. 29, 1914), 210.
16. We suggest, only in passing, that such a focus may also provide a more fruitful approach to a theory of the multiproduct firm.
17. "... industrial differentiation ... has been and remains the type of change characteristically associated with the growth of production. Notable as has been the increase in the complexity of the apparatus of living, as shown by the increase in the variety of goods offered in consumers' markets, the increase in the diversification of intermediate products and of industries manufacturing special products or groups of products has gone even further." Allyn Young, "Increasing Returns and Economic Progress," *Economic Journal* XXXVIII, No. 152 (Dec. 1928), 537.
18. A.P. Usher, *History of Mechanical Inventions* (Cambridge: Harvard University Press, 1954), especially chs. xiii-xv.
19. On this point it is difficult to avoid the conclusion that we are still suffering, in our understanding of technological innovation, from a Schumpeterian blight. For all his profound understanding of the capitalist process, Schumpeter never quite overcame his preoccupation with the charismatic aspects of leadership and its role in instituting changes in the operation of the economic system. As a result, his own towering intellectual leadership in this area has led to an excessive concern with the more dramatic and discontinuous aspects of innovation, with the circumstances surrounding the initial "breakthrough," and to a neglect of the less spectacular dimensions of innovation. We refer, in particular, to two other aspects: (1) the cumulative impact of relatively small innovations (which were of great importance in the design, development, and adaptation of machines), and (2) the determinants of the rate and the area over which an innovation, once made, is eventually diffused. These points will be treated subsequently in this paper.
20. George Stigler, "The Division of Labor is Limited by the Extent of the Market," *The Journal of Political Economy,* LIX, No. 3 (June 1951), 190.
21. Woodbury, "Legend of Eli Whitney." (See n. 7.)
22. Charles Fitch, "Report on the Manufactures of Interchangeable Mechanism," *Tenth Census of the United States* (1880), II, 13-19. (Hereafter cited as Fitch, *Report.*)
23. The application to the last item was apparently too quick, for Blanchard was upheld, as late as 1849, in an infringement suit involving the application of his lathe to turning shoemaker's lasts. *See Journal of the Franklin Institute* XLVII (1849), 259-62.
    The British commission on arms manufacture which visited the United States in 1853 was particularly impressed with the gunstock lathe, at the time apparently still unknown in England. "It is most remarkable that this valuable labour-saving machine should have been so much neglected in England, seeing that it is capable of being applied to so many branches of manufacture, its introduction into the armory will prove a national benefit. *Report of the Committee on Machinery of U.S.* (H.C., 1855), p. 39. The British Government subsequently purchased these machines from the Ames Manufacturing Company of Chicopee, Massachusetts.
24. Fitch reports that die-forging machines were employed at Harper's Ferry as early as 1827. (Fitch, *Report,* p. 20.) The most significant subsequent improvements were achieved at the Colt armory under the guidance of its ingenious superintendent, Elisha K. Root. Although the sewing-machine industry relied more heavily upon casting than on forging, drop forging with dies was introduced into sewing-machine manufacture as early as 1856. (Fitch, *Report,* p. 37.) Compare also W.F. Durfee, "The History and Modern Development of the Art of Interchangeable Construction in Mechanism," *Transactions, American Society of Mechanical Engineering,* XIV (1893), especially 1,250-51.
25. Fitch, *Report,* pp. 22-26; Joseph W. Roe, "History of the First Milling Machine," *American Machinist,* XXXVI (June 27, 1912), 1,037-38; Robert S. Woodbury, *History of the Milling Machine* (Cambridge: M.I.T. Press, 1960), chs. i and ii. Woodbury's book, which is part of a series devoted to the history of machine tools, is an invaluable guide to the detailed technical development of the milling machine.

26. Fitch, *Report,* pp. 3 and 25; Felicia J. Deyrup, *Arms Makers of the Connecticut Valley* (Northampton: Smith College Press, 1948), pp. 153-54.
27. Fitch, *Report,* p. 26. Fitch cites examples for 1880 of armsmaking plants where milling machines constituted between 25 and 30 per cent of the total number of machines in use (p. 22).
28. *Ibid.,* p. 28.
29. E.G. Parkhurst, "Origin of the Turret, or Revolving Head," *American Machinist,* XXIII (May 24, 1900), 489-91. Compare also Guy Hubbard, "Development of Machine Tools in New England," *American Machinist,* LX (Feb. 21, 1924), 272-74.
30. Fitch, *Report,* pp. 27-28; Joseph W. Roe, *English and American Tool Builders,* (New Haven: Yale University Press, 1916), p. 143. The successors of the Robbins and Lawrence Company, the Jones and Lamson Machine Company, remained one of the leaders in turret lathe production for many years. See James Hartness, *Machine Building for Profit* (Springfield, Vt.: Jones & Lamson Machinery Co., 1909). Hartness introduced the flat-turret lathe and was a pioneer in the application of hydraulic feeds to machine tools.
31. Guy Hubbard, "Development of Machine Tools in New England," *American Machinist,* LXI (Aug. 21, 1924), 314; Roe, *Tool Builders,* p. 176.
32. *Manufactures of the United States in 1860* (compiled from the original returns of the Eighth Census), p. clxxxix. The Twelfth Census conveniently collates the basic earlier census data on the sewing machines; *Twelfth Census* (1900), X, 404. An illustration and description of the Singer Sewing Machine, as it appeared when patented in 1851, appears on the front page of *Scientific American,* Vol. VII, Number 7 (November 1, 1851).
33. "In the Exhibition at London in 1851, only two very imperfect sewing machines were exhibited. In 1855, at Paris, there were 14 varieties of sewing machines, some of which were so perfect that little or no material advance has since been made; and in 1862, in the London Exhibition, about 50 different arrangements of machines were shown. . . . In the 'Exposition Universelle,' now open in Paris, there are no less than 87 exhibitors of sewing machines. Their manufacture is now very general in European countries. France has 27 exhibitors, America 21, and England 12. Even so small a principality as Hesse has two exhibitors of sewing machines; the colony of Canada, five." Captain Hichens on "Apparatus for Sewing and Making up Clothing," in *Reports on the Paris Universal Exhibition, 1867. Presented to both Houses of Parliament* (1868); V, 131-32.
34. See the summary of census statistics for 1880 and 1890 for industries in which the sewing machine was employed extensively, in Chauncey M. Depew (ed.), *One Hundred Years of American Commerce* (New York, 1895), II, 538. See also the description of sewing machines for specialized purposes in Frederick A. Paget, "Report on the Machines and Apparatus used in Sewing and Clothing," *Reports on the Philadelphia International Exhibition of 1876. Presented to both Houses of Parliament* (1877); I, 242-43.
35. The sewing-machine industry relied much more heavily than did the firearms industry upon cast iron and was instrumental in bringing about important improvements in foundry operations. The molding press, for example, was introduced by Albert Eames, who was at the time (around 1873), foreman of the foundry at the Wheeler and Wilson Sewing Machine Company and who was earlier employed in firearms manufacture. The molding press played an important role in the production of sewing-machine parts, in hardware generally, and in other industries dependent on casting. Fitch, *Report,* pp. 36-37; also Fitch, "Report on the Manufacture of Hardware, Cutlery, and Edge-tools," *Tenth Census of the United States* (1880); II, 10.
36. Roe, *Tool Builders,* pp. 202-6; Robert S. Woodbury, *History of the Gear-Cutting Machine* (Cambridge: M.I.T. Press, 1958), pp. 80-81. Brown and Sharpe undertook the production of automatic gear cutters in 1877.
37. Woodbury, *History of the Grinding Machine* (Cambridge: M.I.T. Press, 1959), pp. 60-61.
38. *Ibid.* Compare Luther D. Burlingame, "The Universal Milling Machine," *American Machinist,* XXXIV (Jan. 5, 1911), 9.
39. I am grateful to Professor Duncan McDougall for kindly placing at my disposal his data on the sale of machinery output of the Brown and Sharpe Company.

40. Woodbury, *Grinding Machine*, pp. 60-61.
41. *Ibid.*, pp. 61-62; Guy Hubbard, "100 Years of Progress in American Metalworking Equipment," *Automotive Industries*, CXIII, No. 5 (Sept. 1, 1955), p. 315.
42. Woodbury, *Grinding Machine*, pp. 64-71. Brown and Sharpe applied the hydraulic feed technique to the grinding machine in 1902, a technique which makes it easier to operate any machine tool automatically. *Ibid.*, p. 139.
43. Burlingame, "Milling Machine" (cited in n.38).
44. An admirable descriptive survey, profusely illustrated, of the "state of the arts" in machine tools in 1880 may be found in *Tenth Census* (1880), Vol. XXII, "Report on Power and Machinery Employed in Manufactures."
45. It is of more than passing interest as evidence of technological convergence between sewing machines and bicycles to note that, in England as well as in the United States, the earliest bicycles were produced in sewing-machine plants. Colonel Pope in 1878 produced his first "Columbia" in a corner of the Weed Sewing Machine Company plant at Hartford. In England in the late 1860's, the Coventry Sewing Machine Company played a similar role.
See *Twelfth Census* (1900), X, 331; Albert A. Pope, "The Bicycle Industry," in Depew, *One Hundred Years*, II, 550; G.C. Allen, *The Industrial Development of Birmingham and the Black Country*, 1860-1927 (London: G. Allen, 1929), p. 243; J. Clapham, *An Economic History of Great Britain* (Cambridge: The University Press, 1952), II, 96-97.
46. *Special Reports of the Census Office* (1905), Part 4, "Selected Industries," p. 289. Compare *Thirteenth Census of the United States* (1910), X, "Manufactures," 825-28.
47. Some of the earliest bicycles are reported to have weighed over one hundred pounds.
48. *Twelfth Census* (1900), "Manufactures," 385-88. See also Fred Colvin, *Sixty Years with Men and Machines* (New York: Whittlesey House, 1947), 88-89.
49. "The successful ball bearing depends upon having the balls themselves perfectly spherical and all of identical diameter. The ball must run in races perfectly circular, perfectly concentric, and of exact dimensions. Not only must the balls and their races be machined to a fine surface finish, but all these dimensions must be held to close tolerances, and all these parts must be hardened. Only grinding could deal with this problem." Woodbury, *Grinding Machine*, p. 110.
50. *Ibid.*, p. 111.
51. A former Brown and Sharpe employee, who had been initiated earlier into the problems of grinding at the Seth Thomas Clock Company.
52. Woodbury, *Gear-Cutting Machine*, pp. 78-126.
53. "The bicycle impinged upon the locomotive through the medium of machine tools when the new turret lathes, which had been developed to form wheel hubs, were applied to the manufacture of locomotive crankpins. . . ." Colvin, *Sixty Years*, p. 89.
54. In 1894, Pope's bicycle plant in Hartford built ". . . what was then the most up-to-date mill in the country for making cold drawn steel tubing. There was also a research department for testing metals and improving bicycle design." John B. Rae, *American Automobile Manufacturers* (Philadelphia: Temple Press, 1959), p. 9.
55. *Special Reports of the Census Office* (1905), Part 4, "Manufactures," p. 269. In the 1900 Census Reports, the brief treatment of motor vehicles appears in the chapter on locomotives. Most of the 4,192 motor vehicles mentioned in that report were in fact powered by steam or electricity. *Twelfth Census* (1900), "Manufactures," 255-59.
56. *Census of Manufactures* (1914), II, "Reports for Selected Industries," 731-32.
57. Woodbury states that after 1900 the automobile industry became ". . . the largest single customer of the machine tool industry, taking 25 to 30 per cent of the output. . . ." Woodbury, *Grinding Machine*, p. 120.
58. *Ibid.*, p. 99.
59. "When the demand for bicycles decreased some manufacturers turned to the automobile, and many establishments that made only bicycles in 1900 are now [1905] devoted primarily to the manufacture of automobiles, while others make them to a greater or less degree in connection with the manufacture of bicycles. . . ." *Special Reports of the Census Office* (1905), Part 4, "Manufactures," p. 289. See also Rae *American Automobile Manufacturers*, pp. 8-10. The emergence of early automobile

firms out of bicycle firms is a sequence which was reproduced in Great Britain. See S.B. Saul, "The Motor Industry in Britain to 1914," *Business History* V, No. 1 (Dec. 1962), 22-44.

60. H.J. Hinde, "Relation of Power Presses and Dies to the Automobile Industry," *Mechanical Engineering,* XLIII (Aug. 1921), 531.

61. Just as, at an earlier date, the working action of the Colt revolver was incorporated into machines used for *producing* the revolver. ". . . the same arrangement of parts characteristic of the Colt revolver seems to have been carried through the principal machines for its manufacture, the horizontal chucking lathes, cone-seating and screw machines, barrel-boring, profiling, and mortising machines, and even the compound crank-drops, exhibiting the same general arrangement of working parts about a center." Fitch, *Report,* pp. 27-28.

62. S. Einstein, "Machine-Tool Milestones, Past and Future," *Mechanical Engineering,* LII (Nov. 1930), 961; F.K. Hendrickson, "The Influence of the Automobile on the Machine-Tool Industry in General," *Mechanical Engineering,* XLIII (Aug. 1921), 530. In discussing the relationship between the milling machine and the automobile, Einstein concluded: "The automotive industry asked of the machine-tool designer automatic machines, either through modification of standard designs or machines of entirely special design - quick-acting fixtures, either hand-operated or automatically operated - more powerful and stronger machines occupying a minimum amount of floor space. On the other hand, the automotive industry supplied the machine-tool designer with a vast variety of highly successful mechanisms and constructive details, from which he could draw freely such elements and such ideas as could be adapted to the design of machine tools." "Discussion at the Machine Shop Practice Session," *Mechanical Engineering,* XLIII (Aug. 1921), 534.

63. For example, the redesigning of milling cutters at the Cincinnati Milling Machine Company, when it was discovered that ". . . the cutters of the time were not as strong as the machines that were driving them and therefore gave out long before the maximum power of the machine was reached." Woodbury, *Milling Machine,* p. 80.

64. Guy Hubbard, "Metal-Working Plants," *Mechanical Engineering,* LII (Apr. 1930), 411. See also Frederick W. Taylor, *The Art of Cutting Metals* (New York: American Association of Mechanical Engineers, 1907); *Special Reports of the Census Office* (1905), Part 4, "Metal-working Machinery," pp. 232-33; Ralph Flanders, "The Influence of the Automobile on Lathe Practice," *Mechanical Engineering,* XLIII (Aug. 1921), 532; Carl J. Oxford, "One Hundred Years of Metal Cutting Tools," *Centennial of Engineering 1852-1952* (Chicago: American Association of Engineers, 1953), pp. 346-50; S. Einstein, "Machine-Tool Milestones, Past and Future," *Mechanical Engineering,* LII (Nov. 1930), 959-62.

65. Just as it simplified, for example, the development of the typewriter, whose problems remained unsolved until it was placed in the hands of the skilled machinists and technical experts of E. Remington and Sons, gun manufacturers at Ilion, New York. *Twelfth Census* (1900), "Manufactures," 442.

66. *Accuracy for Seventy Years, 1860-1930* (Hartford, Conn.: Pratt & Whitney Co., 1930).

67. Guy Hubbard, "Development of Machine Tools in New England," *American Machinist,* LX (Jan. 31, 1924), 171-73.

68. C.B. Owen, "Organization and Equipment of an Automobile Factory," *Machinery,* XV (Mar. 1909), 493.

69. The government's current role in subsidizing research and development activity has an earlier and interesting parallel in the innovations emerging out of firearms production - an industry where, as we have seen, the government played a role as a major producer as well as "consumer."

2. America's rise to woodworking leadership

1. Furthermore, in a world of high transport costs, especially away from bodies of navigable water, most productive activities were dependent upon local resources within a very small radius. It is important also to emphasize the fact, obvious though it may be,

296 Notes

that wood supplies were not abundant everywhere, and that supply conditions for any particular locality may have altered drastically over time.

2. Tench Coxe claimed, in 1810, that the potash and pearlash derived in clearing the trees of new farm land for cultivation would "nearly compensate the settler" for the expenses thus incurred, at least where the land was "convenient for boat navigation." Tench Coxe, *A Statement of the Arts and Manufactures of the United States of America*, 1810, p. xvii. For a similar earlier statement, see Tench Coxe, *A View of the United States of America*, Philadelphia, 1794, p. 454.
3. Lewis Mumford, *Technics and Civilization*, New York, 1934, pp. 119-20.
4. See *Eighth Census of United States: Manufactures*, pp. 733-42. Value added by manufacture was $54,671,082 for cotton goods and $53,569,942 for lumber.
5. Nathan Rosenberg, "Innovative Responses to Materials Shortages," *American Economic Review Papers and Proceedings*, May 1973.
6. John Richards, *A Treatise on the Construction and Operation of Wood-working Machines*, London, 1872, p. iv.
7. *Ibid.*
8. *Ibid.*, p. 33.
9. Victor Clark, *History of Manufactures in the United States*, New York, 1929, 3 vols., vol. 1, p. 48.
10. James Elliot Defebaugh, *History of the Lumber Industry of America*, Chicago, 1907, 2 vols., vol. 2, p. 9.
11. Roger Burlingame, *March of the Iron Men*, New York, 1938, p. 39.
12. J. Leander Bishop, *A History of American Manufactures from 1608 to 1860*, Philadelphia, 1868, 3 vols., vol. 1, p. 492. See also Albert S. Bolles, *Industrial History of the United States*, Norwich, Conn., 1878, pp. 218-221. The number cited by Bishop is suspiciously high. Bolles refers to a machine which was perfected in 1810 "which was able to make a hundred nails a minute" (p. 220). See also Greville and Dorothy Bathe, *Jacob Perkins*, Philadelphia, 1943, p. 20.
13. Bolles, *op. cit.*, p. 220. Temin points out that, in colonial days, abandoned houses were often burned down in order to recover the nails. Peter Temin, *Iron and Steel in Nineteenth Century America*, Cambridge, 1964, p. 42.
14. Fogel has pointed out, in criticizing Rostow's overemphasis upon the railroad demand for iron, that "in 1849 the domestic production of nails probably exceeded that of rails by over 100 per cent." Robert Fogel, *Railroads and American Economic Growth*, Baltimore, 1964, p. 135.
15. For a later period (1872) one highly knowledgeable observer estimated that three-quarters of all woodworking machinery in America was devoted to the preparation of building materials. Richards, *op. cit., p. 48.*
16. Siegfried Giedion, *Space, Time and Architecture*, Cambridge, 1941, p. 347.
17. *Ibid.*, pp. 347-55.
18. The Census of 1810 reported 2,541 sawmills. In addition to its other deficiencies, however, this census provided only partial coverage including, e.g., no report on sawmills in Connecticut, New Hampshire, New York, North Carolina, and Vermont. All that can be said, therefore, is that the total number of sawmills must have greatly exceeded the number cited.
19. Defebaugh, *op. cit.*, vol. 1, p. 490.
20. Fred H. Gilman, "History of the Development of Saw Mill and Woodworking Machinery," *Mississippi Valley Lumberman*, February 1, 1895, 61.
21. Rodney C. Loehr, "Saving the Kerf: The Introduction of the Band Saw Mill," *Agricultural History*, July 1949, 168-69.
22. *Ibid.*, p. 169.
23. *Niles Weekly Register*, July 19, 1817, p. 336, reported that the saw mill of Stewart and Hill of Baltimore had installed a rapid circular saw for cutting veneers. It pointed out, significantly, that "two boys may attend the machine."
24. Defebaugh, *op. cit.*, vol. 2, p. 442.
25. Richards, *op. cit.*, p. 141.
26. G.L. Molesworth, "On the Conversion of Wood by Machinery," *Proceedings of the*

Institution of Civil Engineers, vol. 17, 1857-58, p. 22. Richards stated that English saws in 1860 were, on the average, half the thickness of American saws. John Richards, "Woodworking Machinery," *Journal of the Franklin Institute,* June 1870, 399. See also M. Powis Bale, *Woodworking Machinery,* London, 1896 (2nd edition), pp. 327-28.

27. For many years large, usable wood scraps simply had been burned. "The slab butts and edgings of boards were carried outside of the mills and board piles, and thrown into a common pile to be burned and which was kept constantly burning, winter and summer. Thus millions of slabs were burned to get rid of them, and the burning did not entirely cease until about 1835 or 1840, although the best of them were cut into lath or were used for other purposes much earlier." Defebaugh, *op. cit.,* vol. 2, p. 443.

28. *Niles Weekly Register,* March 27, 1819, p. 93, and also September 28, 1833.

29. Loehr, *op. cit.,* p. 169.

30. Richards, *op. cit.,* pp. 207-39.

31. *Special Report of Mr. Joseph Whitworth* (1854) as reprinted in Nathan Rosenberg, *The American System of Manufactures,* Edinburgh, 1969, p. 345.

32. Bale, *op. cit.,* pp. 88-89.

33. "Mortises and tenons represent in wood work what bolts and rivets do in metal work, - the mechanical means of connecting the different parts of frames and structures: the analogy is, however, far from complete. Metals are joined by fusion or welding; they are also connected without special reference to fibre, while in wood work all connections or joints must be mechanical, and every piece arranged with strict reference to its fibre. Transversely, it has no capacity for withstanding tensile strain, at least none that need to be practically considered, while parallel with the fibre its coefficient compares with many metals. Its employment, however, to resist tensile strain is rendered impracticable, or nearly so, from the want of some means of connecting its ends, so as to represent a continued strength throughout such joints." Richards, *op. cit.,* p. 239.

34. *Ibid.,* p. 243. The reciprocating mortising machine was an excellent example of American boldness in machinery design. Richards further states: "To develop the reciprocating mortising machine, as it has been done in America, requires three things: highly-skilled labour, a long experience, and very limited amount of engineering knowledge with the builders of the machines. This last condition is rather a curious one, but a skilled engineer, conversant with all the principles of the operation and difficulties to be encountered in making and using such machines, could not conscientiously recommend them, except for the lighter class of works." *Loc. cit.*

35. *Report of the Committee on the Machinery of the United States,* in Rosenberg, *The American System of Manufactures,* p. 171.

36. *Whitworth Report* in *ibid.,* p. 344. The *Report of the Committee on the Machinery of the United States* had observed: "The several parts of a house are got up in separate manufactories, such as stairs and staircases. Here there is every appliance for bending and twisting the wood, and working it under awkward forms. For doors, window-frames, and sashes. For this purpose special tools are employed for mortising, tenoning, and forming the mouldings." *Ibid.,* p. 171.

37. See *ibid.,* pp. 29-36.

38. Dwight Goddard, *Eminent Engineers,* New York, 1905, p. 73. See also the detailed description provided by the British parliamentary committee, which observed them in operation in Springfield in 1854, in Rosenberg, *The American System of Manufactures,* pp. 137-43. Whitworth, in his report, provided a precise breakdown of the total labor time consumed at Springfield Armory in making gunstocks by machinery. He reported that all the operations together took just over 22 minutes of labor time. *Ibid.,* p. 365.

39. Asa Waters, *Biographical Sketch of Thomas Blanchard and his Inventions,* Worcester, 1878, p. 8. It was later claimed that the Blanchard lathe was originally invented in England. See, for example, D.K. Clark, *The Exhibited Machinery in 1862,* London Exhibition, 1864, p. 221. No evidence whatever is provided in support of the assertion. But, in any case, it is clear that if some neglected English mechanical genius had invented a similar machine, he had absolutely no influence on subsequent developments. When British engineers first examined American woodworking machinery in the early 1850's, they acclaimed the Blanchard lathe as a wonderful invention, unknown to them,

and purchased large numbers of them from the Ames Manufacturing Company in Chicopee, Massachusetts for use in the Enfield Arsenal. And their Report had stated: "It is most remarkable that this valuable labor-saving machine should have been so much neglected in England, seeing that it is capable of being applied to so many branches of manufacture, its introduction into the armoury will prove a national benefit." As reprinted in Rosenberg, *The American System of Manufactures,* p. 138.

40. Charles Fitch, "Report on the Manufactures of Interchangeable Mechanism," *Tenth Census of the United States,* 1880, vol. 2, p. 14.

41. An amusing account of European incredulity is provided by Zachariah Allen: "On my way from Brussels to Haerlem to view the national exhibition of the manufactures of Belgium, holden under the auspices of the king and honoured by his presiding at the distribution of the prizes, having accidentally fallen into company in a diligence with a Flemish artist on his way to the same place with some of his new machines, our conversation turned upon the subject of steam navigation, then lately introduced into that country. He inquired if there were any steamboats in America, and was surprized on being informed that they had been in successful operation there nearly twenty years. I took occasion to describe to him several American inventions, among others the machine for cutting and heading nails, which are completely finished and fall from the engine as fast as one can count them. The machine for making weavers reeds or slaies seemed to strike his attention as a wonderful invention, whereby the mechanism is made to draw in the flattened wire from a reel, to insert it between the side pieces, to cut it off at the proper length, and finally to bind each dent firmly in its place with tarred twine, accomplishing the whole operation without the assistance of the attendant, in a more perfect manner than can be performed by the most skillful hand. Although he possessed a good share of intelligence, the complicated operations of these machines, performing processes which he supposed could only be brought about by manual dexterity, appeared to him incomprehensible. But when I proceeded to describe Blanchard's lathe in which gun stocks and shoe lasts are turned exactly to a pattern, his belief seemed somewhat wavering, and on continuing to give him a description of Whitmore's celebrated Card Machine, which draws off the card wire from the reel, cuts it off at a proper length for the teeth, bends it into the form of a staple, punctures the holes in the leather, and inserts the staples of wire into the punctures, and finally crooks the teeth to the desired form - performing all these operations with regularity without the assistance of the human hand to guide or direct it, the credulity of my travelling companion in the diligence would extend no farther, and he evidently began to doubt all the statements I had been making to him, manifesting at the same time some little feeling of irritation at what he appeared to consider an attempt to impose upon him such marvellous accounts. Uttering an emphatic humph! he threw himself back into the corner of the diligence, and declined further conversation during the remainder of our ride upon the subject of mechanics and of the improvements made in Flemish manufactures." Zachariah Allen, *The Science of Mechanics,* Providence, R.I., 1829, pp. 348-49. The incident occurred during Allen's continental tour in 1825.

42. "In those districts of the United States of America that the Committee have visited the working of wood by machinery in almost every branch of industry, is all but universal; and in large establishments the ordinary tools of the carpenter are seldom seen, except in finishing off, after the several parts of the article have been put together.
   "The determination to use labour-saving machinery has divided the class of work usually carried on by carpenters and the other wood trades into special manufactures, where only one kind of article is produced, but in numbers or quantity almost in many cases incredible." *Report of the Committee on the Machinery of the United States,* as reprinted in Rosenberg, *The American System of Manufactures,* p. 167.

43. The extent to which Blanchard's lathe wasted wood may well have been the critical factor in the British failure to show more interest in it until the later, improved models were developed. Thus, an experienced London gunmaker, who was questioned concerning stockmaking machinery by the Parliamentary Select Committee on Small Arms, replied: "The first stock machine that I ever saw was introduced into England 20 years ago, an American invention; it was sent over here, a beautiful working model of it,

and submitted for purchase right of working it to all the gun-makers here, and it was forwarded from London to Birmingham, and there a very eminent manufacturer, and a very extensive one, brought it into active operation, but he found that the guns that cost him 1 s. for stocking by machinery he could do for 7 d. and 8 d. by hand labour, the waste in that machinery is so very great. This working model we had in Pall Mall, it was in the shop of the elder Mr. Wilkinson at the time; there were pieces of wood of about 18 inches long that were fixed in, and you saw the stock turned and worked." "Yet the loss of material was so great that it was discontinued?" "Yes." *Report from the Select Committee on Small Arms,* Parliamentary Papers, 1854, vol. XVIII, Q. 7273 and Q. 7274. See also Q. 7520 and Q. 7521, and Molesworth, *op. cit.,* pp. 22, 45-6.
44. See A. William Hoglund, "Forest Conservation and Stove Inventors - 1789-1859," *Forest History,* Winter 1962. Hoglund states that, in spite of much criticism of stoves, fireplaces had come to be regarded as "old-fashioned" by the 1840's. Fireplaces persisted, of course, on the frontier.
45. Louis Hunter, *Steamboats on the Western Rivers,* Cambridge, 1949, pp. 130-33.
46. "Lumber manufacture, from the log to the finished state, is, in America, characterized by a waste that can truly be called criminal" (Richards, *op. cit.,* p. 141).
47. This trade-off was, to some degree, also influenced by the crudeness of design and manufacture of machinery during this period.

3. Anglo-American wage differences in the 1820's

1. Zachariah Allen, *The Science of Mechanics* (Providence, R.I.: Hutchens and Cory, 1829), p. iv.
2. Of the decade of the 1820's in Britain, Clapham commented: ". . . by 1820-30 the professional purveyor of machines made with the help of other machines, the true mechanical engineer of the modern world, was just coming into existence - in Lancashire and London where the demand was at its maximum." J.H. Clapham, *An Economic History of Modern Britain: The Early Railway Age, 1820-1850* (Cambridge: Cambridge University Press, 1964), p. 152.
3. Victor Clark, *History of Manufactures in the United States* (Washington: McGraw-Hill Book Company, 1929), I, 392.
4. H.J. Habakkuk, *American and British Technology in the 19th Century* (Cambridge: Cambridge University Press, 1962), especially pp. 21-22. See also pp. 151-56. Habakkuk cites the passage in Victor Clark's book which in turn relies on Allen's figures: "Was the differential for skill smaller in America than in England? V.S. Clark, referring apparently to the 1820's, considered that differences in wages between England and America were greater in unskilled than in skilled occupations." *Ibid.,* p. 22.
5. Allen's inability to provide an estimate for this class of French labor is particularly regrettable.
6. See *Report from the Select Committee on Artizans and Machinery,* Parliamentary Papers, 1824, Vol. V, especially columns 299-302, where spinner and power loom manufacturers complain of the inability to employ mechanics for the production of textile machinery, or the repair of old machinery. See also column 566, William Fairbairn's comments on the scarcity of competent millwrights.
7. H.J. Habakkuk, *International Social Science Bulletin,* Vol. VI, No. 2, 1954, p. 196.
8. Some interesting material on regional differentials for this period may be found in Stanley Lebergott, *Manpower in Economic Growth* (New York: McGraw-Hill Book Company, 1964), especially pp. 539, 541, 546, and 547.
9. It is estimated that cotton textile wage earners increased from 12,000 to 55,000 during the 1820's and that domestics increased from 110,000 to 160,000 during this period. Even so, these two groups constituted barely 5 per cent of the total labor force in 1830, when 70 per cent of the labor force was in agriculture. It is estimated that there were 610,000 free (i.e., nonslave) farm laborers in 1830. There are no estimates of nonfarm common labor, but the number of farm laborers alone was almost three times the total of cotton textile workers and domestics in 1830. See Lebergott, *Manpower in Economic Growth,* pp. 510-11.

10. Habakkuk, *American and British Technology*, pp. 151-52.
11. *Ibid.*, p. 154.
12. *Ibid.*, p. 156.
13. *Ibid.*, p. 8.
14. In Manchester in 1824 the wages of mechanics and engineers engaged in millwork was 28 shillings per week, which corresponds very closely to Allen's figure for 1825 of four shillings six pence per day for ordinary machine makers. Whitesmiths earned 27 shillings per week, and fitters, turners, and ironmolders earned 30 shillings per week. Arthur L. Bowley, *Wages in the United Kingdom in the Nineteenth Century* (Cambridge University Press, 1900), table opposite p. 123. Unfortunately, wages for other engineering skill classifications are not available.

4. Problems in the economist's conceptualization of technological innovation

1. See, for example, Murray Brown, *On the Theory and Measurement of Technological Change* (Cambridge University Press, 1966) for a useful survey of this literature.
2. Joseph A. Schumpeter, *History of Economic Analysis* (Oxford University Press, 1954), p. 1027.
3. Theodore W. Schultz, *Investment in Human Capital* (New York: The Free Press, 1971), p. 203. Emphasis Schultz's. For a promising attempt to deal with the problems posed here within the framework of a theory of induced innovation, see Yujiro Hayami and Vernon Ruttan, *Agricultural Development* (Baltimore: The Johns Hopkins Press, 1971).
4. Salter, *op. cit.*, p. 15.
5. *Loc. cit.* Even blueprints literally on the shelf may not be as readily (i.e., costlessly) "available" as might be casually assumed. Arrow has pointed out that "when the British in World War II supplied us with the plans for the jet engine, it took ten months to redraw them to conform to American usage." Kenneth Arrow, "Classificatory Notes on the Production and Transmission of Technological Knowledge," *American Economic Review Papers and Proceedings* (May 1969), p. 34.
6. The view that technological change may best be viewed as a cumulation of small, individually minor improvements was authoritatively stated by A.P. Usher in his *A History of Mechanical Inventions* (Harvard University Press, 1954). See also S.C. Gilfillan's two books, *The Sociology of Invention* (MIT Press, 1970, first published in 1935), and *Inventing the Ship* (Follett Publishing Company, 1935). Some related perspectives are presented in Nathan Rosenberg, "The Direction of Technological Change," *Economic Development and Cultural Change* (October 1969). For two valuable empirical studies of the cumulative quantitative importance of small improvements, see John Enos, "A Measure of the Rate of Technological Progress in the Petroleum Refining Industry," *Journal of Industrial Economics* (June 1958), and Samuel Hollander, *The Sources of Increased Efficiency* (MIT Press, 1965).
7. "(W)e shall impose a restriction on our concept of innovation and henceforth understand by an innovation *a change in some production function which is of the first and not of the second or a still higher order of magnitude.* A number of the propositions which will be read in this book are true only of innovation in this restricted sense." Joseph A. Schumpeter, *Business Cycles*, vol. 1, p. 94. Emphasis Schumpeter's. Elsewhere, Schumpeter states: "This historic and irreversible change in the way of doing things we call 'innovation' and we define: innovations are changes in production functions which cannot be decomposed into infinitesimal steps. Add as many mail-coaches as you please, you will never get a railroad by so doing." Joseph A. Schumpeter, "The Analysis of Economic Change," as reprinted in *Readings in Business Cycle Theory* (Philadelphia: The Blakiston Company, 1944).
8. "It is leadership rather than ownership that matters. The failure to see this and, as a consequence, to visualize clearly entrepreneurial activity as a distinct function *sui generis,* is the common fault of both the economic and the sociological analysis of the classics and of Karl Marx." *Business Cycles*, vol. 1, pp. 103-4. Elsewhere Schumpeter states that "successful innovation . . . is a special case of the social phenomenon of leadership." Joseph A. Schumpeter, "The Instability of Capitalism," reprinted in Nathan Rosenberg (ed.), *The Economics of Technological Change* (Penguin, 1971), pp. 33-4.

9. Schumpeter, "The Analysis of Economic Change," p. 10. Elsewhere Schumpeter had stated: "To undertake such new things is difficult and constitutes a distinct economic function, first, because they lie outside of the routine tasks which everybody understands and, secondly, because the environment resists in many ways that vary, according to special conditions, from simple refusal either to finance or to buy a new thing to physical attack on the man who tries to produce it. To act with confidence beyond the range of familiar beacons and to overcome that resistance requires aptitudes that are present in only a small fraction of the population and that define the entrepreneurial type as well as the entrepreneurial function. This function does not essentially consist in either inventing anything or otherwise creating the conditions which the enterprise exploits. It consists in getting things done." Joseph A. Schumpeter, *Capitalism, Socialism and Democracy* (Unwin University Books, 1966), p. 132.

10. Schumpeter, *Business Cycles*, vol. 1, p. 85. Furthermore, in defining an innovation as a shift in a production function, or a "new combination," Schumpeter was encompassing in his definition much more than the specifically technological innovations with which we are primarily concerned here. "This concept covers the following five cases: (1) The introduction of a new good - that is one with which consumers are not yet familiar - or of a new quality of a good. (2) The introduction of a new method of production, that is one not yet tested by experience in the branch of manufacture concerned, which need by no means be founded upon a discovery scientifically new, and can also exist in a new way of handling a commodity commercially. (3) The opening of a new market, that is a market into which the particular branch of manufacture of the country in question has not previously entered, whether or not this market has existed before. (4) The conquest of a new source of supply of raw materials or half-manufactured goods, again irrespective of whether this source already exists or whether it has first to be created. (5) The carrying out of the new organization of any industry, like the creation of a monopoly position (for example through trustification) or the breaking up of a monopoly position." Joseph A. Schumpeter, *The Theory of Economic Development* (Harvard University Press, 1934), p. 66.

11. John Enos, "Invention and Innovation in the Petroleum Refining Industry," in *The Rate and Direction of Inventive Activity* (Princeton University Press, 1962), pp. 299-321.

12. Enos's findings are used or relied upon in many places. Edwin Mansfield, in his own influential work, has reproduced Enos's data on the lag between invention and innovation in two books. See his *Economics of Technological Change* (Norton, 1968), p. 101, and *Industrial Research and Technological Innovation* (Norton, 1968), p. 110. Others interested in examining this lag have relied upon Mansfield. See, for example, Harvey Leibenstein, "Organizational or Frictional Equilibria, X-Efficiency, and the Rate of Innovation," *Quarterly Journal of Economics* (November 1969), 622. See also Robert Fogel and Stanley Engerman (eds.), *The Reinterpretation of American Economic History* (Harper and Row, 1971), p. 206.

13. *Ibid.*, p. 309.

14. *Ibid.*, p. 308.

15. Edwin Mansfield, *The Economics of Technological Change*, chap. IV.

16. Eli Whitney's methods for the interchangeable manufacture of muskets were extremely crude and imperfect in their early stages and took many years to improve, but they were nevertheless sufficiently superior to alternative hand techniques that they were a commercial success from the start.

17. Jewkes, *op. cit.*, pp. 30-1.

18. *Ibid.*, p. 279.

19. *Ibid.*, pp. 310-11.

20. *Ibid.*, pp. 314-17.

21. *Ibid.*, p. 318.

22. *Ibid.*, p. 339.

23. *Ibid.*, p. 263.

24. For further discussion of the role of technological variables, see Nathan Rosenberg, "Factors Affecting the Diffusion of Technology," *Explorations in Economic History* (Fall 1972), pp. 3-33.

25. A conspicuous exception is Griliches's study of the diffusion of hybrid corn, where it is

explicitly recognized that the seed had to be specifically adapted to the unique requirements of different geographic regions. Zvi Griliches, "Hybrid Corn: An Exploration in the Economics of Technological Change," *Econometrica* (October 1957).

26. See Everett Rogers with F. Floyd Shoemaker, *Communication of Innovations* (The Free Press, 1971), 2 ed., for an exhaustive survey of the literature dealing with such factors.

27. Frank Lynn, "An Investigation of the Rate of Development and Diffusion of Technology in our Modern Industrial Society," in The National Commission on Technology, Automation, and Economic Progress, *The Employment Impact of Technological Change* (U.S.G.P.O., 1966), vol. 2, pp. 55-7.

28. This point is well documented in Harold C. Passer's excellent study, *The Electrical Manufacturers, 1875-1900* (Harvard University Press, 1953).

29. Nathan Rosenberg, "Factors Affecting the Diffusion of Technology," *Explorations in Economic History* (Fall 1972), 3-33.

30. NSF defines development as follows: "Development is the systematic use of knowledge and understanding gained from research and directed to the production of useful materials, devices, systems and methods; such work includes the design, testing and improvement of prototypes and processes. Development is directed to generally predictable and very specific ends." For some apposite remarks on the role of development activities, see Jewkes, *op. cit.,* pp. 28-33.

31. Much of the 19th century progress in iron and steel production, for example, consisted essentially in redesigning the productive process in such a way as to minimize heat loss. Much of the 20th century progress in electric power generation has consisted of thousands of minor modifications of design or substitution of new materials which have continuously and drastically reduced the fuel cost of a kilowatt-hour of electricity.

32. OECD, *Gaps in Technology: General Report,* Paris, 1968, p. 17. See also OECD, *Gaps in Technology: Analytical Report,* Paris, 1970, Book III, chapter 2.

33. Nathan Rosenberg, "Economic Development and the Transfer of Technology: Some Historical Perspectives," *Technology and Culture* (October 1970), pp. 564-5.

34. "The step we commonly call research, advanced development or basic invention, accounts, typically, for less than 10% of the total innovative effort. The other components, which we do not usually associate with the innovative process, account for something like 90% of the total effort and cost. Engineering and designing the product, tooling and manufacturing-engineering, manufacturing start-up expenses, and marketing start-up expenses, are all essential to the total process." U.S. Department of Commerce, *Technological Innovation* (U.S. Department of Commerce, 1967), p. 9.

35. Joseph Ben-David, *Fundamental Research and the Universities,* OECD, Paris, 1968, p. 49.

36. A. Phillips, "The Tableau Economique as a Simple Leontief Model," *Quarterly Journal of Economics* (February 1955).

37. Of course the story is much more complicated - and interrelated - than it is possible to indicate in the present context. For example, theoretical developments exercised a significant influence over the shape of the emerging tools of national income measurement. The Keynesian definitions of savings and investment - their *ex post* equality - were a major influence in the structuring of the national income accounts.

38. See, however, Simon Kuznets's characteristically astute review of the book, *American Economic Review* (June 1940) 250-71. The review is reprinted in Simon Kuznets, *Economic Change* (Norton, 1953), pp. 105-24.

39. Moses Abramovitz, "Resource and Output Trends in the United States since 1870," *American Economic Review Papers and Proceedings* (May 1956), 5-23; Robert Solow, "Technical Change and the Aggregate Production Function," *Review of Economics and Statistics* (August 1957), 312-20. Both articles are reprinted in Nathan Rosenberg (ed.), *The Economics of Technological Change* (Penguin Books Ltd., 1971).

40. See Evsey Domar, "On Total Productivity and All That," *Journal of Political Economy* (December 1962), 597-608, for a perceptive treatment of some of the thorny methodological issues raised by this approach.

5. Neglected dimensions in the analysis of economic change

1. (Abramovitz 1956:11). Broadly similar results were reached, for a different time period and employing a different technique, by Robert Solow (1957).
2. It is possible, of course, to *define* technological change in such a manner as to include all alterations in the relationship between inputs and outputs, but it is difficult to see what would be gained thereby. Definitions ought to assist in the clarification and classification of phenomena, whereas such a definition would serve to blur and obscure important differences in the sources of productivity change, and implications drawn from such a definition could never be disproved.
3. We should include, as part of this consumption, an increase in the availability of leisure time which, historically, has been one of the major "outputs" of a successfully developing economy. This has been one of the most important factors which has transformed the whole pattern of life in high-income countries.
4. In this paper we will use the term "feedback" to refer to any process whereby the future quality of the human agent as a factor input is altered as a result of current participation in economic activity.
5. Even where it dealt with questions of growth this tradition was wedded to the point of view - already discussed - that the process of growth can be explained in terms of changes in the supplies of homogeneous capital and labor inputs.
6. Indolence is not, however, an inborn human trait. As Rotwein points out in his valuable introduction, Hume regards indolence and the pursuit of 'the pleasures of idleness' as a symptom of frustration. (Hume 1955:XLVIII-XLIX).
7. (Hume 1955:50). Smith's position is the same. "In a country which has neither foreign commerce, nor any of the finer manufactures, a great proprietor, having nothing for which he can exchange the greater part of the produce of his lands which is over and above the maintenance of the cultivators, consumes the whole in rustic hospitality at home." (Smith 1937:385). Smith also notes the pernicious influence of court society upon character and industry: "Our ancestors were idle for want of a sufficient encouragement to industry. It is better, says the proverb, to play for nothing, than to work for nothing. In mercantile and manufacturing towns ... they are in general industrious, sober and thriving. ... In those towns which are principally supported by the ... residence of a court ... they are in general idle, dissolute and poor." (Smith 1937:319).
8. (Hume 1955:11. See also p. 22). Adam Smith makes the same point: "We may observe that the greater number of manufacturers there are in any country, agriculture is the more improved, and the causes which prevent the progress of these react, as it were, upon agriculture." (Smith 1956:230).
9. (Hume 1955:14. See also pp. 21-22.) Cf. (Mill 1909:581): "A people may be in a quiescent indolent, uncultivated state, with all their tastes either fully satisfied or entirely undeveloped, and they may fail to put forth the whole of their productive energies for want of any sufficient object of desire. The opening of a foreign trade, by making them acquainted with new objects, or tempting them by the easier acquisition of things which they had not previously thought attainable, sometimes works a sort of industrial revolution in a country whose resources were previously undeveloped for want of energy and ambition in the people: inducing those who were satisfied with scanty comforts and little work, to work harder for the gratification of their new tastes, and even to save, and accumulate capital, for the still more complete satisfaction of those tastes at a future time."
10. (Smith 1937:766). For a more extended discussion see (Rosenberg, December 1960).
11. (Smith 1937:734-35). In his earlier lectures Smith had stated: "... in every commercial nation the low people are exceedingly stupid. The Dutch vulgar are eminently so, and the English are more so than the Scotch. The rule is general; in towns they are not so intelligent as in the country, nor in a rich country as in a poor one." (Smith 1956:256).
12. It is only in very recent years that mechanical techniques for the picking of cotton have been developed, by contrast with the mechanization of grain harvesting which has been

going on for over a century. The nature of the cotton crop made its mechanization inherently a much more difficult technical problem. It is interesting to speculate on the final outcome of the slavery issue had mechanized cotton-picking developed along with mechanized grain harvesting in the 1840's.

13. The extreme sensitivity of midwestern farmers in transferring resources from one use to another in response to market forces is embodied in the so-called corn-hog cycle.

14. See the succinct presentation of comparative figures in (Campbell 1960:70-77).

15. It has recently been suggested that Asian collectivization may come to grief because of its inability to cope with the extreme delicacy of the rice plant. The great care and personal effort and attention involved in its cultivation make it a recalcitrant candidate for the rigors and centralized discipline of collectivization. "Plants . . . stand in no awe of Communism. In the whole of the vegetable kingdom, it is the rice plant which possesses the most 'bourgeois background' and which is proving the most 'reactionary' in the face of the policy of collectivization which has been applied in Asia during the past few years. Rice may well be the factor which brings about the downfall of Asian Communism." (Chi 1962:104).

16. See the highly suggestive comparison of Britain and the United States in (Habakkuk 1962).

17. Even though Adam Smith had long ago included in his definition of fixed capital ". . . the acquired and useful abilities of all the inhabitants or members of the society. The acquisition of such talents, by the maintenance of the acquirer during his education, study, or apprenticeship, always costs a real expense, which is a capital fixed and realized, as it were in his person. Those talents, as they make a part of his fortune, so do they likewise of that of the society to which he belongs. The improved dexterity of a workman may be considered in the same light as a machine or instrument of trade which facilitates and abridges labour, and which, though it costs a certain expense, repays that expense with a profit." (Smith 1937:265-66).

18. See the important work on this subject by Theodore Schultz (1959, 1960 and 1961). See also (The Journal of Political Economy, 1962).

19. (Oshima 1957, Kuznets 1962:6, and Martin and Lewis 1956). For the historical experience of the United States see (Fabricant 1952 and Bator 1960).

20. See the data on the historical diffusion of the automobile in eight industrial countries in (Rostow 1960:84-85 and 168-71).

21. It appears that, in recent years, the automobile has become too cumbersome and complicated and now requires equipment and facilities too elaborate to serve as extensively as it once did as a learning device. The growing interest in "hot-rods" may reflect an attempt to freeze the technical development of the automobile at a stage where it is still manageable without highly expensive and complex garage facilities. Alternatively, the motors of such devices as power-driven lawn mowers, particularly in suburbia, may provide a partial learning substitute.

6. The direction of technological change: inducement mechanisms and focusing devices

1. Albert O. Hirschman, *The Strategy of Economic Development* (New Haven, Conn.: Yale University Press, 1958), pp. 24-28; see also chap. 8.

2. J.R. Hicks, *The Theory of Wages* (New York: Macmillan Co., 1932), pp. 124-25.

3. W.E.G. Salter, *Productivity and Technical Change* (New York: Cambridge University Press, 1960), pp. 43-44. Cf. Fellner's succinct statement that "in the event of purely competitive factor hiring, with no factor rationing to individual firms, the market provides no incentive to seek for any given factor inputs one rather than another distribution of the factor-saving effects" (W. Fellner, "Does the Market Direct the Relative Factor-saving Effects of Technological Progress?" in Universities - National Bureau Committee for Economic Research, *The Rate and Direction of Inventive Activity* [Princeton N.J.: Princeton University Press, 1962], p. 177). See also Paul Samuelson, "A Theory of Induced Innovation along Kennedy-Weisacker Lines," *Review of Economics and Statistics* (November 1965).

4. In general, we would expect individual countries or regions to excel in innovations

which constitute technical solutions to specific problems of special importance to them. Thus it is hardly surprising that pioneer America made major contributions to the improvement of the ax, or that contemporary Britain should assume world leadership in the development of techniques for the fully automatic landing of airplanes in dense fog.

5. See the interesting article by Paul David, "The Mechanization of Reaping in the Ante-Bellum Midwest," in *Industrialization in Two Systems,* ed. Henry Rosovsky (New York: John Wiley & Sons, 1966), pp. 3-39.

6. Albert O. Hirschman and Charles E. Lindblom, "Economic Development, Research and Development, Policy Making: Some Converging Views," *Behavioral Science* (April 1962), p. 214.

7. Paul Mantoux, *The Industrial Revolution in the Eighteenth Century* (London, 1948), pp. 211-13, 244-51; H.J. Habakkuk, *American and British Technology in the Nineteenth Century* (New York: Cambridge University Press, 1962), p. 134. For a recent more rigorous formulation by an economic historian, see J.R.T. Hughes, "Foreign Trade and Balanced Growth: The Historical Framework," *American Economic Review Papers and Proceedings* (May 1959), pp. 330-37. See also Tibor Scitovsky, "Growth - Balanced or Unbalanced?" in *The Allocation of Economic Resources,* by M. Abramovitz et al. (Stanford, Calif.: Stanford University Press, 1959), esp. pp. 215-16.

8. The incredible improvisations which Watt resorted to, before availing himself of Wilkinson's machinery, are suggested in the following: "The close fitting of the piston in the cylinder, which did not so greatly matter in the Newcomen engine, was essential to the proper working of Watt's. For want of an accurate boring-machine, the cylinder of his first engine was made of tin and hammered to shape against a hard-wood block; the gaps between piston and cylinder were sealed as far as possible with felt, paper, oiled rags, and the like" (K.R. Gilbert, "Machine Tools," in *A History of Technology,* ed. Charles Singer et al. (London: Oxford University Press, 1958), 4:421.

9. From this point of view the critical question is why some economies have been capable of a creative response to this sort of demand while others have not been so capable. On the role of the machine tool industry as a creator and diffusor of new techniques, see Paul Strassmann, "Interrelated Industries and the Rate of Technological Change," *Review of Economic Studies* (October 1959), pp. 16-22; and Nathan Rosenberg, "Technological Change in the Machine Tool Industry, 1840-1910," *Journal of Economic History* (December 1963), pp. 414-43. Some of the historical examples in the next section have been drawn from the latter paper.

10. Neither the forming tool nor the oil-tube drill (which had previously been used in drilling gun barrels) had originated in the bicycle industry. However, both had previously been used for only very limited purposes, and their application to new kinds of uses in bicycle production was directly responsible for their much wider industrial application. The shaping of bicycle hubs was one of the first, if not in fact the first, use of the forming tool on hardened metals (see Rosenberg, pp. 435, 438-39; *Twelfth Census of the United States* [1900], 10:385-88; Fred Colvin, *Sixty Years with Men and Machines* [New York: McGraw-Hill Book Co., 1947], pp. 88-89).

11. Peter Temin, *Iron and Steel in Nineteenth Century America* (Cambridge, Mass.: M.I.T. Press, 1964), pp. 135-36.

12. *Sir Henry Bessemer, F.R.S.: An Autobiography* (London: Offices of "Engineering," 1905), pp. 131-36. Referring to his experimental tests of the gun with the cooperation of the French government at Vincennes, Bessemer concludes: "The experiments at Vincennes took place on or about the 22nd December, 1854, and before the close of that year I found myself once more at Baxter House, busy with plans for the production of an improved metal for the manufacture of guns, which improvement in the quality of the iron I proposed to effect by the fusion of steel in a bath of molten pig-iron in a reverberatory furnace. I soon determined on the form of furnace, and applied for a patent for my "Improvements in the Manufacture of Iron and Steel," which was dated as early as January 10th, 1855 - that is, within three weeks after the experiments in the Polygon at Vincennes" (ibid., pp. 136-37).

13. "Tool steels of the older type retained their hardness up to only $350°$F, whereas high-speed steels show no appreciable softening up to $1100°$" (Carl J. Oxford, "One

Hundred Years of Metal Cutting Tools," *Centennial of Engineering 1852-1952* [Chicago: Museum of Science and Industry, 1953], p. 346).

14. "Metal-working Machinery," *Special Reports of the Census Office* (1905, Part IV), p. 232.

15. Guy Hubbard, "Metal-working Plants," *Mechanical Engineering* 52 (1930): 411. Cf. also H.I. Brackenbury, "High-Speed Tools and Machines to Fit Them," *Proceedings, Institution of Mechanical Engineers* 63 (1910): 929-51.

16. A similar set of imbalances and compulsive sequences, with very similar results, has occurred more recently in connection with the introduction of cemented carbide tools and ceramic tools. See, e.g., The American Society of Mechanical Engineers, *Manual on Cutting of Metals: Single-point Lathe Tools* (New York: ASME, 1939), pp. 24-25, and the numerous articles on ceramic tools in the *Tool Engineer,* beginning in 1955.

17. *Special Reports of the Census Office* (1905), p. 233. In an important article Einstein stated: "Meanwhile high-speed steel had appeared on the market, which revolutionized the machine-tool industry. Cutting speeds of the various tools could materially be increased with this new material, and at the same time higher rates of traverse of tool or work could be used. This resulted in the development of the so-called single-pulley machine which adopted the positive-gear-change mechanism of the feedbox and used it for the drive of the machine. These new types of machines, heavier in structure, with a positive drive to the work and the tool, successfully took up the battle with high-speed steel, and in a large number of cases new shapes of tools had to be developed to utilize successfully the full cutting capacity of these new machines" (S. Einstein, "Machine-Tool Milestones, Past and Future," *Mechanical Engineering* [November 1930], p. 960).

18. Einstein, p. 959.

19. Robert S. Woodbury, *History of the Milling Machine* (Cambridge, Mass.: M.I.T. Press, 1960), p. 80.

20. Robert S. Woodbury, *History of the Grinding Machine* (Cambridge, Mass.: M.I.T. Press, 1959), esp. pp. 73-108; Rosenberg, pp. 435-38.

21. See, e.g., Edward J. Reed and Edward Simpson, *Modern Ships of War* (New York, 1888); or the article "Battleships" in *Encyclopedia Britannica*.

22. "Ordnance versus Armour," *Mechanics Magazine,* n.s. 17 (March 15, 1867): 157. A recent book summarizes the resulting changes in the British Navy as follows: "The guns used at Inkerman and Balaclava did not differ greatly from those which had been used at Torres Vedras or Waterloo; and in the fleet which sailed to the Crimea, the standard armament of the line-of-battle ship was a 32-pounder smoothbore, firing a shot with a velocity of 1,600 feet per second, mounted upon a crude wooden carriage with a recoil controlled by the friction of large wooden axles, and served by a crew so large that they practically hid their unwieldy little weapon from sight. . . . By the late Eighties all this was changed. The guns were changed; as early as the Seventies Britain was building a 16-inch 80-ton gun - an almost incredible weapon. The ships to carry such guns had in turn to be changed out of recognition. By the late Eighties we have the *Hood,* 'a turret ship with eight redoubts' mounting four 67-ton guns and ten 6-inch quick-firing guns, eighteen smaller quick-firing guns and eight machine guns. The powerful guns, in turn, had called for thicker armour, 'till in 1881 the *Inflexible* carried, in places, 24-inch plates.' Such weights of armour could obviously not be carried along the entire length of a ship, and so had, in turn, led to the collecting of all the vital parts of warships into a 'citadel'; it was these massive citadels which gave the new kind of warship its characteristic appearance. In the years between 1858 and 1888 the wooden-wall broadside navy had gone for ever: the Navy of 1858 was still the navy that had fought at Trafalgar; the Navy of 1888 was already the navy that fought at Jutland" (J.D. Scott, *Vickers: A History* [London: Weidenfeld & Nicolson, 1962], p. 23).

23. Lynn White, Jr., "The Act of Invention," in *The Technological Order,* ed. Carl Stover (Detroit: Wayne University Press, 1963), p. 111. Elsewhere White has observed that "the new violence of the lance at rest called for heavier armour, which in turn produced the crossbow as an 'anti-tank gun' " (Lynn White, Jr., in *Scientific Change,* ed. Alistair C. Crombie [New York: Basic Books, 1963]. p. 278). The imbalance relationship, of course, sometimes terminates with the complete triumph of one component in the

relationship over the others. In the sixteenth century, e.g., the armorer fought a frequently ingenious and often highly decorative but inevitably losing battle with firearms, since significant protection against such weapons required intolerably heavy armor. It is reported that, even without the leg pieces, the armor made for the duc de Guise in 1588 weighed over 100 pounds. "By the mid-seventeenth century infantry commonly wore nothing more formidable than a leather jerkin, and cavalry only an iron cuirass" (A.R. Hall "Military Technology," in *A History of Technology,* ed. Charles Singer et al. [London: Oxford University Press, 1957], 3:353).

24. Why, one might be inclined to ask, have such disequilibria been most important in areas dominated by a fairly sophisticated mechanical technology? Why have they not been equally important, say, in agriculture? This is a big question. Part of the answer, I would suggest, is that the person receiving such signals in agriculture - the farmer - is not, himself, competent by training or occupational experience to evaluate them in a creative way. In the industrial sector there has been a continuous and direct confrontation between technical problems and personnel with the competence to solve them, whereas many of the problems of agriculture have had to await the development of specialized institutions (e.g., land-grant colleges) and specialized professions with the requisite skills - genetics, soil chemistry, and, of course, mechanical technology as well.

25. Karl Marx, *The Poverty of Philosophy* (Moscow, n.d.), p. 161. Earlier in the same work Marx had stated: ". . . from 1825 onwards, almost all the new inventions were the result of collisions between the worker and the employer who sought at all costs to depreciate the worker's specialized ability. After each new strike of any importance, there appeared a new machine" (p. 134).

26. Karl Marx, *Capital* (Moscow, 1961), 1:436.

27. Ibid.

28. Samuel Smiles, *Industrial Biography* (London: John Murray, 1908), pp. 267-70. Smiles called the self-acting mule "one of the most elaborate and beautiful pieces of machinery ever contrived" (p. 267).

29. Andrew Ure, *Philosophy of Manufactures* (London, 1835), p. 368. Cf. Gaskell, who stated: "These unions, which have ramified throughout the entire factory labourers, are sources of great embarrassment to the master, whose energy, activity, and capital, are more or less at their mercy. He cannot, however, dissolve them by force; and the knowledge which the men have of the *cheapening* power of machinery keeps them together. If the master cannot overcome them by force, he has another and a sure, though gradual agent of destruction in the workshop of the machinist; and it is this that he is wielding with silent but irresistible force, which will free him from his adult and combined labourers. . . . It is impossible to deny that the men have accelerated the advance of the mighty agent which is to destroy them" (P. Gaskell, *Artisans and Machinery* [London, 1836], p. 349). Nasmyth, who commented frequently on the dilatoriness of labor, wrote enthusiastically of self-acting tools: "They were always ready for work, and never required a Saint Monday" (*James Nasmyth: An Autobiography,* edited by Samuel Smiles [London, 1883], p. 307). See also H.W. Dickinson, "Richard Roberts, His Life and Invention," *Transactions of the Newcomen Society* 25 (1945-47): 127. A sales agent for the McCormick reaper in the 1850s stated that it was the purpose of his work to "place the farmer beyond the power of a set of drinking Harvest Hands with which we have been greatly annoyed" (William T. Hutchinson, *Cyrus Hall McCormick,* vol. 1: *Seed-Time, 1809-1856* [New York: Century Co., 1930], p. 470). As quoted by Paul David, "Mechanization of Reaping in the Ante-Bellum Midwest," in *Industrialization in Two Systems,* ed. H. Rosovsky (New York: John Wiley & Sons, 1966), p. 18.

30. "The origin of this invention was somewhat similar to that of the self-acting mule. The contractors for the Conway Tubular Bridge while under construction, in 1848, were greatly hampered by combinations amongst the workmen, and they despaired of being able to finish the girders within the time specified in the contract. The punching of the iron plates by hand was a tedious and expensive as well as an inaccurate process; and the work was proceeding so slowly that the contractors found it absolutely necessary to adopt some new method of punching if they were to finish the work in time. In their

emergency they appealed to Mr. Roberts, and endeavored to persuade him to take the matter up. He at length consented to do so, and evolved the machine in question during his evening's leisure - for the most part while quietly sipping his tea. The machine was produced, the contractors were enabled to proceed with the punching of the plates independent of the refractory men, and the work was executed with a despatch, accuracy, and excellence that would not otherwise have been possible" (Smiles, *Industrial Biography*, pp. 271-72).

31. Marx, *Capital*, p. 436.
32. *The Life of Sir William Fairbairn, Bart.*, edited and completed by William Pole (London, 1877), p. 163.
33. Ibid., p. 164. In 1873, shortly before his death, Fairbairn wrote: "The introduction of new machinery and the self-acting principle owed much of their efficacy and ingenuity to the system of strikes, which compelled the employers of labour to fall back upon their own resources, and to execute, by machinery and new inventions, work which was formerly done by hand" (ibid., pp. 419-20). Cf. also p. 46, and *Mechanics Magazine* 32 (1840): 190. An illustration as well as description of the machine appeared in *Official Description and Illustrated Catalogue of the Great Exhibition of the Works of Industry Of All Nations* (London, 1851), pp. 286-87.
34. Charles Babbage, *On the Economy of Machinery and Manufactures* (London, 1835), pp. 298-300. Cf. Richard Prosser, *Birmingham Inventors and Inventions* (Birmingham, 1881), p. 86. Goodman states: "The invention of making gun barrels by means of grooved rolls is due to a Birmingham manufacturer of the name of Osborne. It was on the occasion of a strike of the barrel welders that he was led to make the experiment. He was not allowed to introduce his system without opposition, for no sooner were his rolls set to work than twelve hundred barrel welders, each armed with his forge hammer, proceeded to the private residence of Mr. Osborne, in the Stratford Road, threatening its destruction. The military were called out before the disturbance could be quelled, and for many days afterwards a guard was placed over the mill in which the work was carried on" (John D. Goodman, "The Birmingham Small Gun Trade," in *Birmingham and the Midland Hardware District,* ed. S. Timmins [London, 1866], p. 389).
35. "Report of the Committee on the Machinery of the U.S.A.," *Parliamentary Papers* (1854-55), vol. 50. The history of the British government's troubled relationship with the gun trade is exhaustively ventilated in "Report of the Select Committee on Small Arms," *Parliamentary Papers* (1854), vol. 18. In its final report, the committee conveniently summarized Nasmyth's views as follows: "Mr. Nasmyth, the well-known inventor of the steam-hammer, considers the systematic introduction of machinery into the gun trade to be highly desirable. . . . He considers that in the gun trade, the masters are in a state of dependence upon the skilled workmen, from which they can only be emancipated by the substitution of machinery for hand labour" (p. vi).
36. Thomas Brassey, *Work and Wages* (New York, 1883), pp. 129-30. See also the testimony of John Anderson, "Report from the Select Committee on Small Arms," Q. 356. Brassey was strongly committed to the proposition that the high price of labor is a major stimulus to inventive activity. Chapter 5 of *Work and Wages* is titled "Dear Labour Stimulates Invention." Conversely "the cheap labour at the command of our competitors seems to exercise the same enervating influence as the delights of Capua on the soldiers of Hannibal" (ibid., p. 142).
37. Smiles, *Industrial Biography*, pp. 294-95.
38. Such as "the adoption of the Fourdrinier machine after 1800 by English paper manufacturers was due largely to their desire to break the power of skilled labour in the industry" (Habakkuk [see n. 7 above], p. 153). See also D.C. Coleman, *The British Paper Industry* (London: Oxford University Press, 1958), pp. 258-59.
39. The point about uncertainties is, of course, *in principle* a general one. Under nineteenth-century conditions, it was the disruption in their labor supply which seems to have constituted the most serious threat to entrepreneurs who were concerned with maintaining the continuity of their productive operations. But uncertainties about power failure may persuade an entrepreneur to avoid an electricity-using process, or uncertainties over the supply of water may militate against a reliance upon water power.

In fact, there were many instances in the nineteenth century when the erratic variations in water supply (partly, of course, a seasonal phenomenon) were crucial in inducing mill owners to introduce steam-power.

40. David Landes, "Technological Change and Development in Western Europe, 1750-1914," in *The Cambridge Economic History of Europe,* ed. M.M. Postan and H.J. Habakkuk (London: Cambridge University Press, 1965), 6: 340-41: See also L.F. Haber, *The Chemical Industry during the Nineteenth Century* (London: Oxford University Press, 1958), chaps. 1 and 2; T.C. Barker, R. Dickinson, and D.W.F. Hardie, "The Origins of the Synthetic Alkali Industry in Britain," *Economica* (May 1956), pp. 158-71.

41. *Life of Robert Owen,* p. 171, as quoted in Habakkuk, p. 135.

42. Marx, *Capital,* 1:433-36. Writing only a few years after the events, Marx exclaimed: "But who, in 1860, the Zenith year of the English cotton industry, would have dreamt of the galloping improvements in machinery, and the corresponding displacement of working people, called into being in the following 3 years, under the stimulus of the American Civil War?" (p. 433). The examples Marx subsequently cited include techniques which obviously economized on capital and labor inputs more than they did on cotton (cf. Habakkuk, p. 159). Figures on cotton prices and the volume and sources of cotton imports immediately before, during, and after the cotton famine may be found in Thomas Ellison, *The Cotton Trade of Great Britain* (London, 1886), p. 91.

43. Jesse Markham, *The Fertilizer Industry* (Nashville, Tenn: Vanderbilt University Press, 1958), p. 98.

44. George W. Stocking and Myron W. Watkins, *Cartels in Action* (New York: Twentieth Century Fund, 1946), chap. 9.

45. Although wartime situations and military requirements have thus been a most important historical force, one should nevertheless avoid overstatement. The following quotation from Bernal is, surely, a gross exaggeration: "Science and warfare have always been most closely linked; in fact, except for a certain portion of the nineteenth century it may fairly be claimed that the majority of significant technical and scientific advances owe their origin directly to military or naval requirements" (J.D. Bernal, *The Social Function of Science* [New York: Macmillan Co., 1939], p. 165). For a forceful statement of an opposing view, see John Nef, *War and Human Progress* (New York: Russell & Russell, 1950).

46. William Haynes, *This Chemical Age* (New York: A.A. Knopf, 1945), pp. 164-67.

47. B.S. Keirstead, *The Theory of Economic Change* (Toronto: Macmillan Co., 1948), p. 172.

48. H.W. Dickinson and Rhys Jenkins, *James Watt and the Steam Engine* (London: Oxford University Press, 1927), pp. 149-56; H.W. Dickinson, *A Short History of the Steam Engine* (London: Cass, 1963), pp. 79-82; Samuel Smiles, *Lives of the Engineers: Boulton and Watt* (London: John Murray, 1904), chap. 12.

49. See William Fairbairn, *An Account of the Construction of the Britannia and Conway Tubular Bridges* (London, 1849); William Fairbairn, *Useful Information for Engineers* (London, 1860), 2:223-28, 244-64, 268-81, 282-92.

50. Similarly, a heart attack or other symptom of ill health may in fact be life-prolonging provided one (1) survives the initial attack and (2) interprets the medical signals correctly and makes appropriate modifications in future behavior. Thus, the ill health of David Ricardo's brother Moses was frequently referred to in Ricardo's correspondence. Moses gave up his medical practice at an early age and went to live in retirement at Brighton. Whereas poor David Ricardo, whose health had given no particular cause for alarm, died at the age of fifty-one, his "sickly" brother lived into his ninetieth year, nursing his frail constitution to the very end (Piero Sraffa, ed., *The Works and Correspondence of David Ricardo* [London: Cambridge University Press, 1955], 10:56).

51. Having suggested avenues of exploration as obviously heretical as the foregoing, it is some consolation to be able to invoke the voice of Marshall (although admittedly an elderly Marshall) in an analogous line of reasoning. In accounting for Holland's earlier commercial supremacy, Marshall committed to print the following statement: "Holland seemed poor in physical resources: but her poverty was a part of her strength: for it led

her to give her whole energies to developing those resources which she possessed" (Alfred Marshall, *Industry and Trade* [London: Macmillan Co., 1927], p. 692). Essentially the same view has been stated more recently by a distinguished economic historian: "Flanders owes its industry as much to its geographical limitations as to its geographical facilities" (M.M. Postan, in *the Cambridge Economic History of Europe,* ed. M.M. Postan and H.J. Habakkuk [London: Cambridge University Press, 1952], 2: 183). Similarly, Pirenne attributed Venice's early rise to commercial leadership to her lack of agricultural resources: "This ineluctable necessity was imposed on Venice from its very foundation on the sandy islets of a lagoon, on which nothing would grow. In order to procure a livelihood its first inhabitants had been forced to exchange salt and fish with their continental neighbours for the corn, wine and meat which they could not have obtained otherwise. But this primitive exchange inevitably developed, as commerce made the town richer and more populous, and at the same time increased its demands and sharpened its enterprise" (Henri Pirenne, *Economic and Social History of Medieval Europe* [London: Routledge, 1949], pp. 26-27). It is perhaps needless to add that Marshall, Postan, and Pirenne were well aware of the locational advantages which geography had conferred upon the places mentioned. No one seriously believes that poverty of resources, by itself, is sufficient to generate economic growth. For an analogous argument in accounting for the growth of inventiveness in New England, see Daniel Boorstin, *The Americans: The National Experience* (New York: Random House, 1965), p. 22. For the suggestion that the origins of Mesopotamian civilization owed something to a regional shortage of resources, see Frank Hole, "Investigating the Origins of Mesopotamian Civilization," *Science* 5 (August 1966): 608.

52. The earliest advertisements for automobiles in weekly magazines frequently showed a team of runaway horses bearing down on oblivious bystanders (usually children at play) with indignant captions such as, "These horrible accidents must be stopped!" Next to this awful scene of impending disaster would be an illustration of a horseless carriage.

53. The Firth of Forth bridge, e.g., incorporated improvements in design gleaned from the disastrous collapse of the Tay railway bridge in 1879 (see T.K. Derry and Trevor I. Williams, *A Short History of Technology* [London: Oxford University Press, 1960], pp. 449 and 457). The failure of bridges - especially railway bridges in the early days of railway construction - was an unfortunately common phenomenon. These dramatic accidents led to highly valuable public investigations (see, e.g., "Report of the Commissioners Appointed to Inquire into the Application of Iron to Railway Structures," *Parliamentary Papers* [1849], vol. 29). See also Fairbairn, *The Life of Sir William Fairbairn, Bart.,* p. 187. On boiler explosions, see ibid., chap. 16.

54. "It is noteworthy that sensational accidents or unusual occurrences are followed immediately by a large influx of applications for patents connected in some way with them, as for example, devices to prevent fires at sea and for the protection of property following upon the L'Atlantique disaster and the epidemic of 'smash and grab' raids respectively" (H.W. Dickinson, "The Evolution of Invention," *Proceedings, Institution of Mechanical Engineers* [January 1934], p. 8).

55. Joseph Rossman, *Industrial Creativity* (New York: Universe Books, 1964), p. 128.

56. John Anderson, *The Strength of Materials and Structures* (London, 1872), pp. 280-82; see also the *Journal of the Society of Arts* 17 (August 13, 1869): 746; and Edwin Clark, *The Britannia and Conway Tubular Bridges* (London, 1850), 2:690-94.

57. Lebergott has argued that in the context of the American economy in the nineteenth century, entrepreneurial expectations concerning the trend of future wages justified a capital-intensive bias. "Given (1) a long run upward pressure on wages, and (2) an uncertain future course for the price of capital, the entrepreneur's wisest choice, time and again, proved to be: Adopt techniques that were not labor-intensive" (Stanley Lebergott, *Manpower in Economic Growth* [New York: McGraw-Hill Book Co., 1964], pp. 230-31). Lebergott also makes the interesting point that entrepreneurs who had recently immigrated to the United States regarded wages as "high" in some absolute sense because they implicitly retained European wage-level standards. This tended to reinforce the capital-intensive bias (ibid., p. 231).

58. Somewhat analogously, Babbage asserted that depressed business conditions provided a

strong stimulus to technical innovation (Babbage, p. 176). More recently it has been argued that machine tool producers introduce tools of new design during periods of depression in an attempt to offset the decline in the demand for their products (William Brown, "Innovation in the Machine Tool Industry," *Quarterly Journal of Economics* [1957], pp. 406-25). The argument is interesting and suggestive, if not entirely persuasive. Rostow at one point has combined the deterioration of business conditions with the downward inflexibility of wages as an explanation of labor-saving inventions. Writing of the British economy during the 1870s, he suggests that "machinery was sought as a means of escaping the tyranny of money wages that could not be reduced. Everywhere 'the growing depression stimulated invention of labour-saving devices' " (W.W. Rostow, *British Economy of the Nineteenth Century* [New York: Oxford University Press, 1948], p. 75). Landes has recently argued that the loss of nearby markets for Swiss textile products at the end of the Napoleonic Wars "incited her textile industry to lower costs by mechanization and seek compensatory outlets in distant lands." Landes goes on to broaden his case by stating that "there is good reason to believe that adversity is at least as often a stimulus to achievement as a deterrent. It is no accident that techniques often achieve their highest development after the appearance of more efficient equipment or methods have rendered them obsolescent: recall the resistance of the sailing ship to the competition of steam, of puddling to that of the convertor and the open hearth, of coaching to the railway" (David Landes, "Factor Costs and Demand: Determinants of Economic Growth," *Business History* [January 1965], p. 31).

59. This signaling mechanism has its parallels, of course, in political processes. In discussing the problems of Brazil's underdeveloped Northeast province, Hirschman points out that "a bad drought usually jolts the government into a major new effort" (Albert Hirschman, *Journeys toward Progress* [New York: Twentieth Century Fund, 1963], p. 18). As a signaling device, the threat to the established political order which is implicit in a failure to take a remedial - or at least amelioratory - action is similar to the threat to a business firm which fails to respond to a direct threat to its profit position. For a fascinating account of some of the other dimensions of the political problem in the context of recent Brazilian history, see ibid., chaps. 1 and 4. Dorfman has exhorted the economics profession to accept a view of the business firm which is very much in harmony with the approach suggested here: "We must recognize the firm for what operations research has disclosed it to be: often fumbling, sluggish, timid, uncertain and perplexed by unsolvable problems. . . . It reacts in familiar ways to the familiar and avoids the novel as long as it dares" (Robert Dorfman, "Operations Research," *American Economic Review* [September 1960], p. 622).

60. James G. March and Herbert A. Simon, *Organizations* (New York: John Wiley & Sons, 1958), p. 185. One might alternatively restate this as a short-run companion to Keynes's dictum: "In the long run we are all dead." It could then read: "In the short run we are all preoccupied."

7. Karl Marx on the economic role of science

1. "The bourgeoisie cannot exist without constantly revolutionising the instruments of production, and thereby the relations of production, and with them the whole relations of society. Conservation of the old modes of production in unaltered form, was, on the contrary, the first condition of existence for all earlier industrial classes" (Marx and Engels 1951, 1:36).

2. Marx 1906, pp. 382-83. Engels states: "Like all other sciences, mathematics arose out of the *needs* of men; from the measurement of land and of the content of vessels; from the computation of time and mechanics" (Engels 1939, p. 46; emphasis Engels's. Cf. Marx 1906, p. 564).

3. Marx 1906, p. 411. He adds: "In the same way the irregularity caused by the motive power in mills that were put in motion by pushing and pulling a lever, led to the theory, and the application, of the flywheel, which afterwards plays so important a part in

Modern Industry. In this way, during the manufacturing period, were developed the first scientific and technical elements of Modern Mechanical Industry."
4. Engels 1954, p. 247. Earlier in the paragraph, he had stated: "The successive development of the separate branches of natural science should be studied. First of all, astronomy, which, if only on account of the seasons, was absolutely indispensable for pastoral and agricultural peoples. Astronomy can only develop with the aid of mathematics. Hence this also had to be tackled. Further, at a certain stage of agriculture and in certain regions (raising of water for irrigation in Egypt), and especially with the origin of towns, big building structures and the development of handicrafts, mechanics also arose. This was soon needed also for navigation and war. Moreover, it requires the aid of mathematics and so promoted the latter's development."
5. Ibid., p. 248. The editor of Engels's unfinished manuscript points out that Engels had written in the margin of the manuscript opposite this paragraph: "Hitherto, what has been boasted of is what production owes to science, but science owes infinitely more to production."
6. Since the subsequent discussion turns directly upon the Marxian periodization scheme, it is important to remind the reader of the meaning which Marx attaches to the terms "handicraft," "manufacture," and "modern industry." Engels expressed Marx's meanings succinctly as follows: "We divide the history of industrial production since the Middle Ages into three periods: (1) handicraft, small master craftsmen with a few journeymen and apprentices, where each laborer produces the complete article; (2) manufacture, where greater numbers of workmen, grouped in one large establishment, produce the complete article on the principle of division of labor, each workman performing only one partial operation, so that the product is complete only after having passed successively through the hands of all; (3) modern industry, where the product is produced by machinery driven by power, and where the work of the laborer is limited to superintending and correcting the performances of the mechanical agent" (Engels 1910, pp. 12-13).
7. Actually, Marx's use of the term "science" was sufficiently broad that it included bodies of systematized knowledge far beyond what we ordinarily mean when we speak today of pure or even applied science - e.g., engineering and machine building. It was not a term which he attempted to use with precision. In *Theories of Surplus Value*, for instance, he refers to science simply as "the product of mental labour" (Marx 1963, pt. 1, p. 353).
8. "With regard to the mode of production itself, manufacture, in its strict meaning, is hardly to be distinguished, in its earliest stages, from the handicraft trades of the guilds, otherwise than by the greater number of workmen simultaneously employed by one and the same individual capital. The workshop of the medieval master handicraftsman is simply enlarged" (Marx 1906, p. 353. Cf. Marx and Engels 1947, pp. 12-13).
9. "With manufacture was given simultaneously a changed relationship between worker and employer. In the guilds the patriarchal relationship between journeyman and master maintained itself; in manufacture its place was taken by the monetary relation between worker and capitalist - a relationship which in the countryside and in small towns retained a patriarchal tinge, but in the larger, the real manufacturing towns, quite early lost almost all patriarchal complexion" (Marx and Engels 1947, p. 52).
10. Machinery had sometimes been employed in earlier periods, but Marx clearly regarded these instances as exceptional. "Early in the manufacturing period the principle of lessening the necessary labour-time in the production of commodities, was accepted and formulated: and the use of machines, especially for certain simple first processes that have to be conducted on a very large scale, and with the application of great force, sprang up here and there. Thus, at an early period in paper manufacture, the tearing up of the rags was done by paper mills; and in metal works, the pounding of the ores was effected by stamping mills. The Roman Empire had handed down the elementary form of all machinery in the water-wheel" (Marx 1906, p. 382). In a footnote Marx makes the extremely interesting observation that "the whole history of the development of machinery can be traced in the history of the corn mill" (ibid., p. 382, n. 3).
11. "The needlemaker of the Nuremberg Guild was the cornerstone on which the English

needle manufacture was raised. But while in Nuremberg that single artificer performed a series of perhaps 20 operations one after another, in England it was not long before there were 20 needlemakers side by side, each performing one alone of those 20 operations; and in consequence of further experience, each of those 20 operations was again split up, isolated, and made the exclusive function of a separate workman" (ibid., pp. 370-71).

12. "While simple co-operation leaves the mode of working by the individual for the most part unchanged, manufacture thoroughly revolutionises it, and seizes labour-power by its very roots. It converts the labourer into a crippled monstrosity, by forcing his detail dexterity at the expense of a world of productive capabilities and instincts; just as in the States of La Plata they butcher a whole beast for the sake of his hide or his tallow" (ibid., p. 396).

13. "For a proper understanding of the division of labour in manufacture, it is essential that the following points be firmly grasped. First, the decomposition of a process of production into its various successive steps coincides, here, strictly with the resolution of a handicraft into its successive manual operations. Whether complex or simple, each operation has to be done by hand, retains the character of a handicraft, and is therefore dependent on the strength, skill, quickness, and sureness, of the individual workman in handling his tools. The handicraft continues to be the basis. This narrow technical basis excludes a really scientific analysis of any definite process of industrial production, since it is still a condition that each detail process gone through by the product must be capable of being done by hand and of forming, in its way, a separate handicraft. It is just because handicraft skill continues, in this way, to be the foundation of the process of production that each workman becomes exclusively assigned to a partial function, and that for the rest of his life, his labour-power is turned into the organ of this detail function" (ibid., pp. 371-72).

14. Ibid., p. 408; see also p. 410. In his early work, *The Poverty of Philosophy,* Marx had stated: "The machine is a unification of the instruments of labour, and by no means a combination of different operations for the worker himself. 'When, by the division of labour, each particular operation has been simplified to the use of a single instrument, the linking-up of all these instruments, set in motion by a single engine, constitutes - a machine.' (Babbage, *Traité sur l'Economie des Machines,* etc., Paris 1833). Simple tools; accumulation of tools; composite tools; setting in motion of a composite tool by a single hand engine, by men; setting in motion of these instruments by natural forces, machines; system of machines having one motor; system of machines having one automatic motor - this is the progress of machinery" (Marx, n.d., pp. 132-33. This book was first published in 1847).

15. There is an important learning experience at the technological level before this can be done well. "It is only after considerable development of the science of mechanics, and accumulated practical experience, that the form of a machine becomes settled entirely in accordance with mechanical principles, and emancipated from the traditional form of the tool that gave rise to it" (Marx 1906, p. 418, n. 1). A typical aspect of the innovation process, therefore, is that machines go through a substantial process of modification after their first introduction (see ibid., p. 442).

16. Ibid., pp. 414-15. Later, Marx adds: "The implements of labour, in the form of machinery, necessitate the substitution of natural forces for human force, and the conscious application of science, instead of rule of thumb. In Manufacture, the organization of the social labour-process is purely subjective; it is a combination of detail labourers; in its machinery system, Modern Industry has a productive organism that is purely objective, in which the labourer becomes a mere appendage to an already existing material condition of production" (p. 421).

17. "As soon as a machine executes, without man's help, all the movements requisite to elaborate the raw material, needing only attendance from him, we have an automatic system of machinery, and one that is susceptible of constant improvement in its details" (ibid., p. 416).

18. In a valuable article, "Karl Marx and the Industrial Revolution," Paul Sweezy argues that many of the important differences between Marx and his classical predecessors

reduced to the fact that the classical economists "took as their model an economy based on manufacture, which is an essentially conservative and change-resistant economic order; while Marx, recognizing and making full allowance for the profound transformation effected by the industrial revolution, took as his model an economy based on modern machine industry" (Sweezy 1968, p. 115).

19. Marx 1906, p. 504. The manufacturing stage needs to be seen as an essential step in the introduction of science into the productive process. The application of science required that productive activity be broken down into a series of separately analyzable steps. The manufacturing system, even though it continued to rely upon human skills, accomplished precisely this when it replaced the handicraftsman with a number of detail laborers. In this important sense it "set the stage" for the advent of modern industry.

20. Ibid., p. 532. Marx (1959) examines the vast possibilities for capital-saving innovations and improvements in an advanced capitalist economy in *Capital*, vol. 3, chaps. 4 and 5.

21. "As inventions increased in number, and the demand for the newly discovered machines grew larger, the machine-making industry split up, more and more, into numerous independent branches, and division of labour in these manufactures was more and more developed. Here, then, we see in Manufacture the immediate technical foundation of Modern Industry. Manufacture produced the machinery, by means of which Modern Industry abolished the handicraft and manufacturing systems in those spheres of production that it first seized upon" (Marx 1906, p. 417).

22. Ibid., pp. 417-18. Marx saw the improvements in the means of communication and transportation as particularly significant in pushing the productive process beyond the limitations inherent in the manufacturing system. "The means of communication and transport became gradually adapted to the modes of production of mechanical industry, by the creation of a system of river steamers, railways, ocean steamers, and telegraphs. But the huge masses of iron that had now to be forged, to be welded, to be cut, to be bored, and to be shaped, demanded, on their part, cyclopean machines, for the construction of which the methods of the manufacturing period were utterly inadequate" (pp. 419-20).

23. Ibid., p. 420. Marx saw this process as culminating during his own time. "It is only during the last 15 years (i.e., since about 1850), that a constantly increasing portion of these machine tools have been made in England by machinery, and that not by the same manufacturers who make the machines" (p. 408).

24. "If we now fix our attention on that portion of the machinery employed in the construction of machines, which constitutes the operating tool, we find the manual implements reappearing, but on a cyclopean scale. The operating part of the boring machine is an immense drill driven by a steam-engine; without this machine, on the other hand, the cylinders of large steam-engines and of hydraulic presses could not be made. The mechanical lathe is only a cyclopean reproduction of the ordinary footlathe; the planing machine, an iron carpenter, that works on iron with the same tools that the human carpenter employs on wood; the instrument that, on the London wharves, cuts the veneers, is a gigantic razor; the tool of the shearing machine, which shears iron as easily as a tailor's scissors cut cloth, is a monster pair of scissors; and the steam hammer works with an ordinary hammer head, but of such a weight that not Thor himself could wield it. These steam hammers are an invention of Nasmyth, and there is one that weighs over 6 tons and strikes with a vertical fall of 7 feet, on an anvil weighing 36 tons. It is mere child's play for it to crush a block of granite into powder, yet it is not less capable of driving, with a succession of light taps, a nail into a piece of soft wood" (ibid., p. 421; see also pp. 492-93).

25. At one point Marx presents what one might be tempted to call a Toynbeean "challenge-response" mechanism to account for the emergence of high productivity societies. It is not true, he says, "that the most fruitful soil is the most fitted for the growth of the capitalist mode of production. This mode is based on the dominion of man over nature. Where nature is too lavish, she 'keeps him in hand, like a child in leading-strings.' She does not impose upon him any necessity to develop himself. It is not the tropics with their luxuriant vegetation, but the temperate zone, that is the mother country of capital. It is not the mere fertility of the soil, but the differentiation

of the soil, the variety of its natural products, the changes of the seasons, which form the physical basis for the social division of labour, and which, by changes in the natural surroundings, spur man on to the multiplication of his wants, his capabilities, his means and modes of labor. It is the necessity of bringing a natural force under the control of society, of economising, of appropriating or subduing it on a large scale by the work of man's hand, that first plays the decisive part in the history of industry" (ibid., pp. 563-64).

26. In this light, there is no necessary conflict between Marx's materialist conception of history and his treatment of science as a productive force under advanced capitalism. I therefore disagree with the following statement of Bober: "Marx intends to offer a materialistic conception of history. Yet he frequently stresses the power of science as a component of modern technique and production. The incorporation of science in the foundation of his theory is no more defensible than the inclusion of all other nonmaterial phenomena" (Bober 1965, p. 21).

27. Marx 1968, pt. 2, p. 110. In *The German Ideology* Marx and Engels stated that "the science of mechanics perfected by Newton was altogether the most popular science in France and England in the eighteenth century" (Marx and Engels 1947, p. 56).

28. "Hegel's division (the original one) into mechanics, chemics, and organics, fully adequate for the time. Mechanics: the movement of masses. Chemics: molecular (for physics is also included in this and, indeed, both - physics as well as chemistry - belong to the same order) motion and atomic motion. Organics: the motion of bodies in which the two are inseparable. For the organism is certainly *the higher unity which within itself unites mechanics, physics, and chemistry into a whole* where the trinity can no longer be separated. In the organism, mechanical motion is effected directly by physical and chemical change, in the form of nutrition, respiration, secretion, etc., just as much as pure muscular movement" (Engels 1954, pp. 331-32; emphasis Engels's). For Engels's entire treatment of the subject, see ibid., pp. 322-408. In his book, *Herr Eugen Duhring's Revolution in Science,* Engels draws a sharp distinction between the sciences concerned with inanimate nature and those concerned with living organisms. The former group of sciences (mathematics, astronomy, mechanics, physics, chemistry) are susceptible to mathematical treatment "to a greater or less degree." No such precision is possible in the sciences concerned with living organisms. "In this field there is such a multitude of reciprocal relations and casualties that not only does the solution of each question give rise to a host of other questions, but each separate problem can usually only be solved piecemeal, through a series of investigations which often requires centuries to complete; and even then the need for a systematic presentation of the interrelations makes it necessary again and again to surround the final and ultimate truths with a luxuriant growth of hypotheses" (Engels 1939, pp. 97-99).

29. "Classification of the sciences, each of which analyzes a single form of motion, or a series of forms of motion that belong together and pass into one another, is therefore the classification, the arrangement, of these forms of motion themselves according to their inherent sequence, and herein lies its importance" (Engels 1954, p. 330; see also Zvorikine 1963, pp. 59-74).

30. See Engels 1954, "Preface."

31. The most ambitious attempt to fill this void is the fascinating but seriously flawed four-volume work by the late J.D. Bernal, *Science in History* (1971). His *Science and Industry in the Nineteenth Century* (London, 1953) is more restricted in scope and far more consistently persuasive. Nevertheless, *Science in History* displays an immense erudition, and all but the most remarkably well-informed readers will learn much from, and be greatly stimulated by, its contents.

8. Capital goods, technology, and economic growth

1. J.R. Hicks, *The Theory of Wages,* The MacMillan Company, London, 1932, pp. 124-5. Similar statements may be found in numerous other places, such as N. Kaldor, *Essays on Economic Stability and Growth,* Duckworth, London, 1960, p. 229, and K. Rothschild, *The Theory of Wages,* Blackwell, Oxford, 1956, p. 118.

2. This is a point which Hicks has recently emphasized in a discussion of the problems of underdeveloped countries: "... a market may be large enough to call forth all possible economies of scale in the production of final consumers' goods, but may not be large enough to do the same in the production of the capital goods which are to make those consumer goods, or in many ancillary industries. All the stages of production must be taken into account. . . . It is especially important . . . that many of the most 'advanced' capital goods are among the things which are most affected by economies of large-scale production." J.R. Hicks, *Essays in World Economics,* Oxford, at the Clarendon Press, 1959, pp. 184-5. And later, with respect to Ceylon: "It is only the simplest sorts of capital goods (building materials being the obviously important case) which can expect to command a market within Ceylon sufficient to enable their production to be carried on at an efficient size. One has only to consider that there are plenty of countries that can produce textiles efficiently; but there are very few countries which can keep a textile machinery industry going without considerable reliance on an export market. This is the kind of situation which repeats itself with one sort of specialized capital good after another." Ibid., p. 205.

3. In 1954 the machine-tool industry consisted of 639 companies, the 4 largest of which accounted for 18 per cent., and the 20 largest companies for 49 per cent., of the value of all industry shipments. Of the industry's 81,000 workers, only 15,500 were employed by firms with over 2,500 employees. See M. Brown and N. Rosenberg, "Patents and Other Factors in the Machine Tool Industry," *The Patent, Trademark, and Copyright Journal of Research and Education* (Spring, 1960), pp. 45-46.

4. Cf. Marvin Frankel, "Producers' Goods, Consumer Goods and Acceleration of Growth," *Economic Journal* (Mar. 1961), p. 2.

5. W. Paton and R. Dixon, *Make-or-buy Decisions in Tooling for Mass Production,* Bureau of Business Research, School of Business Administration, University of Michigan, Ann Arbor, 1961, pp. 1-4.

6. C. Carter and B. Williams, *Industry and Technical Progress,* Oxford University Press, London, 1957, p. 155.

7. See, for example, R.S. Eckaus, "The Factor Proportions Problem in Underdeveloped Areas," *American Economic Review,* Sept. 1955, pp. 539-65; H. Singer, "Problems of Industrialization of Underdeveloped Countries," *International Social Science Bulletin,* vol. vi, no. 2 (1954), pp. 217-23; B. Higgins, *Economic Development,* W.W. Norton, New York, 1959, chap. 14.

8. In fact, late developers have never adopted Western technology wholesale and indiscriminately, and it may be suggested that the proponents of the technological dualism hypothesis have been preoccupied with the emergence of particular industries - such as oil extraction and refining - where the elasticity of substitution between labor and capital is very low. Japanese development is a case in point. In the early years after the Meiji Restoration (1868) the adoption of advanced Western techniques was highly selective. In some industries - e.g. textiles - Western methods were introduced at selected stages and processes while old-fashioned cottage industry techniques survived elsewhere. Moreover, where Western machinery was introduced it was both operated and serviced more intensively than was the practice in the West. See G. Ranis, "Factor Proportions in Japanese Economic Development," *American Economic Review* (Sept. 1957), pp. 594-607. In the Russian case, Granick has argued that the development of Soviet metal-working industries was characterized, not by adoption of the most advanced, highest labor productivity technology, but by an attempt to minimize the capital-output ratio, and that the substitution of labor for capital was undertaken wherever possible. Granick cites such examples as the persistence, until fairly recently, in the use of general-purpose as opposed to special-purpose equipment, and the failure to substitute mechanized techniques where possible in individual processes such as hand-scouring, manual moulding, and hand assembly tools. Furthermore, throughout a wide range of auxiliary operations - materials-handling, inspection, repair work, clerical and bookkeeping work - very little substitution of capital for labour has taken place. David Granick, "Economic Development and Productivity Analysis: The Case of Soviet Metalworking," *Quarterly Journal of Economics* (May 1957), pp. 205-33.

9. C. Kennedy, "Technical Progress and Investment," *Economic Journal* (June 1961), p. 294; J. Robinson, *The Accumulation of Capital,* Richard D. Irwin, Homewood, Illinois, 1956, p. 169; Habakkuk, op. cit., p. 168.
10. Karl Marx, *Capital,* vol. i, Random House, New York, 1936, p. 417.
11. George Stigler, "The Division of Labor is Limited by the Extent of the Market," *The Journal of Political Economy* (June 1951), p. 190.
12. Cf. Habakkuk, op. cit., pp. 167-8.
13. Cf. Habakkuk, op cit., pp. 163-4.
14. N. Rosenberg, "Capital Formation in Underdeveloped Countries," *American Economic Review* (Sept. 1960), pp. 706-15.
15. The obvious alternative, of course, is to import capital goods from the low-cost foreign producers of such goods, and this has been a common practice. But, in addition to the fact that this alternative largely deprives the underdeveloped country of many of the external economies resulting from the possession of a well-developed domestic capital goods industry, it has also posed serious problems due to the following related points: (1) the capital goods of the industrial countries are highly capital-intensive in their use, and therefore not appropriate to the resource endowment of the importing country; (2) the imported capital goods frequently embody a high degree of complementarity to skilled labor rather than to unskilled labor and are therefore difficult to operate successfully; (3) the problems of adequate servicing and replacement of parts are often difficult to handle satisfactorily when dealing with a foreign supplier.
16. Simon Kuznets, "Quantitative Aspects of the Economic Growth of Nations: The Share and Structure of Consumption," *Economic Development and Cultural Change* (July 1960), Part II, pp. 23-24.

9. Economic development and the transfer of technology: some historical perspectives

1. For a more formal treatment of the problems of technological leadership and "followership," see Edward Ames and Nathan Rosenberg, "Changing Technological Leadership and Economic Growth," *Economic Journal,* vol. 73, no. 289 (March 1963). A recent study of the diffusion of new technologies in six West European countries suggested that, for five important new processes, "countries which are pioneers tend to have *slower* speeds of diffusion. This result is consistent with the hypothesis that the pioneer faces all sorts of teething troubles, new problems associated with the new technique; these are likely to be - partly and gradually - solved by the time others adopt it. It is therefore not necessarily desirable to be the first to introduce a new technique" (G.F. Ray, "The Diffusion of New Technology: A Study of Ten Processes in Nine Industries," *National Institute Economic Review* [May 1969], p. 82).
2. See Alexander Gerschenkron, "Economic Backwardness in Historical Perspective," in *The Progress of Underdeveloped Areas,* ed. B.F. Hoselitz (Chicago, 1952).
3. See A.P. Usher, *A History of Mechanical Inventions,* rev. ed. (Cambridge, Mass., 1954), esp. chap. 4; Nathan Rosenberg, "Technological Change in the Machine Tool Industry, 1840-1910," *Journal of Economic History* 23, no. 4 (December 1963): 414-46; Nathan Rosenberg, ed., *The American System of Manufactures* (Edinburgh, 1969).
4. Joseph W. Roe, *English and American Tool Builders* (New Haven, Conn., 1916), in fact contains several such genealogies.
5. See the illuminating account of a 16th-century transfer in Warren C. Scoville, "Minority Migrations and the Diffusion of Technology," *Journal of Economic History* 11, no. 4 (Fall 1951): 347-60. A.R. Hall has concluded: "It seems fairly clear that in most cases in the 16th century - and indeed long afterwards - the diffusion of technology was chiefly effected by persuading skilled workers to emigrate to regions where their skills were not yet plentiful" (A.R. Hall, "Early Modern Technology, to 1600," in *Technology in Western Civilization,* ed. Melvin Kranzberg and Carroll W. Pursell, Jr. [New York, 1967], 1:85). For a similar conclusion with respect to mining and metallurgy in an earlier period, see Bertrand Gille, "Technological Developments in Europe: 1100 to 1400," in *The Evolution of Science,* ed. Guy S. Metraux and Francois Crouzet (New York, 1963), p. 201.

6. These episodes are described at length in the introductory essay in Rosenberg, ed., *The American System of Manufactures.*
7. David Landes, *The Unbound Prometheus* (Cambridge, 1969), pp. 148-49. Landes adds: "Perhaps the greatest contribution of these immigrants was not what they did but what they taught. . . . The growing technological independence of the Continent resulted largely from man-to-man transmission of skills on the job" (p. 150). For a more detailed discussion of this subject, see W.O. Henderson, *Britain and Industrial Europe, 1750-1870* (Liverpool, 1954).
8. I. Svennilson, in *Economic Development with Special Reference to East Asia,* ed. K. Berrill (New York, 1964), p. 407.
9. Some of the recent work of Derek Price lends indirect support to this position. Price argues that there exist powerful professional incentives for the scientist to publish the results of his work quickly and authoritatively, whereas the incentive structure of the working technologist is one which (with some exceptions) encourages concealment and a reluctance to publish. Science is *papyrocentric,* whereas technology is *papyrophobic* (see Derek J. De Solla Price, "Is Technology Historically Independent of Science? A Study in Statistical Historiography," *Technology and Culture* 6, no. 4 [Fall 1965]: 553-68).
10. Lucius F. Ellsworth, "A Directory of Artifact Collections," in *Technology in Early America,* ed. Brooke Hindle (Chapel Hill, N.C., 1966), p. 96.
11. This paragraph and the three following are taken, with minor modification, from Nathan Rosenberg, "Technological Change in the Machine Tool Industry, 1840-1910."
12. For more recent evidence on such trends from the point of view of the input-output structure of the economy, see Ann P. Carter, "The Economics of Technological Change," *Scientific American* 214, no. 4 (April 1966): 25-31.
13. In a study of the world market in chemical processing plants, Freeman has pointed out: "The main European 'design-engineering' offices of the large contractors typically employ from 300 to 700 total office-based staff. They are concentrated mainly in London (where half a dozen of them have such offices) and to a lesser extent in The Hague, Paris, and Frankfurt. London is the principal European base of the American contractors and the largest world centre for chemical 'design-engineering' work. But the large American contractors show great ingenuity and flexibility in switching work from one office to another anywhere in the world, in order to achieve an optimum loading of their total resources. The London office, even though employing 500 or fewer, is able to take on several very large contracts because it is part of a world-wide organization and can call upon process design and know-how from the parent company, as well as specialists to deal with crises or bottlenecks. A particular individual, a scale model, or a set of drawings may be flown out at a few hours' notice" (C. Freeman, "Chemical Process Plant: Innovation and the World Market," *National Institute Economic Review* [August 1968], p. 37).
14. G.I.H. Lloyd, *The Cutlery Trades* (London, 1913), pp. 394-95.
15. As quoted in *A Treatise on the Progressive Improvement and Present State of the Manufactures in Metal* (London, 1833), 2:105.
16. Siegfried Giedion, *Mechanization Takes Command* (New York, 1948), p. 90.
17. At the turn of the 20th century, some evidence of change began to appear. A British observer in 1902 made the following highly revealing comments: "Some few years ago no machine-tool maker ventured to offer advice to a manufacturing engineer as to the tools he might buy or the methods he should follow. He would have been told to mind his own business. Today a good deal of help and advice is often requested and the tool maker and user, as they should be, are often in consultation as to tools, methods and organization. Machine tool making in America has been, I take it, considered a reputable business for some years, whereas here for a time it certainly was but little considered" (William H. Booth, "An English View of American Tools," *American Machinist,* no. 45, November 6, 1902, p. 1580).
18. See, for example, Daniel H. Calhoun, *The American Civil Engineer: Origin and Conflict* (Cambridge, Mass., 1960), esp. chap. 6.
19. *Autocar,* September 21, 1912, as quoted in S.B. Saul, "The Motor Industry in Britain

to 1914," *Business History 5*, no. 1 (December 1962): 41. The belief long persisted in British industry that high quality was incompatible with mass production. Much evidence on this point for the late 1920s may be found in Committee on Industry and Trade, *Survey of Metal Industries*, pt. 4 (London, 1928), pp. 227-28, 220-21, and passim.

20. See Burton H. Klein, "A Radical Proposal for R. and D.," *Fortune 57*, no. 5 (May 1958): 112 ff., for a statement of the case for the desirability of parallel research efforts under conditions of uncertainty.

21. It is interesting to note that, as early as the 1840s, American engineers were playing a vital role in bringing the railroad to Russia. See the account of the work of Major George Washington Whistler in Albert Parry, *Whistler's Father* (Indianapolis, 1939). Of Major Whistler, the awed muzhiks are supposed to have exclaimed: "How smart the foreign gentleman, who has harnessed the samovar and made it run." The book is unfortunately thin on technological history. It does, however, provide a minimal account of Whistler's role in directing the construction of the Saint Petersburg to Moscow railway and of his establishment of a factory at Alexandrovsky for building locomotives and other railroad equipment.

22. J.H. Clapham, *An Economic History of Modern Britain*, 3 vols. (Cambridge, 1938), 2:76. English orientation toward export markets doubtless increased the difficulties of standardization of locomotive design (see *Survey of Metal Industries*, pp. 168-72). Nevertheless, the high degree of diversification had developed early in the history of the industry when it was certainly preoccupied with the needs of domestic users.

23. J. Stephen Jeans, ed., *American Industrial Conditions and Competition*, Reports of the Commissioners Appointed by the British Iron Trade Association To Enquire into the Iron, Steel and Allied Industries of the United States (London, 1902), esp. pp. 255-59.

24. *Survey of Metal Industries*, p. 172.

25. H.J. Habakkuk, *American and British Technology in the 19th Century* (Cambridge, 1962), p. 203.

26. A good deal of valuable material on Anglo-American differences in engineering methods at the beginning of the 20th century will be found in the paper by H.F.L. Orcutt, "Modern Machine Methods," and the subsequent discussion, in *Proceedings of the Institution of Mechanical Engineers* (1902), pts. 1-2, pp. 9-112. The greater degree of specialization of both American firms and American machinery as compared with their English counterparts received much attention. J.R. Richardson, who operated a general engineering workshop, justified his unwillingness to purchase American molding machines in the following way: "It was not that English engineers did not understand American methods, but that Americans did not as a rule understand the conditions which obtained in large engineering works in England having a big general practice. There must be a large run of work. Even the most enthusiastic Americans had told him that a large quantity was not needed, that it could be done perfectly well with a dozen, but very often a dozen was a large quantity. Not only was it necessary to have on his catalogue 500 different types and sizes of steam-engines, but an infinite variety of mining and general machinery; and in addition his firm was expected to do anything required, and had to do it even if it only had to be done once" (pp. 72-73). The important unanswered - indeed, unposed - question is why the general engineering firm persisted as it did in England (see S.B. Saul, "The American Impact upon British Industry," *Business History 3* [1960]: 21-27).

27. America's superiority in the contriving of highly specialized machinery was clearly recognized as early as the 1850s. A parliamentary committee of distinguished engineers which visited America in 1853 stated: "As regards the class of machinery usually employed by engineers and machine makers, they are upon the whole behind those of England, but in the adaptation of special apparatus to a single operation in almost all branches of industry, the Americans display an amount of ingenuity, combined with undaunted energy, which as a nation we would do well to imitate, if we mean to hold our present position in the great market of the world" (*Report of the Committee on the Machinery of the U.S.*, as reprinted in Rosenberg, ed., *The American System of Manufactures*, pp. 128-29).

28. As quoted in Habakkuk, p. 203.
29. Freeman, p. 49. Earlier, Freeman had stated: "We found in our case studies that a number of successful process innovations arose from intimate technical cooperation between chemical firms and contractors or chemical firms and component makers, or all three. This is particularly evident in the United States; much less so in Europe" (p. 48). In an article on the plastics industry, Freeman states, in connection with the machinery for making plastics: "Generally speaking a technically advanced and progressive machine industry will principally benefit the country in which it is located, because material-suppliers, machine-makers and fabricators can more easily co-operate there in experiment, development and design. This tripartite co-operation appears to be closer and more satisfactory in Germany than in Britain" (C. Freeman, "The Plastics Industry: A Comparative Study of Research and Innovation," *National Institute Economic Review* [November 1963], pp. 42-43; see also p. 46).
30. C. Freeman, "Research and Development in Electronic Capital Goods," *National Institute Economic Review* (November 1965), p. 63.
31. Paul Strassmann, *Technological Change and Economic Development* (Ithaca, N.Y., 1968), p. 168.
32. Ibid., p. 274.
33. David Granick, *Soviet Metal-fabricating and Economic Development* (Madison, Wis., 1967), chap. 5.
34. Strassmann makes the interesting point that "face-to-face contact to establish trust appears to be a need throughout the technical transfer network. Lack of full trust invariably means some concealment of information which impairs perception of needs at subsequent stations. Moreover, a sense of insecurity may blur the transmitter's vision as well, making him incapable of stating his problems, even on paper, and of reaching decisions. To see faces is to see confusion, satisfaction, or antagonism; to gauge whether status is being, or should be, granted or withheld. The conversational setting from time to time allows the release of tensions through small talk, jokes, and laughter. The urge to conceal as well as the fear of concealment weakens" (Strassmann, pp. 32-33). For a highly suggestive treatment of the spatial diffusion of innovations which turns, in large measure, on the role of person-to-person contacts and personal communications between pairs of individuals, see Torsten Hagerstrand, "Quantitative Techniques for Analysis of the Spread of Information and Technology," in *Education and Economic Development,* ed. C. Arnold Anderson and Mary Jean Bowman (Chicago, 1965), chap. 12.
35. Karl Marx, *Capital* (New York, 1936), p. 417.
36. See Nathan Rosenberg, "Capital Goods, Technology and Economic Growth," *Oxford Economic Papers,* n.s. 15, no. 3 (November 1963): 217-27.
37. In the shaping of technology, many more kinds of adjustments are involved than the factor-proportions adjustment, with which economists are typically preoccupied. It may be possible, for example, to incorporate certain kinds of features into a machine which will compensate for (or simply bypass) deficiencies on the part of the labor force. If workers cannot be relied upon to perform appropriate maintenance procedures on expensive machinery, machines can be designed which *require* less maintenance. This occurred in the United States early in the 20th century when it was found that immigrant labor could not be relied upon to attend to the numerous separate lubrication points of their machinery, thus causing frequent breakdowns. This problem was solved by redesigning machine tools to incorporate the centralized, self-acting lubrication system of the automobile. Since this system went into operation automatically when the machine was activated, reliance upon the worker was, in this respect, bypassed (S. Einstein, "Machine-Tool Milestones, Past and Future," *Mechanical Engineering* 52, no. 11 [November 1930]: 961).
38. As Vernon Ruttan has pointed out: "The ratio of purchased inputs to total output has risen steadily over the last several decades in American agriculture. The farm supply industries have played an important role in channeling technological advances into agriculture. Adoption of new technology has usually increased agriculture's dependence on inputs produced in the nonfarm sectors of the economy and has lessened

agriculture's dependence upon land inputs" (Vernon Ruttan, "Research on the Economics of Technological Change in American Agriculture," *Journal of Farm Economics* 42, No. 4 [November 1960]: 740).

39. For an absorbing account of the Soviet Union's heavy reliance upon American technical personnel in her attempt to achieve a rapid mechanization of agriculture, see Dana G. Dalrymple, "The American Tractor Comes to Soviet Agriculture: The Transfer of a Technology," *Technology and Culture* 6, no. 2 (Spring 1964): 191-214. The nature and extent of early Soviet dependence upon the importation of foreign skills is explored in sector-by-sector detail in Anthony C. Sutton, *Western Technology and Soviet Economic Development, 1917 to 1930* (Stanford, Calif., 1968).

40. On the other hand, the "feedback" role of extension services may also have an important effect upon the productivity of agricultural research. Colleges of agriculture and agricultural experiment stations have frequently suffered from the absence of direct extension-work responsibilities because this has isolated them from a valuable flow of feedback originating at the farm level. (I am grateful to Professor Vernon Ruttan for this observation.)

41. Nor should one ignore social changes which can accelerate the rate of diffusion of existing "best-practice" techniques. Thomas C. Smith (*Agrarian Origins of Modern Japan* [Stanford, Calif., 1959], esp. chap. 7) has pointed to the numerous improvements in farming methods which had been developed under the Tokugawa Shogunate. However, the spread of these techniques had been seriously inhibited by feudal restrictions upon travel and communication. One of the important achievements of the Meiji Restoration was the elimination of barriers to the spread of the best techniques. Consequently, agricultural productivity was raised by narrowing the gap between "best-practice" and "average-practice" techniques. See also Saburo Yamada, "Changes in Output and in Conventional and Nonconventional Inputs in Japanese Agriculture since 1880," *Food Research Institute Studies* 7, no. 3 (1967): 371-413.

42. "Organization for research and development in the farm supply industries varies widely. At one extreme are such industries as hybrid seed corn, where a substantial share of research and development has been conducted at USDA and land-grant college experiment stations. At the other extreme is the farm equipment industry where public contributions have apparently been considerably smaller relative to the contributions by private industry. A third variation is represented by the fertilizer industry where the TVA has provided what is in effect an industry research institute for the fertilizer industry with a research program ranging from fundamental chemical and biological research to applied engineering and economics" (Ruttan, p. 741).

43. Similarly, the identification of obstacles to economic development is a hazardous enterprise. For a valuable and provocative discussion, see Albert O. Hirschman, "Obstacles to Development: A Classification and a Quasi-vanishing Act," *Economic Development and Cultural Change* 13, no. 4, pt. 1 (July 1965): 385-93.

10. Selection and adaptation in the transfer of technology: steam and iron in America, 1800-1870

1. Not, however, by the most astute practitioners of the historian's craft. See, for example, Marc Bloch's admirable "Avenement et conquetes du moulin à eau," *Annales d'histoire économique et sociale,* vii, 1935, pp. 538-63. Bloch provides a masterly analysis, turning primarily on changing legal and economic conditions as they affected the availability of servile labor, of the lag of an entire millennium between the invention of the watermill and its widespread adoption.

2. H.W. Dickinson, *A Short History of the Steam Engine,* London, 1938, p. 89. For evidence on the limited impact of Watt's steam engine on the broad range of British industries by 1800, a full quarter century after first practical success had been achieved, see John Lord, *Capital and Steam Power,* London, 1932, chapter 8. However, since Boulton and Watt's engine was widely pirated and Newcomen engines remained in use, Lord's figures should not be taken as representing the total amount of steam power in

use in Great Britain. See A.E. Musson and E. Robinson, "The Early Growth of Steam Power," *Economic History Review,* 1959 pp. 418-439, and J.R. Harris, "The Employment of Steam Power in the Eighteenth Century," *History,* 1967, pp. 133-148. On the later diffusion of steam power in Britain, Habakkuk has stated that ". . . steam did not begin to play an important part in powering the British economy until the 1830's and 40's, and was not massively applied until the 1870's and 80's. Even as late as 1870 less than a million horse-power was generated by steam in the factories and workshops of Great Britain." H.J. Habakkuk, *American and British Technology in the 19th Century,* Cambridge, 1962, pp. 184-85. Habakkuk's statement is presumably based upon M.G. Mulhall, *The Dictionary of Statistics,* London, 1892, p. 545. Mulhall's figures, although of questionable reliability, are still, surprisingly, the only ones available. This is a classic example of what I have in mind about the neglect of the diffusion process. We know a great deal about the purely *technical* sequence of events concerning the development and subsequent improvement of the steam engine. We know exceedingly little in detail concerning the industrial diffusion of the steam engine in England in the nineteenth century.

3. Even in cotton textiles, the most fully mechanized of the textile branches, full mechanization was achieved only in the second quarter of the nineteenth century. The protracted agony of the hand-loom weavers was being acted out in the 1830's and 1840's as the improved power loom was widely adopted, and it was only in the latter decade that the number of power loom weavers exceeded the number of hand-loom weavers. Even so, there were an estimated 40,000 cotton hand-looms still at work in 1850. In woolen textiles, which was slower to mechanize, hand-loom weavers persisted in and around Leeds, the most highly mechanized woolen center, during the 1850's. J.H. Clapham, *An Economic History of Modern Britain,* vol. I, Cambridge University Press, 1962, chapter XIV; Phyllis Deane and W.A. Cole, *British Economic Growth 1688-1959,* Cambridge University Press, 1962. See also Maurice Dobb, *Studies in the Development of Capitalism,* London, 1946, pp. 263-65.

4. Until well into the nineteenth century, the British government had laws restricting the export of certain kinds of machinery and the emigration of certain kinds of skilled labor.

5. Allen H. Fenichel, "Growth and Diffusion of Power in Manufacturing, 1839-1919" in Dorothy Brady (ed.), *Output, Employment and Productivity in the United States after 1800,* National Bureau of Economic Research, Studies in Income and Wealth no. 30, New York, 1966, Appendix B. For 1879 the proportions were: steam power 64.1% and water power 35.9%. In 1889 steam power had risen to 78.4% and water power had declined to 21.5%. Fenichel also reports that "steam-power capacity grew from 36,100 horse-power in 1839 to 1,216,000 horse-power in 1869 to 13,840,000 horsepower in 1919; and water-power capacity grew from 1,130,000 to 1,765,000 horsepower between 1869 and 1919." *Ibid.,* p. 444.

It is also worth pointing out that the mechanization of agriculture in the U.S. took place with no major changes in traditional power sources. Most of the new machines which transformed farm practices - the reaper, cultivator, steel plow - relied primarily upon the power of the horse and, to a lesser extent, the mule and the ox. There was indeed a substitution of animal power for manpower, and of the horse for the ox, which moved too slowly, but steam power had only a limited application and it was not until the introduction of the internal combustion tractor in the second decade of the 20th century that a major new power source was introduced into agriculture. The draft animal population on U.S. farms reached its all-time peak in the 1920's and declined sharply thereafter. For the role of steam in American agriculture see Reynold M. Wik, *Steam Power on the American Farm,* Philadelphia, 1953.

6. The impact of transportation improvements is, of course, a complex issue. Reductions in transport costs also expand the markets available to commodities which are produced by water power, and may thereby retard the introduction of steam.

7. U.S. Congress, House, *Report on the Steam Engines in the United States,* H. Doc. No. 21, 25th Congress, 3d Session, 1839, p. 376. The data in this report have recently been exploited by Peter Temin, "Steam and Waterpower in the Early Nineteenth Century,"

Journal of Economic History, June 1966, pp. 187-2 Q5. See also Carroll Pursell, Jr., *Early Stationary Steam Engines in America,* Washington, D.C., 1969. It is possible that, at least in the earlier years, the oft-cited greater reliability of steam power over water power was not so decisive as is commonly supposed. Although the supply of water power was likely to be erratic due to seasonal variations resulting from drought, frost or flood, the early steam engines were also subject to frequent breakdowns and extensive maintenance requirements which may not have been readily catered for.

8. Temin, *op. cit.,* pp. 188-189. High pressure steam engines were cheaper to construct than low pressure engines but were profligate in their utilization of fuel. These characteristics made such engines attractive in the resource-abundant environment of the U.S. where it was worthwhile, in effect, to "trade off" relatively large amounts of natural resource inputs for a reduction in fixed capital costs.
9. Victor S. Clark, *History of Manufactures in the United States,* 1929, New York, 3 vols., vol. I, p. 406.
10. *Ibid.,* pp. 406-8. France's contribution to improving the technology of hydraulic engines - particularly the pathbreaking work of Poncelet and Fourneyron - should be seen against the background of her relative poverty of coal supplies.
11. Fenichel, *op. cit.,* p. 456.
12. At least it did after the Louisiana Purchase in 1803. An important motive in consummating that substantial real estate transaction was the fact that the only feasible route for the produce of the Mississippi Valley at the time lay down river by way of New Orleans. It is interesting to note that Robert Fulton's business partner was none other than Robert Livingston, Thomas Jefferson's minister to France, who had earlier negotiated the Louisiana Purchase.
13. After the successful maiden voyage of the Clermont, the *American Citizen,* a New York newspaper, wrote of "Mr. Fulton's Ingenious Steam-Boat, invented with a view to the navigation of the Mississippi from New Orleans upwards . . ." Fulton himself stated, at the same time, that the steamboat ". . . will give a cheap and quick conveyance to the merchandise on the Mississippi, Missouri, and other great rivers, which are now laying open their treasures to the enterprise of our countrymen." As quoted in Louis Hunter, *Steamboats on the Western Rivers,* Cambridge, Mass. 1949, p. 8.
14. *Ibid.,* p. 34.
15. *Report on Steam Engines, op. cit.,* p. 10. The figures, mostly reported but partly estimated, were:

Steamboats	57,019 horsepower
Railroads	6,980
Others	36,319
Total	100,318 horsepower

16. Hunter, *op. cit.,* pp. 122-23.
17. The story of this transformation is beautifully told in Hunter, *op. cit.,* chapter 2.
18. *Ibid.,* pp. 130-33.
19. Albert Fishlow, *American Railroads and the Transformation of the Ante-Bellum Economy,* Cambridge, Mass., 1965, p. 149. Fishlow also reports that "For other rolling stock there is no evidence of reliance upon foreign products."
20. Fishlow, *op. cit.,* pp. 137-38.
21. Peter Temin, *Iron and Steel in 19th Century America,* M.I.T. Press, Cambridge, Mass., 1964, Appendix C. Table C6. By the late 1870's the rails were mostly steel rails.
22. One of the serious problems in adapting the locomotive to the American scene was to design engines which would perform effectively on the short curves prevalent on American railroads. Baldwin's solution was the flexible beam truck, for which he received a patent in 1842. *History of the Baldwin Locomotive Works 1832-1913,* no author, no date, pp. 30-34.
23. In 1867-8 Abram Hewitt said that "In the Welsh iron works . . . it was humiliating to find that the vilest trash which would be dignified by the name of iron went universally by the name of the American rail." Though the Welsh ironmasters greatly preferred to turn out good work, their American customers had the "stupidity and reckless

extravagance" to insist on buying at the lowest prices, irrespective of quality. (British railways at that time usually bought under a guarantee of seven years life for which they paid a substantial premium). J. C. Carr and W. Taplin, *History of the British Steel Industry*, Cambridge, 1962, p. 60, fn. 3.

24. Albert Fishlow, "Productivity and Technological Change in the Railroad Sector, 1840-1910," in Dorothy Brady (ed.), *Output, Employment, and Productivity in the United States after 1800*, Studies in Income and Wealth No. 30, New York, 1966, p. 619.

25. The engineer was Major George Washington Whistler, father of the artist. Whistler also constructed locomotives and their equipment in a factory which had been established for that purpose. See Albert Parry, *Whistler's Father*, Bobbs-Merrill Company, Indianapolis, 1939. "How smart this foreign gentleman," the awed mouzhiks are reported to have exclaimed, "who has harnessed the samovar and made it run." *Ibid.* p. xvi.

The Baldwin Locomotive Works in Philadelphia filled orders for large numbers of locomotives for Russian railways in the 1870's and 1890's. *History of the Baldwin Locomotive Works, op. cit.,* pp. 67-68, 73, 85.

26. See Temin, *op. cit.,* chapters 1 and 5. It is interesting to note that, on the European continent as well, mineral fuel was introduced more rapidly at the refining than at the smelting stage. Landes points out: "In contrast to Britain, where puddling came more than half a century after coke smelting, the continental countries learned to refine with coal first. This is the normal sequence technologically: the fuel and ore economies are greater in refining; the absence of direct contact between fuel and metal excludes some of the most serious difficulties associated with the chemical composition of the materials employed; and the initial cost of the shift to coal in refining is much less than in smelting." David Landes, *The Unbound Prometheus*, Cambridge, 1969, pp. 175-76.

27. T.S. Ashton, *Iron and Steel in the Industrial Revolution*, Manchester, 1924, chapter 2 (reprinted in 1963 with a valuable bibliographical note by W.H. Chaloner); M. W. Flinn, "Abraham Darby and the coke-smelting process," *Economica*, February 1959, p.. 54-59).

28. Ashton states: "Already in 1788 there were 59 coke furnaces in blast, as against 26 charcoal furnaces, but by 1806 the coke furnaces numbered 162, and only 11 charcoal furnaces were to be found in the whole of Great Britain." Ashton, *op. cit.,* p. 99.

29. Temin, *op. cit.,* p. 82 and Appendix C, Table C.3.

30. Ashton, *op. cit.,* pp. 33-34.

31. Only later was it established that the chemical composition of the iron exerted a decisive effect upon the quality of the final product. Indeed, iron containing more than 0.1% phosphorus was unsuitable for the original (acid) Bessemer process. The introduction of the basic lining by the two English chemists Gilchrist and Thomas, in the late 1870's, made it possible to use phosphoric ores. The basic Bessemer process, however, in turn *required* phosphoric iron and could not be used unless the iron contained *more* than 1.5 per cent phosphorus. Since the U.S. contains only very limited amounts of ores which are suitable for exploitation by the basic Bessemer process, the technique was never employed there on a large scale.

32. See Walter Isard, "Some Locational Factors in the Iron and Steel Industry Since the Early 19th Century," *Journal of Political Economy*, June 1948, pp. 203-217.

33. In this age of the jet airplane and high-speed turnpike travel it takes perhaps a stretch of the imagination to recall that, until the construction of the railroads in the 1850's, the Alleghenies constituted an almost impenetrable barrier to the movement of bulky freight between Pittsburgh and Philadelphia.

34. The greater resistance of anthracite to combustion may also have been responsible for the delay in the introduction of the important practice of "hard driving" into the U.S. See Temin, *op. cit.,* pp. 157-163.

35. *Ibid.,* pp. 57-62.

36. *Ibid.,* pp. 57-59.

37. *Ibid.,* Appendix C, Table C.3.

38. *Loc. cit.*

39. In Germany there was a similar delay due to the comparatively late discovery of the richness of the coal deposits of the Ruhr.

40. For some evidence on the supply of skilled labor in the U.S. in the 1820's see Nathan Rosenberg, "Anglo-American Wage Differences in the 1820's," *Journal of Economic History*, June 1967, pp. 221-229.

41. See Richard A. Easterlin, "A Note on the Evidence of History" in C. Arnold Anderson and Mary Jean Bowman (eds.), *Education and Economic Development*, Aldine Publishing Company, 1965, especially Table 1, p. 426. The qualifications in the last sentence of the text seem important. Not all education is equally conducive to technological innovation; indeed, it may be plausibly argued that certain kinds of education have probably retarded this process. Moreover, Britain herself did not rank terribly high in the first half of the 19th century in a ranking of countries by percentage of the total population enrolled in school (Easterlin, *loc. cit.*). Yet it was Britain which *initiated* the new technology - a technology which was, moreover, rapidly diffused within that country. Acquisition of mechanical skills was, to a very large degree, an on-the-job process. The precise role of the sort of skills which are acquired through formal education is still not clearly understood.

42. H. W. Dickinson, *op. cit.*, p. 90. Their use continued primarily at the coal pits where wastefulness of coal was less important. For an early, authoritative treatment of the subject of steam engines, see J. Farey, *Treatise on Steam Engines*, 1827.

43. Hunter, *op. cit.*, p. 55.

44. The number of windmills in Kent employed in the grinding of corn grew from 95 to 239 between 1769 and the 1840's. William Coles Finch, *Watermills and Windmills*, London, 1933, pp. 136-37.

45. Although total U.S. tonnage of sailing vessels peaked in 1861, steam tonnage did not exceed sail tonnage until 1893. *Historical Statistics of the U.S., op. cit.*, pp. 444-45. In Britain, sailing tonnage peaked in 1865 but was not exceeded by steam tonnage until 1883. B. R. Mitchell, *Abstract of British Historical Statistics*, Cambridge, 1962, p. 218. Of course it must be remembered that a ton of steamship did far more work than a ton of sail-boat. The point is that there was no Schumpeterian "gale of creative destruction." For an interesting appraisal of Schumpeter's thesis, as expounded in his book *Capitalism, Socialism and Democracy*, see W. Paul Strassmann, "Creative Destruction and Partial Obsolescence in American Economic Development," *Journal of Economic History*, September 1959, pp. 335-349.

11. Factors affecting the diffusion of technology

1. For an admirable earlier study, see Marc Bloch's "Avenement et conquetes du moulin a eau," *Annales d'histoire economique et sociale*, 7 (1935), 538-563. Bloch provides a masterly analysis, turning primarily on changing legal and economic conditions as they affected the availability of servile labor, of the lag of an entire millennium between the invention of the watermill and its widespread adoption.

2. H. W. Dickinson, *A Short History of the Steam Engine* (Cambridge: Cambridge University Press, 1938), p. 89.

3. Even in cotton textiles, the most fully mechanized of the textile branches, full mechanization was achieved only in the second quarter of the nineteenth century. The protracted agony of the hand-loom weavers was being acted out in the 1830s and 1840s as the improved power loom was widely adopted, and it was only in the latter decade that the number of power-loom weavers exceeded the number of hand-loom weavers - although the proportion of *output* accounted for by the hand-loom weavers was, of course, far smaller than the proportion which they constituted of the weaving labor force. Even so, there were an estimated 40,000 cotton hand-looms still at work in 1850. In woolen textiles, which was slower to mechanize, hand-loom weavers persisted in and around Leeds, the most highly mechanized woolen center, during the 1850s. J. H. Clapham, *An Economic History of Modern Britain*, (Cambridge: Cambridge University Press, 1962), I Ch. XIV; Phyllis Deane and W. A. Cole, *British Economic Growth*

1688-1959 (Cambridge: Cambridge University Press, 1962), pp. 191-192. See also Maurice Dobb, *Studies in the Development of Capitalism* (London: Routledge and Kegan Paul, 1946), pp. 263-265.

4. For evidence on the limited impact of Watt's steam engine on the broad range of British industries by 1800, a full quarter century after first practical success had been achieved, see John Lord, *Capital and Steam Power* (London: P. S. King and Son, 1923), Ch. 8. However, since Boulton and Watt's engine was widely pirated and Newcomen engines remained in use, Lord's figures should not be taken as representing the total amount of steam power in use in Great Britain. See A. E. Musson and E. Robinson, "The Early Growth of Steam Power," *Economic History Review,* 2nd series, 9 (April 1959), 418-439, and J. R. Harris, "The Employment of Steam Power in the 18th Century," *History,* 52 (1967), 133-148.

5. H. J. Habakkuk, *American and British Technology in the 19th Century* (Cambridge: Cambridge University Press, 1962), pp. 184-185. See also A. E. Musson and Eric Robinson, *Science and Technology in the Industrial Revolution* (Manchester: University of Manchester Press, 1969), p. 72, and David Landes, *The Unbound Prometheus* (Cambridge: Cambridge University Press, 1969), p. 221.

6. M. G. Mulhall, *The Dictionary of Statistics* (London: G. Routledge and Sons, Ltd., 1892), p. 545.

7. See, for example, E. Mansfield, "Technical Change and the Rate of Imitation," *Econometrica,* 29 (October 1961,), 741-766.

8. J. R. Harris, "Employment of Steam Power," p. 147.

9. John L. Enos, "Invention and Innovation in the Petroleum Refining Industry," in *The Rate and Direction of Inventive Activity* (Princeton: University Press, 1962), p. 309. Most of Enos's information is drawn from J. Jewkes, D. Sawers, and R. Stillerman, *The Sources of Invention* (London: St. Martin's Press, 1958).

10. *Ibid.,* p. 308.

11. Jewkes et al., *Sources,* p. 283.

12. Enos, *Inventive Activity,* p. 307.

13. John Nef, "The Progress of Technology and the Growth of Large-Scale Industry in Great Britain, 1540-1640," *Economic History Review,* 5 (1934), 15-18.

14. Beer made of malt that had been dried by raw coal was, apparently, practically undrinkable.

15. T. S. Ashton, *Iron and Steel in the Industrial Revolution* (Manchester: University of Manchester Press, 1924), p. 35. Furthermore, as Ashton points out: "It seems very probable that the production of sound coke-iron was not a sudden creation but the result of many trials in which failure and success alternated. The inventory taken at Coalbrookdale in 1718 shows that there was an accumulation of 'sculls,' or defective iron, which were sold at a low price to a neighboring forgemaster. Such sculls may have been produced frequently in the early days of the process, and though it is beyond question that marketable iron was produced every year, it might have been difficult for Darby himself to say exactly when the problem had reached a final solution" (p. 33).

16. The frequent inferiority and, therefore, slow diffusion of new inventions in their early stages of development is strikingly apparent in military history. The English long-bow swept the field at Agincourt in 1415, long after the first introduction of gunpowder and cannon into Europe. Although Europeans sought to develop an effective field artillery in the fifteenth century, they did not succeed in overcoming the technical problems involved until the first half of the seventeenth century. Until that time such weapons were of very limited effectiveness, aside from their use in sieges. Their limited mobility and slow rate of fire enabled them to be easily overcome by massed charges. See C. Cipolla, *Guns, Sails and Empires* (New York: Random House, 1965), Ch. 1 and Epilogue.

The Texas Rangers, in spite of all the guns and powder in their armories, could not establish a decisive superiority over the fierce Comanches under the special circumstances of mounted combat until the availability of the six shooter. For an absorbing account, see W. P. Webb, *The Great Plains* (Boston: Ginn and Co., 1931), pp. 167-179 Before the availability of multi-shot weapons, Webb concludes that the Indians were at a distinct advantage. "In the first place, the Texans carried at most three shots; the Comanche carried twoscore or more arrows. It took the Texan a minute to reload his weapon; the Indian could in that time ride three hundred yards and discharge twenty arrows. The Texan had to dismount in order to use his rifle effectively at all, and it was his most reliable weapon; the Indian remained mounted throughout the combat. Apparently the one advantage possessed by the white man was a weapon of longer range and more deadly accuracy than the Indian's bow, but the agility of the Indian and the rapidity of his movements did much to offset this advantage" (p. 169).

17. This is not to suggest that a continuous rate of improvement in a technology implies some continuous rate of its diffusion. When a technology is still at a very primitive stage in its development, even substantial reductions in cost may have little effect upon diffusion. On the other hand as a new technology reaches the cost levels of competing methods, relatively small additional cost reductions may bring it below critical threshold levels and thus lead to rapidly accelerating rates of diffusion. For a rigorous application of a threshold function in the study of diffusion, see Paul David, "The Mechanization of Reaping in the Ante-Bellum Midwest," in *Industrialization in Two Systems,* ed., H. Rosovsky (New York: John Wiley and Sons, 1966), pp. 3-28. David's threshold function, it should be noted, relates to farm size, and he demonstrates how the rising relative cost of harvest labor lowered that farm size, leading finally to the rapid introduction of the reaper in the Midwest during the 1850s.

18. James Mak and Gary Walton, "Steamboats and the Great Productivity Surge in River Transportation," *Journal of Economic History,* 32 (September 1972), p. 625. They add: "We qualify this conclusion to the extent that the productivity gains from steam power, which are reflected in the fall in rates, 1815-20, may have been understated somewhat because of slow entry or limited competition in this early period. Nevertheless, there were 17 vessels in operation on Western rivers in 1817, and 69 were in operation by 1820. It seems reasonable to assume that the initial impact of the steam engine had occurred by 1820" (footnote 15, p. 625).

19. A cumulation of minor design changes on the steamboat had the effect of substantially increasing the length of the navigation season for each steamboat size class. By steadily reducing the draft in relation to tonnage and cargo carrying capacity, steamboat designers and builders brought about major improvements in the productivity of capital by enabling steamboats to operate a longer portion of the year. Thus, Mak and Walton state that, as a rough average, "The navigation season was extended from approximately six months, before 1830, to about nine months, during the last half of the ante-bellum period" (Mak and Walton, "Steamboats," p. 634). See also Louis Hunter, *Steamboats on the Western Rivers* (Cambridge: Harvard University Press, 1949), Ch. 2 and pp. 219-225.

20. John L. Enos, "A Measure of the Rate of Technological Progress in the Petroleum Refining Industry," *Journal of Industrial Economics,* 6 (June 1958), 180. See also the same author's "Invention and Innovation in the Petroleum Refining Industry," in *The Rate and Direction of Inventive Activity* (Princeton: Princeton University Press, 1962), pp. 299-321.

21. This limited conception of the potential of the radio in turn slowed the pace at which it was developed because, as a result, "most of the original research on the development of wireless communication was initially oriented toward the relatively simple task of transmitting impulses for telegraphic communications." Frank Lynn, *An Investigation of the Rate of Development and Diffusion of Technology in Our Modern Industrial Society,* in *The Employment Impact of Technological Change,* 6 vols. (Washington, D.C.: U.S.G.P.O., 1966), II, p. 68.

22. In a closely related context, Simon Kuznets has pointed to the almost congenital pessimism of professional judgments on the possibilities for technological change over the years. "Experts are usually specialists skilled in, and hence bound to, traditional views; and they are, because of their knowledge of one field, likely to be cautious and unduly conservative. Hertz, a great physicist, denied the practical importance of shortwaves, and others at the end of the nineteenth century reached the conclusion that little more could be done on the structure of matter. Malthus, Ricardo, and Marx, great economists, made incorrect prognoses of technological change at the very time that the

scientific bases for these changes were evolving. On the other hand, imaginative tyros like Jules Verne and H. G. Wells seemed to sense the potentialities of technological change. It is well to take cognizance of this consistently conservative bias of experts in evaluating the hypothesis of an unlimited effective increase in the stock of knowledge and in the corresponding potential of economic growth" (Simon Kuznets, *Economic Growth and Structure* [New York: W. W. Norton and Co., 1965], p. 89).

23. Marx saw this point clearly: "To what an extent the old forms of the instruments of production influenced their new forms at first starting, is shown by, amongst other things, the most superficial comparison of the present powerloom with the old one, of the modern blowing apparatus of a blast-furnace with the first inefficient mechanical reproduction of the ordinary bellows, and perhaps more strikingly than in any other way, by the attempts before the invention of the present locomotive, to construct a locomotive that actually had two feet, which after the fashion of a horse, it raised alternately from the ground. It is only after considerable development of the science of mechanics, and accumulated practical experience, that the form of a machine becomes settled entirely in accordance with mechanical principles, and emancipated from the traditional form of the tool that gave rise to it" (Karl Marx, *Capital* [New York: Modern Library Edition, n.d.], p. 418). Similarly on water, abortive attempts were made to imitate nature. One such attempt - Lord Stanhope's ill-fated paddle steamer - has been preserved in the sad lines of the poet, T. Baker (fl. 1837-1857):

> Lord Stanhope hit upon a novel plan
> Of bringing forth this vast Leviathan
> (This notion first Genevois' genius struck);
> His frame was made to emulate the duck;
>
> Webb'd feet had he, in Ocean's brine to play;
> With whale-like might he whirl'd aloft the spray;
> But made with all this splash but little speed;
> Alas! the duck was doom'd not to succeed!

The Steam-Engine, Canto IV

As reprinted in D. B. Wyndham Lewis and Charles Lee, *The Stuffed Owl: An Anthology of Bad Verse* (London: J. M. Dent and Sons 1952), pp. 193-194.

24. In this sense, the *sequence* of events in history becomes very important in explaining the experiences of individual industries. The problems encountered and solved in industry A often turn out to provide valuable externalities in the form of knowledge, techniques and labor skills which become available to industries B, C, and D. Thus, Usher has argued that the industrial revolution in England owed much to the technical skills which had been developed by generations of craftsmen in the production of clocks and watches. Machines which had been developed in this trade "stand out as the most conspicuous examples of instruments of precision. The lessons learned by the craftsmen of these trades formed the basis for the development of the engineering sciences in the late eighteenth century and the early nineteenth century. These timekeepers presented a substantial array of notable devices for the control of motion. These devices involved all the primary problems of geared mechanisms. The marine chronometer required delicate adjustments to the expansion and contraction of metals during small changes of temperature. The pendulum clocks presented important problems in the theory of dynamics. The development of the pendulum clocks rested upon a complete mathematical treatment of the forces operating in a pendulum. The escapements of both clocks and watches called for considerable refinements in the design of gear teeth, and the problems received full mathematical treatment in the course of the eighteenth century. Much of the work done for Arkwright on the spinning machine was entrusted to a clockmaker. George Stephenson learned much of his mechanics by repairing and studying clocks. The rapid development of the engineering sciences after Watt's inventions was largely due to the extensive mathematical treatment of the problems of dynamics involved in the construction of these small instruments of precision" (W. Bowden, M. Karpovich and A. P. Usher, *An Economic History of Europe Since 1750* [New York: American Book Co., 1937], p. 308). See also A. E. Musson and E.

Robinson, "The Origins of Engineering in Lancashire," *Journal of Economic History,* 20 (June 1960), especially 219-222. Musson and Robinson conclude "that clock-, if not watch-, makers, and above all clock-tool makers, were in very great demand for textile-machine making and contributed materially to the early growth of engineering" (p. 222).

25. A. Alchian, "Reliability of Progress Curves in Airframe Production," *Econometrica,* 31 (October 1963), 679 692; Werner Hirsch, "Firm Progress Ratios," *Econometrica,* 24 (April 1956), 136-143; Leonard Rapping, "Learning and World War II Production Functions," *Review of Economics and Statistics,* 47 (1965), 81-86; Paul David, "Learning by Doing and Tariff Protection: A Reconsideration of the Case of the Ante-Bellum U.S. Cotton Textile Industry," *Journal of Economic History,* 30, (September 1970), 521-601; Kenneth Arrow, "The Economic Implications of Learning by Doing," *Review of Economic Studies,* 29 (June 1962), 155-173. According to Hirsch, the U.S. Air Force "for quite some time had recognized that the direct labor input per airframe declined substantially as cumulative airframe output went up. The Stanford Research Institute and the RAND Corporation initiated extensive studies in the late forties, and the early conclusions were that, in so far as World War II airframe data were concerned, doubling cumulative airframe output was accompanied by an average reduction in direct labor requirements of about 20%. This meant that the average labor requirement after doubling quantities of output was about 80% of what it had been before. Soon the aircraft industry began talking about the 'eighty per cent curve' " (Hirsch, p. 136). It is possible, of course, that cost reductions which have been attributed to learning by doing have actually been due to other factors which have not been correctly identified, especially in cases where learning by doing has been defined as a residual. See John Chipman, "Induced Technical Change and Patterns of International Trade," in *The Technology Factor in International Trade,* ed. Raymond Vernon (New York: National Bureau of Economic Research, 1970), pp. 95-98.

26. Charles K. Hyde, "Technological Change in the British Iron Industry, 1700-1860," unpublished doctoral dissertation, University of Wisconsin, pp. 112-113. See also Landes, p. 92.

27. Loss of life was especially fearful aboard steamboats on western rivers, and this loss of life led to an early assertion of the investigatory and regulatory activities of the federal government. See John G. Burke, "Bursting Boilers and Federal Power," *Technology and Culture,* 7 (Winter 1966).

28. Nathan Rosenberg, "Economic Development and the Transfer of Technology: Some Historical Perspectives," *Technology and Culture,* 11 (October 1970), 550-575.

29. K. R. Gilbert, "Machine Tools," in Charles Singer et al., *A History of Technology* (London: Oxford University Press, 1958). IV, p. 421. See also S. Smiles, *Industrial Biography* (London: John Murray 1908), pp. 178-182. As Landes points out about the separate condenser, Watt "saved the energy that had previously been dissipated in reheating the cylinder at each stroke. This was the decisive breakthrough to an 'age of steam' not only because of the immediate economy of fuel (consumption per output was about a fourth that of the Newcomen machine) but even more because this improvement opened the way to continuing advances in efficiency that eventually brought the steam-engine within reach of all branches of the economy and made of it a universal prime mover" (Landes, *Unbound Prometheus,* p. 102).

30. At a much earlier date, the cannon makers had borrowed important techniques, especially that of casting in bronze, from the makers of church bells. Such are the vagaries - and ironies - of the industrial learning process. Cipolla, pp. 23, 25.

31. S. H. Hollingdale and G. C. Tootill, *Electronic Computers* (London: Pelican, 1965), p. 46 and Ch. 2 and 3. Babbage, it is interesting to note, had borrowed the system of punched cards from the Jacquard loom - where they had been used to control the introduction of threads in weaving brocade. See Charles Babbage, *Passages from the Life of a Philosopher* (London: Longman, Green, Longman, Roberts, and Green, 1864), pp. 116-118.

32. Nathan Rosenberg, "Technological Change in the Machine Tool Industry, 1840-1910," *Journal of Economic History,* 23 (December 1963).

33. A. Fishlow, "Productivity and Technological Change in the Railroad Sector,

1840-1910," in Studies in Income and Wealth, no. 30, *Output, Employment and Productivity in the United States after 1800* (New York: National Bureau of Economic Research 1966), pp. 635,641.

34. As Fishlow points out, the main advantage of the air brake and automatic coupler was increased speed. "Greater speed in itself is not an unmixed blessing, however. Unless engine capacity is not being fully utilized, higher speeds can be attained only by the sacrifice of load. What the air brake and automatic coupler really did, therefore, was to allow a greater element of choice in train operation, permitting higher speed when it was more desirable than larger loads." *Ibid.*, p. 636.

35. *Ibid.*, pp. 635, 639-640.

36. "Report on the Commissioners Appointed to Inquire into the Application of Iron to Railway Structures," *Parliamentary Papers*, 1849, Vol. 29; William Fairbairn, *Useful Information for Engineers*, London, 1860, vol. II, pp. 223-228.

37. The contemporary experience with the introduction of new, high-yielding rice and wheat varieties in Southeast Asia forcefully exemplifies the argument advanced here. It is currently being discovered that the adoption of the high-yielding rice varieties generates a whole new series of requirements with respect to fertilizer use, water management, harvesting, processing (drying, storing, milling, grading), and disease and insect control. Although many of these problems can be dealt with by conventional means, others cannot, e.g., double-cropping calls for low-cost harvesting, threshing, and other machinery of a kind not existing ten years ago. Much attention is currently being devoted to the development of such machinery at the International Rice Research Institute at Los Banos, The Philippines, where the new rice varieties were first developed.
The really distinctive feature of the new rice varieties is that they have been genetically designed so that they are highly fertilizer-responsive. Indeed, without large doses of fertilizer the new varieties are no more productive than the old ones. A continued diffusion of the new varieties would seem to call for cost-reducing innovations in the provision of fertilizer and the entire range of complementary inputs. Better still would be the development of new seed varieties not *requiring* these complementary inputs.

38. William Pole, *The Life of Sir William Fairbairn, Bart* (London: Longmans, Green and Co., 1877), and William Fairbairn, *Treatise on Mills and Millwork* (London: Longman; Part I, 1861 and Part II, 1863).

39. R. L. Hills, *Power in the Industrial Revolution* (Manchester: Manchester University Press 1970), Ch. 8. Hills also points out that the waterwheel was long preferred to the steam engine in spinning because "it was essential to have as regular speed as possible. Until some method of automatically controlling the revolutions had been found, the waterwheel with its greater steadiness was preferable" (*Ibid.*, p. 175).

40. Victor Clark, *History of Manufactures in U.S.* (New York: McGraw-Hill, 1929), II, pp. 407-408. Although of French origin, much of the practical development of the turbine was performed in America.

41. Allen H. Fenichel, "Growth and Diffusion of Power in Manufacturing, 1839-1910," in Studies in Income and Wealth no. 30, *Output, Employment and Productivity in the United States after 1800* (New York: National Bureau of Economic Research, 1966), pp. 443-478, Appendix B. In absolute terms waterpower capacity continued to grow through the first decade of the twentieth century (Table A-1).

42. William Coles-Finch, *Watermills and Windmills* (London: C. W. Daniel, 1933), pp. 136-137.

43. W. S. Jevons, *The Coal Question* (London: Macmillan and Co., 1865), p. 136.

44. "The proportion of pig iron made with charcoal declined from close to 100 per cent in 1840 to about 45 per cent in the middle 1850s . . . and to about 25 per cent at the close of the Civil War" (Peter Temin, *Iron and Steel in 19th Century America* [Cambridge: MIT Press, 1964], p. 82).

45. *Ibid.*, Table C-2.

46. Robert Fogel and Stanley Engerman, "A Model for the Explanation of Industrial Expansion During the 19th Century: With an Application to the American Iron

Industry," as reprinted in *The Reinterpretation of American Economic History*, Robert Fogel and Stanley Engerman, (New York: Harper and Row, 1971), pp. 159-162.

47. C. S. Graham, "The Ascendancy of the Sailing Ship. 1850-1885," *Economic History Review*, 2nd series, 9 (August 1956), 81. For a careful examination of the economic impact of technological changes in the steamship, see Charles K. Harley, "The Shift from Sailing Ships to Steamships, 1850-1890: a Study in Technological Change and its Diffusion," in *Essays on a Mature Economy: Britain after 1840*, ed. Donald N. McCloskey (London: Methuen, 1971), pp. 215-231.

48. *Ibid.,* p. 84.

49. S.C. Gilfillan, *Inventing the Ship* (Chicago: Follett Publishing Co., 1935), p. 157. It is not unusual for the "new" technology to extend the life of the "old" by providing it with some form of externality. Thus, the arrival of the steamboat on western rivers brought about significant reductions in labor costs in flatboat operation by providing flatboatmen with a speedy form of upriver transportation - a trip which had previously been both very slow and costly. See E. Haites and J. Mak, "Ohio and Mississippi River Transportation, 1810-1860," *Explorations in Economic History*, 8 (Winter 1970-1971), fn. 36, and Mak and Walton, "Steamboats," p. 19.

50. See Douglass North, "Ocean Freight Rates and Economic Development, 1750-1913," *Journal of Economic History*, 18, (December 1958); Douglass North, *Growth and Welfare in the American Past* (Englewood Cliffs, N.J.: Prentice-Hall, 1966), Ch. 9; and Gary Walton, "Productivity Change in Ocean Shipping After 1870: A Comment," *Journal of Economic History*, 30 (June 1970). North states: "Although the steamship substituted for the sailing ship in passenger travel as early as 1850, it did not substitute for the sailing ship in the carriage of bulk goods in ocean transportation until much later. Indeed, as late as 1880 most of the goods carried in ocean transportation were going by sail, and the changeover from sail to steam did not occur in most of the long-haul routes in the world until the very end of the nineteenth century. . . ." (*Growth and Welfare*, p. 110).

51. The sailing ship adapted to its more specialized role as a long-distance carrier of bulk cargoes by modifying its sails and rigging so as to reduce crew requirements. In part this was done by "abolishing the lightest sails, broadening the upper yards, and cutting in two the largest sails until furling could replace almost all reefing" (S. C. Gilfillan, p. 160). The American merchant fleet was slower than the British in substituting steam for sail. As late as 1913 almost 20% of the gross tonnage of American merchant vessels still consisted of sailing vessels. *Historical Statistics of the U.S.., Colonial Times to 1957* (Washington, D.C.: U.S. Printing Office, 1960), p. 444.

52. North threw out the following challenge several years ago: "I would hazard the speculation that if we ever did the research necessary to get some crude idea of the magnitudes involved, we would discover that improved economic organization was as important as technological change in the development of the Western world between 1500 and 1830. I mean by this, improvements in the factor and product markets, reduction in impediments to efficient resource allocation, and economies of scale. Moreover, the complementarity between physical and human capital in the development, application, and spread of technological change requires equal analytical attention before we can begin to make sense on this subject. Clearly, we need to overhaul our view of the whole process by which the Western world developed in the last five or six centuries" (Douglass North, "The State of Economic History," *American Economic Review, Papers and Proceedings*, 55 [May 1965], 87-88.

53. *Journal of Political Economy*, 76 (September-October 1968), 953-970.

54. *Ibid.,* p. 953.

55. *Ibid.,* pp. 960-963.

56. *Ibid.,* p. 964.

57. Thus Walton states, with respect to colonial shipping: "As the obstacles of piracy and similar hazards were eliminated, specialized cargo-carrying vessels possessing the input characteristics of the flyboat were adopted. In the process, the costs of shipping were substantially reduced, which had a favorable impact on the development of a trading

Atlantic community" (Gary Walton, "Obstacles to Technical Diffusion in Ocean Shipping, 1675-1775," *Explorations in Economic History*, 8 (Winter 1970-1971), p. 136.

58. North, "Sources of Productivity Change," p. 967.

59. Robert Fogel and Stanley Engerman, *The Reinterpretation of American Economic History* (New York: Harper and Row, 1971), p. 206, emphasis added.

60. *Ibid.*, p. 5, emphasis added. Also, p. 100: "North argues that most of these changes were due to improved organization rather than new equipment. . . . He holds that no *new* technological knowledge was required for the switch from fleets of predominantly small ships to fleets of predominantly large ones. . . . What then explains the dominance of the small over the large ship in the seventeenth and eighteenth centuries and then the rapid shift toward large ships between 1800 and 1860? North again finds the answer not in *new* technological knowledge, but in institutional change. He argues that the elimination of piracy made it feasible to build large, light vessels for the exclusive purpose of carrying cargoes." [Emphasis added.]

12. Technology and the environment: an economic exploration

1. See the excellent cost-benefit analysis of the projected supersonic transport in Stephen Enke, "Government-Industry Development of a Commercial Supersonic Transport," *American Economic Review Papers and Proceedings* 57, no. 2 (May 1967): 78-79. Enke points out that "about 85 percent of U.S. residents have never flown, those who do fly do not always take long-haul flights, and perhaps less than 5 percent of all Americans will ever fly SST's at their higher fares. Private, nonexpense account, long-haul passengers will mostly continue to fly subsonically. . . . Further, American SST passengers will tend to travel to and from a few areas, such as New York, Chicago, Los Angeles, San Francisco, Seattle, Washington, D.C., and Miami. Americans living elsewhere may never use an SST except on international flights. But 100 million Americans may find themselves subjected daily to sonic booms if overland SST flights are permitted."

2. Merrill Eisenbud, "Environmental Protection in the City of New York," *Science* 170, no. 3959 (November 13, 1970): 711.

3. Tastes should be defined here to include preferences with respect to family size. Indeed, other people's tastes concerning family size may be the most important externality of all, in view of the intimate relationship between many externalities and population densities. From my point of view, the most significant decisions which "the rest of the world" makes affecting my welfare are those which, collectively, determine the size of the total population of which I find myself a member. Although much thinking on this thorny and perplexing matter treats the question of population control as if it were subject to a purely technological solution, the problem is, more deeply, one of social and cultural values. The beginning of wisdom here is the recognition that large families are, for the most part, the product of couples who *want* numerous offspring, as attitude studies in different countries have repeatedly demonstrated. A recent survey in India, for example, showed a mean desired family size of 4.7 children, and, in the Philippines, no less than 6.8 children. The whole formidable arsenal of Western contraceptive technology will not bring a drastic reduction in rates of population growth unless and until attitudes concerning desired family size are also drastically revised.

4. See Eugene Smolensky, "The Past and Present Poor," in *The Concept of Poverty* (Washington, D.C., 1965); and "Investment in the Education of the Poor: A Pessimistic Report," *American Economic Review Papers and Proceedings*, vol. 56, no. 2 (May 1966). See also Robert Lampman, "The Low Income Population and Economic Growth," Study Paper no. 13, prepared for consideration by the Joint Economic Committee, U.S. Congress, 86th Cong., 1st sess. (Washington, D.C., 1959).

5. As Lynn White, jr., has recently pointed out, "Ever since man became a numerous species he has affected his environment notably. The hypothesis that his firedrive method of hunting created the world's greatest grasslands and helped to exterminate the

monster mammals of the Pleistocene from much of the globe is plausible, if not proved. For 6 millennia at least, the banks of the lower Nile have been a human artifact rather than the swampy African jungle which nature, apart from man, would have made it. The Aswan Dam, flooding 5000 square miles, is only the latest stage in a long process" ("The Historical Roots of Our Ecologic Crisis," *Science* 155, no. 3767 [March 10, 1967]: 1203).

6. For a detailed treatment of the effects of westward expansion upon wildlife in America, see Peter Mathiessen, *Wildlife in America* (New York, 1959).

7. Allen V. Kneese, "Research Goals and Progress toward Them," in Henry Jarrett, ed., *Environmental Quality in a Growing Economy* (Baltimore, 1966), p. 71.

8. If anyone wants to add that we can have *both* the pure water and these other things by reducing yet *other* expenditures - for example, military operations in southeast Asia - I can only add my enthusiastic endorsement to the suggestion. However, I would also endorse the termination of this particular activity even if the resources so liberated were to be applied to no useful purpose whatever. That is, I regard the activity in question as producing a "bad" and not a "good." The general point is the necessity for choice when confronted, as we are, with limited resources and multiple goals.

9. "In most large cities . . . the electric utilities consume up to half of all fuel burned. Most utilities have made reasonable efforts to reduce the emission of soot and fly ash; virtually all new power plants, and many old ones, are now equipped with devices capable of removing a large fraction of such emissions. Utilities, however, are still under pressure, both from the public and from supervising agencies, to use the cheapest fuels available. This means that in New York and other eastern-seaboard cities the utilities burn large volumes of residual fuel oil imported from abroad, which happens to contain between 2.5 and 3 percent of sulfur, compared with only about 1.7 percent for domestic fuel oil. When the oil is burned, sulfur dioxide is released. Recent studies show that the level of sulfur dioxide in New York City air is almost twice that found in other large cities" (Abel Wolman, "The Metabolism of Cities," *Scientific American* 213, no. 3 [September 1965]: 187-88).

10. Older cars are also, incidentally, usually the main offenders against clean air.

11. "The process of vegetative degeneration is accelerated by the cultivators' habit of firing the land before planting. Fire is a useful and practical method of clearing land. Firing also gives an initial beneficial effect to the soil, which Jack states is due to a temporary increase in bacteria numbers in the soil. It has also the added advantage of making the soil friable, and so renders digging unneccessary. But fire at the same time destroys a large amount of wood, leaves, and other vegetative matter which are the raw materials of humus. As the fallow period is shortened and fires repeatedly invade the same clearing, a stage will be reached when, instead of the forest trees, there will be expanses of grassland made up mainly of the obnoxious weeds - *Imperata arundinacae* and *Imperata cylindrica* ('lalang' in Malaya, 'alang-alang' in Indonesia, and 'cogen' in the Philippines). Once established, these grasses tend to exclude all other vegetation. The rhizomes of *Imperata* extend deep underground and are not affected by fires on the surface. The fires kill off other non-resistant plants but leave the grasses intact and dominant, so that the regeneration of the forest can only occur if these fires are prevented. Once the vegetation has been converted to grasses and weeds, the land is almost sterile. As cattle fodder *Imperata* grasses are of low nutritive value, and cattle suffer from scour and sore mouths when they are forced to eat them. Only the young shoots are reasonably palatable. African herdsmen periodically burn the savannah to obtain a good crop of young shoots for their cattle. Land given over to *Imperata* is a good indication of past destructive agricultural practices, as well as being a sign of heavy pressure of population on the land" (Ooi Jin-bee, "Rural Development in Tropical Areas, with Special Reference to Malaya," *Journal of Tropical Geography* [March 1959], p. 94).

12. For example, although we are all much concerned currently over the forms of pollution arising out of an industrial society heavily dependent upon fossil fuels as an energy source, I suffer from a recurrent nightmare about the pollution problems in a world dependent upon nuclear power plants. I have the distinct impression (which I do not *think* can be attributed to incipient paranoia) that the experts have been continuously

revising in a downward direction their estimates of radiation dosages to which human beings can be "safely" exposed. When I combine this with what must be the unique problems of the disposal of atomic waste materials, not to mention accidents, I wonder whether we may not now be living in what some future generation will one day regard as a golden age, almost totally innocent of *really* serious pollution problems.

13. Just as technological events have repeatedly falsified conservationist predictions of the impending exhaustion of essential inputs of natural resources. A major thrust of the sciences of modern materials has been to broaden the range of substitutes, especially by learning to exploit abundant materials in place of scarce ones. This has been of particular significance with respect to the utilization of low-concentrate resources, a development which (so long as cheap power is available) holds open an almost limitless potential for the future (for a perceptive treatment of this subject see Harold J. Barnett and Chandler Morse, *Scarcity and Growth* [Baltimore, 1963]). To the conservationist's preoccupation with the shrinking heritage of natural resources and the need to preserve them for future generations, Barnett and Morse retort, "In the United States . . . the economic magnitude of the estate each generation passes on - the income per capita the next generation enjoys - has been approximately double that which it received, over the period for which data exist. Resource reservation to protect the interest of future generations is therefore unneccessary. There is no need for a future-oriented ethical principle to replace or supplement the economic calculations that lead modern man to accumulate primarily for the benefit of those now living. The reason, of course, is that the legacy of economically valuable assets which each generation passes on consists only in part of the natural environment. The more important components of the inheritance are knowledge, technology, capital instruments, economic institutions. These far more than natural resources, are the determinants of real income per capita" (pp. 247-48).

14. We might want to label this behavior the "Sputnik syndrome," in recognition of an event which brought a vast outpouring of federal funds for research and education in America.

15. Nevertheless, the recent heated controversy in the Soviet Union over the impending pollution of Lake Baikal by the effluvia of a new paper mill sharply underlines the fact that certain painful social choices are independent of economic systems, in spite of the efforts of several decades of Marxist critics to pin the tail of pollution and environmental destruction firmly upon a capitalist donkey. Recent reports from the Soviet Union even indicate that the growing pollution of the Caspian Sea by industrial and urban sewage, petroleum products, and discharge of waste and ballast by ships constitutes an imminent threat to the sturgeon population and that caviar, as a result, is already in very short supply. Perhaps this will generate an appropriate sense of outrage and decisive action in the appropriate ruling circles! Indeed, all evidence suggests that the Soviet Union is moving into the same range of environmental problems as those with which we are currently contending. Her lag is primarily a function of lower income levels and population densities. One wonders, for example, how successful the Soviet regime will be in suppressing privately owned automobiles as the demand for them increases with rising incomes.

16. Of course DDT has made a massive contribution to malaria control as well. The use of DDT spray in the years after the Second World War reduced the reported incidence of malaria in Ceylon from 2.8 million cases in 1946 to 110 cases in 1961. Thereafter, the spraying was discontinued in the belief that it was no longer necessary. As a result, according to a recent statement by the World Health Organization, 2.5 million Ceylonese were afflicted with the disease in 1968 and 1969.

17. The term "miracle rice" is most unfortunate insofar as it suggests that large increases in output will be both easy and cheap to attain when the new varieties are introduced. In fact, the new strains require costly complementary inputs in order to be successful: fertilizer, insecticides, and dependable water control systems.

18. John Stuart Mill, *Principles of Political Economy* (London, 1909), p. 750.

13. Technological innovation and natural resources: the niggardliness of nature reconsidered

1. Furthermore, the simplifying assumption was usually employed that capital and labor

were applied in doses of fixed proportion, thus collapsing a three variable model to one of two variables. Note that this procedure also has the effect of ignoring the possibilities for offsetting by capital formation the decline in worker productivity due to the deteriorating resources/man ratio.

2. For an incisive analysis of the Malthusian theory, see K. Davis, "Malthus and the Theory of Population," in P. Lazarsfeld (ed.), *The Language of Social Research*, Glencoe, Illinois, 1955. See also Mark Blaug, *Economic Theory in Retrospect*, Homewood, Illinois, 1968, revised edition, chapter 3.

3. John Bates Clark, *The Distribution of Wealth*, New York, 1899, p. 56.

4. *Ibid.*, p. 55-6.

5. See *The Coal Question*, chapter 7, "Of Supposed Substitutes for Coal."

6. *Ibid.*, p. 215. Emphasis Jevons's.

7. Samuel P. Hays, *Conservation and the Gospel of Efficiency: The Progressive Conservation Movement, 1890-1920*, Harvard University Press, 1959.

8. Clifford Pinchot, *The Fight For Conservation*, 1910, pp. 43, 123-124. As quoted in H. Barnett and C. Morse, *Scarcity and Growth*, Baltimore, 1963, p. 76. Emphasis Barnett and Morse.

9. See, e.g., R. Nelson, "A Theory of the Low-Level Equilibrium Trap in Underdeveloped Countries," *American Economic Review*, December 1956; Harvey Leibenstein, *Economic Backwardness and Economic Growth*, John Wiley and Sons, 1957, chapter 10.

10. D. H. Meadows et al., *The Limits to Growth*, New York, 1972.

11. George Stigler, "The Ricardian Theory of Value and Distribution," *Journal of Political Economy*, June 1952. Mark Blaug has made the same point: "If Malthus' theory were indeed a theory, we would want to ask: What would happen if the theory were not true? The answer is, or ought to be, that income per head would rise, not fall, with increasing population. The history of Western countries, therefore, disproves Malthus' theory. The defenders of Malthus reply: But what of India today? No one denies that India is overpopulated and poor. It is overpopulated because the death rate was lowered by the introduction of Western medicine, thus divorcing population growth from the current level of income. It follows that India would be better off if she could also "Westernize" her birth rate. But what has this piece of advice to do with Malthusian theory of population?" Blaug, *op. cit.*, pp. 79-80.

12. Simon Kuznets, *Modern Economic Growth*, New Haven, 1966, chapter 2.

13. Kuznets, *op. cit.*, chapter 3.

14. T. W. Schultz, "The Declining Economic Importance of Agricultural Land," *Economic Journal*, December 1951, p. 727. Schultz presents quantitative evidence to support his generalization, which he holds ". . . to be empirically valid for such technically advanced communities as France, the United Kingdom, and the United States." (p. 739).

15. Zvi Griliches, "Research Costs and Social Returns: Hybrid Corn and Related Innovations," *Journal of Political Economy*, October 1958.

16. William Parker and Judith Klein, "Productivity Growth in Grain Production in the United States, 1840-60 and 1900-10," in Dorothy Brady, (ed.), *Output, Employment and Productivity in the United States after 1800*, New York, 1966.

17. See A. J. Youngson, "The Opening of New Territories," chapter 3 in M. M. Postan and H. J. Habakkuk (eds.), *The Cambridge Economic History of Europe*, vol. VI: *The Industrial Revolutions and After*, Cambridge, 1965.

18. Paul David, "The Mechanization of Reaping in the Ante-Bellum Midwest," in H. Rosovsky (ed.), *Industrialization in Two Systems*, New York, 1966.

19. Z. Griliches, "Hybrid Corn: An Exploration in the Economics of Technological Change," *Econometrica*, October 1957.

20. Hayami and Ruttan, *Agricultural Development*, pp. 53-54.

21. *Ibid.*, p. 84.

22. For further treatments of technological change in agriculture with particular emphasis on their impact upon agricultural productivity, see: V. Ruttan, "Research on the Economics of Technological Change in American Agriculture," *Journal of Farm Economics*, November 1960, and the numerous citations contained therein; R. Loomis and G. Barton, *Productivity in Agriculture: United States, 1870-1958*, U.S. Dept. of Agriculture Technical Bulletin No. 1238, Washington, D.C., 1961; Z. Griliches,

"Productivity and Technology," appearing under "Agriculture" in *International Encyclopedia of the Social Sciences;* Z. Griliches, "The Sources of Measured Productivity Growth: U.S. Agriculture, 1940-60," *Journal of Political Economy,* August 1963; Irwin Feller, "Inventive Activity in Agriculture, 1837-1900," *Journal of Economic History,* December 1962; W. Rasmussen, "The Impact of Technological Change in American Agriculture, 1862-1962," *Journal of Economic History,* December 1962; W. Parker, "Agriculture," Chapter 11 in L. Davis, R. Easterlin, W. Parker et al., *American Economic Growth,* New York, 1972. For an informative, wide-ranging survey of worldwide food prospects and the possibilities for their augmentation by use of scientific knowledge, see *The World Food Problem,* vol. II, A Report of the President's Science Advisory Committee, U.S.G.P.O., 1967.

23. The material in this paragraph is drawn from chapter 8 of Barnett and Morse's *Scarcity and Growth:* "The Unit Cost of Extractive Products."

24. As Carter points out: "There is also research directed specifically toward the use of low-priced materials, which tends to blur even more the operational distinction between technological development and price substitution." (p. 85)

25. *Ibid.* For empirical evidence to support this proposition, see chapter 6. ·

26. The degree of substitutability, the number of such substitutes, and the quantity in which they are available are, of course, critical to long-run arguments over the possibility for avoiding rising costs imposed by the increasing scarcity of particular mineral deposits. On this issue, Barnett and Morse point out that ". . . near-perfect substitutes are better scarcity mitigators than imperfect ones. If consumers are virtually indifferent to whether they obtain Btu for heating, cooling, and lighting from gas, coal, or electricity, a change in the proportions in which these commodities are used will not, in itself, affect welfare. The same is true of rayon or silk, cinder-block or wood-frame construction, margarine or butter, and so on. The point is relevant to the present discussion because, by and large, scarcity-induced substitutions often involve commodities that can be employed or consumed in widely varying proportions without making much difference in welfare terms. Preferences for different forms of heating fuel, construction, clothing, or even foods are largely matters of socially conditioned taste, and are therefore quite flexible over time. Long-run elasticities of substitution are thus greater than short-run elasticities." Barnett and Morse, *op. cit.,* p. 130.

27. Joseph Fisher and Edward Boorstein, *The Adequacy of Resources for Economic Growth in the United States,* Study Paper no. 13, materials prepared in connection with the Study of Employment, Growth, and Price Levels, U.S.G.P.O., Washington, D.C., 1959, p. 42.

28. *Ibid.,* p. 43.

29. The basic empirical justification for this statement is the very low level of American imports expressed as a percentage of GNP. See U.S. Dept. of Commerce, *Historical Statistics of the U.S.: Colonial Times to 1957,* Washington, D.C., 1960, p. 542. For a careful study of the changing natural resource composition of U.S. foreign trade, see J. Vanek, *The Natural Resource Content of the U.S. Foreign Trade, 1870-1955,* Cambridge, 1963. See also the detailed assessments for individual materials in President's Materials Policy Commission, *Resources for Freedom,* Washington, D.C., U.S.G.P.O., 1952.

30. N. Potter and F. T. Christy, Jr., *Trends in Natural Resources Commodities: Statistics of Prices, Output, Consumption, Foreign Trade, and Employment in the U.S., 1870-1957,* Baltimore, 1962.

31. Anthony Scott, "The Development of the Extractive Industries," *The Canadian Journal of Economics and Political Science,* February 1962.

32. *Ibid.,* p. 81. For an interpretation of the British industrial revolution as revolving around the substitution of abundant, inorganic materials for increasingly scarce organic ones, see E.A. Wrigley, "The Supply of Raw Materials in the Industrial Revolution," *Economic History Review,* August 1962.

33. "The Declining Economic Importance of Agricultural Land," *op. cit.,* p. 725. Schultz cites the following statement from page 20 of Harrod's book: ". . . I propose to discard the law of diminishing returns from the land as a primary determinant in a progressive

economy. . . . I discard it only because in our particular context it appears that its influence may be quantitatively unimportant."

34. *Natural Resources and Economic Growth,* p. 172.
35. Barnett and Morse, *op. cit.,* pp. 247-8.
36. Improvements in the technology of resource *discovery,* for example, have been an important factor in continually falsifying the pessimistic predictions of imminent exhaustion of oil reserves periodically issued by the U.S. Geological Survey. See Hans Landsberg and Sam Schurr, *Energy in the United States,* New York, 1968, p. 98.
37. Nathan Rosenberg, *Technology and American Economic Growth,* New York, 1972, pp. 19-20.
38. J. F. Dewhurst and Associates, *America's Needs and Resources,* New York, 1955, pp. 765-6. Chapter 31 of this book contains much useful data on individual materials.
39. Fisher and Boorstein, *op. cit.,* p. 59.
40. Similarly, a sufficient rise in the price of fresh water would activate techniques for the desalination of ocean water which are already established but are, at present, prohibitively costly. Even our *present* state of technological knowledge provides us with many such options which, in effect, place an upper limit to the prospective rise in raw material prices.
41. V. Ruttan and J. Callahan, "Resource Inputs and Output Growth: Comparisons between Agriculture and Forestry," *Forestry Science,* March 1962, p. 78.
42. *Why* this environmental concern should have emerged when it did is, itself, an interesting sociological question. One interpretation is that it is primarily a demand phenomenon associated with rising incomes. "(I)n relatively high-income economies the income elasticity of demand for commodities and services related to sustenance is low and declines as income continues to rise, while the income elasticity of demand for more effective disposal of residuals and for environmental amenities is high and continues to rise. This is in sharp contrast to the situation in poor countries where the income elasticity of demand is high for sustenance and low for environmental amenities. The sense of environmental crisis in the relatively affluent countries at this time stems primarily from the dramatic growth in demand for environmental amenities." V. Ruttan, "Technology and the Environment," *American Journal of Agricultural Economics,* December 1971, pp. 707-8.
43. "Forrester's model is unsound in basic structure. He fails to build in feedbacks whereby societies utilize knowledge to overcome resource depletion and pollution. When specific mineral resource deposits begin to run out in modern societies, the prospect of rising costs feeds back signals. Man then undertakes to find new ones or substitutes, and he develops technology to utilize formerly uneconomic deposits. In response to feedback signals, he counters pressures upon agricultural land by devising hybrid plants, high yield varieties, fertilizers, pesticides, transportation, and refrigeration. Responding to feed-back, he overcomes anticipated energy shortages by exploiting deeper or new resources of great volume. In modern nations he has been successful in devising new technologies not merely to maintain real output per capita but to increase it. The real costs per unit of output have declined greatly in the face of population increases. In the U.S. the unit cost of agricultural products, in terms of labor and capital, has fallen by more than half and of mineral products by three quarters since the Civil War, and the rate of decline has accelerated since World War I." Harold Barnett, review of Jay Forrester, *World Dynamics,* Cambridge, 1971, in *Journal of Economic Literature,* September 1972, p. 853.
44. A well-known feature of the price mechanism as a signalling device is that it fails to provide socially correct signals where externalities are present, as is notoriously the case with pollution phenomena. In large measure, this is due to the peculiarities of our property rights system which allows private individuals to make unrestricted use of common property resources in the discharge of their pollutants. For an intelligent discussion of a range of policy alternatives for dealing with the problem of pollution of common property resources, see J. H. Dales, *Pollution, Property and Prices,* Toronto, 1968.
45. For a searching and suggestive analysis of the demographic response over the course of

American history, see R. A. Easterlin, "Does Human Fertility Adjust to the Environment?" *American Economic Review Papers and Proceedings,* May 1971, pp. 339-407. Easterlin's conclusion is ". . . that both theory and the empirical research done so far on historical American fertility suggest that human fertility responds voluntarily to environmental conditions. If this is so - and it seems hard to ignore the evidence - then the nature of what is called 'the population problem' takes on a radically different guise. The question is not one of human beings breeding themselves into growing misery. Rather, the problem is whether the voluntary response of fertility to environmental pressures results in a socially optimal adjustment. In thinking about this, it seems useful to distinguish between the potential for population adjustment and the actual degree of adjustment. The staggering change in American reproductive behavior over the past century and a half clearly demonstrates the immense potential for adjustment. Whether, currently, the degree of adjustment is socially optimal remains a matter for research." (p. 407). See also the same author's book, *Population, Labor Force, and Long Swings in Economic Growth: The American Experience,* New York, 1968.

46. S. Kuznets, "The Economic Growth of Small Nations," in E. A. G. Robinson (ed.), *Economic Consequences of the Size of Nations,* pp. 27-28. Elsewhere Kuznets stated: "(A)ny emphasis on relative scarcity of irreproducible resources, as a factor in determining low levels of economic performance extending over a long period, must be countered with the question why no successful effort has been made by the victim of such scarcity to overcome it by changes in technology. To be retained, the hypothesis must, therefore, be rephrased: the have-not societies are poor because they have not succeeded in overcoming scarcity of natural resources by appropriate changes in technology, not because the scarcity of resources is an inexorable factor for which there is no remedy. And obviously human societies with low levels of economic performance are least able to overcome any scarcities of irreproducible resources by changes in technology; but this is a matter of social organization and not of bountifulness or niggardliness of nature." S. Kuznets, *Economic Change,* New York, 1953, p. 230.

14. Innovative responses to materials shortages

1. All numbers in this paragraph are drawn from J. Fisher and E. Boorstein, p. 43.
2. The basic empirical justification for this statement is the very low level of imports expressed as a percentage of American *GNP.* See U.S. Dept. of Commerce, p. 542. For a careful study of the changing natural resource composition of U.S. foreign trade, see J. Vanek, especially Ch. 5.
3. It is worth noting here that many of our environmental problems are exacerbated by the *cheapness* of materials inputs as compared to the prices of labor and capital. Thus, we may observe that two of the major concerns of environmentalists are really based upon conflicting assumptions about relative resource prices. The pollution problem is often based upon the cheapness of raw material inputs, making recovery uneconomic. The resource exhaustion problem, on the other hand, is based upon the assumption of inexorably rising prices of resources. But, if materials prices were much higher, we could be saving our old newspapers, collecting discarded beer cans, and hauling off to the junkyards the tens of thousands of automobiles which are abandoned on our roads and highways every year - or at least it would be worth *someone's* effort to provide these services for us. An interesting implication, therefore, is that a rise in raw material costs may be expected to *reduce* the severity of some of our pollution problems.
4. On the output side, one can also (a) redesign final products and (b) alter the *mix* of final products to consume smaller quantities of resources - smaller cars or buses, smaller and/or better insulated houses, more blankets and less fuel oil, more cereal products and less dairy products. Products may even be designed for longer life - admittedly an alien and somewhat novel idea!
5. One of the principal reasons for the limited application of nuclear power in commercial uses has been the continued improvements in energy production with conventional fuels.

15. Science, invention, and economic growth

1. Zvi Griliches, "Hybrid Corn: An Exploration in the Economics of Technological Change," *Econometrica,* October 1957.
2. The issue is not just whether the scientific and technological spheres are autonomous or not, although that has been a much-debated issue. Even if one were satisfied, for example, that the scientific realm is an autonomous sphere, it need not follow that events in that sphere are unpredictable. They may not be directly influenced by economic variables, but they may be moving subject to an internal logic or an external set of forces which can be identified and then used, by economists, to explain sequences of inventive activity.
3. Schmookler draws the implication from his data on interindustry variations in capital goods invention that "... *inventive activity with respect to capital goods tends to be distributed among industries about in proportion to the distribution of investment.* To state the matter in other terms, *a 1 per cent increase in investment tends to induce a 1 percent increase in capital goods invention."* Schmookler, *op. cit.,* p. 144. Emphasis Schmookler's. It is important to note that Schmookler's results "... depend critically on the fact that our capital goods inventions were classified according to the industry that will use them, not according to the industry that will manufacture the new product or the intellectual discipline from which the inventions arise." *Ibid.,* p. 164. See also p. 166.
4. Schmookler, *op. cit.,* pp. 180-1. Of course Schmookler is well aware that consumer expenditure on particular classes of goods is not entirely a function of prices and incomes. Such factors as age structure of a population, climate, geography, and extent of urbanisation, will also play an important role.
5. *Ibid.,* pp. 210-11.
6. *Ibid.,* p. 218. Emphasis Schmookler's.
7. "Thus, independently of the motives of scientists themselves and with due recognition of the fact that anticipated practical uses of scientific discoveries still unmade are often vague, it seems reasonable to suggest - without taking joy in the suggestion - that the demand for science (and, of course, engineering) is and for a long time has been derived largely from the demand for conventional economic goods. Without the expectation, increasingly confirmed by experience, of 'useful' applications, those branches of science and engineering that have grown the most in modern times and have contributed most dramatically to technological change - electricity, electronics, chemistry and nucleonics - would have grown far less than they have. If this view is approximately correct, then even if we choose to regard the demand for new knowledge for its own sake as a non-economic phenomenon, the growth of modern science and engineering is still primarily a part of the economic process." *Ibid.,* p. 177.
8. See *ibid.,* chapter 2, for a searching examination of the problems involved in using patent statistics as a surrogate for inventions and also for Schmookler's justification for his belief that the deficiencies in the patent data and the problems posed by vast qualitative differences in inventions are less than is generally supposed. For a careful discussion of the measurement problems involved in the economics of inventive activity, see Simon Kuznets, "Inventive Activity: Problems of Definition and Measurement," in R. R. Nelson (ed.), *The Rate and Direction of Inventive Activity,* Princeton, 1962, pp. 19-43.
9. Schmookler, *op. cit.,* p. 163. See also p. 208, footnote 1.
10. It is, of course tautologically true to say, as Schmookler does, that "A given percentage improvement in productivity is more valuable in a large than in a small industry." *Ibid.,* p. 91.
11. *Ibid.,* p. 172. Emphasis Schmookler's. See also pp. 209 and 212.
12. *Ibid.,* p. 184.
13. This is the judgment recently delivered by medical historians. See Thomas McKeown and R. G. Brown, "Medical Evidence Related to English Population Changes in the 18th Century," *Population Studies,* 1955-66, pp. 119-41, and Thomas McKeown and R. G. Record, "Reasons for the Decline of Mortality in England and Wales During the 19th Century," *Population Studies,* 1962, pp. 94-122.

14. See T. S. Kuhn, *The Structure of Scientific Revolutions,* Chicago, 1962, chapter VIII and the same author's article, "The History of Science," in the *International Encyclopedia of the Social Sciences.*

15. The great nineteenth century breakthroughs in organic chemistry in turn laid the basis for the subsequent twentieth century revolution in biology. As Bernal points out: "The new organic chemistry had another essential part to play in the history of science - it was to lead to a fuller understanding of biological processes. In fact, the beginning of any deeper understanding than the microscope could provide was totally impossible without a knowledge of the laws of combination and the types of structure actually to be met with in biological systems. The nineteenth-century development of organic chemistry had to precede logically any attempt to formulate a fundamental biology." J. D. Bernal, *Science in History,* Cambridge, Mass., 1971, 4 vols., vol. 2, p. 633.

16. On the great inherent complexity of biological studies Bernal makes the following interesting observations: ". . . (T)he same degree of complexity of even the simplest forms of life is something of an entirely different order from that dealt with by physics or chemistry. What we had admired before in the external aspects of life, in the symmetry and beauty of plants and flowers, or in the form and motion of the higher organisms, now appear, in the light of our wider knowledge, relatively superficial expressions of a far greater internal complexity. That internal complexity is itself a consequence of the long evolutionary history through which living organisms have raised themselves to their present state." *Ibid.,* vol. 3, p. 868. The notion that scientific progress has moved in an orderly sequence from the less complex to the progressively more complex aspects of the physical universe is clearly expressed in Frederick Engels, *The Dialectics of Nature,* Moscow, 1954.

17. For a brief but highly perceptive treatment of some of the underlying problems, see William N. Parker, "Economic Development in Historical Perspective," *Economic Development and Cultural Change,* October 1961, pp. 1-7.

18. The complexity and costliness of water management methods in the growing of rice is a major reason why the new wheat varieties have so often been introduced more rapidly and with greater success than the new rice varieties. This has been the case, for example, in India.

19. "We know from experience in the U.S. that the rapid introduction and widespread use of new crop varieties accelerates the biological dynamics of crop disease - host plant relationships." Albert H. Moseman, *Building Agricultural Research Systems in the Developing Nations,* N.Y., Agricultural Development Council, 1970, p. 97.

20. See T. S. Ashton, *Iron and Steel in the Industrial Revolution,* Manchester 1924; John Nef, "The Progress of Technology and the Growth of Large-Scale Industry in Great Britain, 1540-1640," *Economic History Review,* 1934, pp. 3-24; John Nef, "Coal Mining and Utilization," in Charles Singer *et al., A History of Technology,* London, 1957, 5 vols., vol. 3, pp. 72-88; E. A. Wrigley, "The Supply of Raw Materials in the Industrial Revolution," *Economic History Review,* August 1962, pp. 1-16.

21. It is interesting to note that the historic links between coal and the iron and steel industry persist even today, in spite of extensive attempts to sever the links. As a matter of fact, one of the reasons for the relatively large size of the coal industry today in the face of strong competition from other fuels has been the inability thus far, in spite of prolonged exploration, to develop a satisfactory technique for producing iron without the use of high-grade coal. Although other fuels have been readily substituted for coal in many uses, the substitution in metallurgical processes poses unique and so far intractable difficulties.

22. David Landes, *The Unbound Prometheus,* Cambridge, 1969, p. 83. Landes also points out that, even after machinery was introduced into the wool industry, the machines could be operated only much more slowly than in cotton. *Ibid.,* pp. 87-8.

23. William N. Parker, "Agriculture," in Lance Davis *et al., American Economic Growth,* New York, Harper and Row, 1971, p. 385.

24. It is worth mentioning here that our lack of interest in the study of failures may also have contributed in an important way to an underestimation of the costs of invention. In our preoccupation with success stories we inevitably ignore the substantial

commitment of resources to unsuccessful inventive efforts, and recognize only those which were connected with a successful outcome.

25. Note that my emphasis upon supply and cost considerations does not imply any sort of scientific or technological determinism. More costly inventions can always precede less costly ones in time if demand conditions are sufficiently strong.

Epilogue

1. Charles Wilson, speaking of the early nineteenth century, has stated: "The most striking contrast between Britain and France at this period is the higher rate at which, in Britain, inventions were adopted, developed, and passed into application. No nation in the world showed more vivid inventive genius than the French, but a high proportion of their inventive talent proved abortive or was put to profitable use elsewhere - notably in England and Scotland." Charles Wilson, "Technology and Industrial Organization," in Charles Singer et al. (eds.), *A History of Technology,* Oxford University Press, 1958, vol. v, p. 800.

2. "How good is Japanese science? . . . Perhaps the most reliable guide is a 'self-evaluation' published in the English-language magazine *Technical Japan.* Based on answers to a questionnaire sent to various Japanese science institutes and societies, the survey concluded that Japan excels in subjects where the object of research is either peculiar to or abundant in Japan, such as volcanoes, encephalitis, rice blight, and astronomy; in subjects related to state-run enterprises, such as train operation, magnetometry, and microwave research; in subjects related to industries peculiar to Japan, such as silkworm genetics and zymogenic microorganisms; and in studies related to Japanese industries which compete in international markets, such as shipbuilding, textiles and vitamin synthesis." "Japan (I): On the Threshold of an Age of Big Science?" *Science,* 2 January 1970, p. 32.

3. O.E.C.D., *The Conditions for Success in Technological Innovation,* Paris, 1971, p. 20.

4. Joseph Needham, *Science and Civilization in China* (seven volumes in twelve parts) Cambridge University Press, 1954-. See also Joseph Needham, *The Grand Titration: Science and Society in East and West,* University of Toronto Press, 1969.

5. Francis Bacon, *The New Organon,* The Bobbs-Merrill Co., Indianapolis, 1960, p. 118.

6. Some of the problems of natural-resource scarcity may even be, to a degree, self-solving in a regime of continued economic growth. The growing demand for personal services which has been associated with rising incomes is, in effect, a growing demand for outputs which are less natural-resource-intensive than commodity output generally. The implications for resource requirements of a growing personal services sector merit far more attention than they have so far received.

7. There is strong evidence that it exists in the less developed countries as well. Contrary to the uniformly dreary picture which is often painted of "explosive" population growth, there is in fact widespread recent evidence of reductions in fertility in those less developed countries which are achieving significant socioeconomic advance. On the basis of evidence drawn from the 1960s, Dudley Kirk suggests that a new demographic transition is under way, and that "If progress in modernization continues, notably in the larger countries, the demographic transition in the less developed world will probably be completed much more rapidly than it was in Europe." Dudley Kirk, "A New Demographic Transition?" in *Rapid Population Growth,* prepared by a Study Committee of the Office of Foreign Secretary, National Academy of Sciences, The Johns Hopkins Press, Baltimore, 1971, p. 146. Needless to say, even if Kirk is correct, problems of immense proportions will remain, and it is well to add the cautionary note on which Kirk closes: "On any assumptions concerning the reduction of fertility that may occur with socioeconomic progress, it still follows that one may anticipate and must accommodate an enormous increase in the world population and that these increases will be greatest precisely in those countries economically least well-equipped to absorb the increase in numbers." *Ibid.,* p. 146.

8. See R. A. Easterlin, "Does Human Fertility Adjust to the Environment?" *American Economic Review, Papers and Proceedings,* May 1971, pp. 399-407.

Index

Abramovitz, Moses, 1, 9, 83, 86, 105, 207, 290, 302, 303, 305
abrasives, 24
Academie des Sciences, 283
Acheson, Edward G., 25
Africa, 333
agriculture, 78, 88, 95, 111, 247, 250, 252, 254, 282, 296, 303, 304, 310, 312, 320, 321, 322, 330, 333-337 *passim*, 340; the laziness of farmers, 89-90; varieties of, 97-98; Marx on, 128, 136, 137; diffusion of agricultural technology, 151, 168-171, 202; and the environment, 218-228; the niggardliness of nature, 229, 232, 235-238, 242-243; and science, 260, 264, 271, 274-275
aircraft industry, 68, 73, 163, 300, 324
Alchian, A., 329
Allen, G.C., 294
Allen, Zachariah, 50, 54, 55, 56, 57, 58, 298, 299
aluminum industry, 33
American Machinist, 14, 292
American Society of Mechanical Engineers, 306
"American System of Manufacturing", 120; *see also* mass production technology
Ames, E., 317
Ames Manufacturing Company, 19, 292, 298
Amoskeag Manufacturing Company, 13
Anderson, C.A., 320, 325
Anderson, J., 291, 308, 310
anthracite coal, *see* coal
Arkwright, R., 328
Armstrong, W., 116, 154
Arrow, K., 300, 329
Ashton, T.S., 324, 326, 340
Asia, 304
astronomy, 312
Aswan Dam, 333
atomic energy, 74, 257, 280, 333, 334, 338
Atomic Energy Commission, 171, 223
Austria, 155

automobiles, 104, 111, 144, 280, 304, 320; machine tools important to, 13, 18, 22-25 *passim,* 26-28, 30, 31, 116, 294, 295; mass production of, 159-161; and the environment, 215, 216, 220, 333, 334, 338

Babbage, C., 200, 276, 308, 310, 311, 313, 329
"baby boom," *see* demographic shift
Bach, J.S., 192
backward linkage, 112
Bacon, F., 286, 341
Baker, T., 328
Balaclava, 306
Baldwin, G.B., 70
Baldwin, M., 291
Baldwin Locomotive Works, 13, 161, 180, 324
Bale, M.P., 297
balloon-frame housing, 38, 253
Barker, T.C., 309
Barnett, H.J., 238, 242, 334, 335, 336, 337
Barnum, P.T., 30
Barton, G., 335
Bathe, G. and D., 296
Bator, F., 104, 105
Baumol, W.J., 82
bauxite, 257
bearings, 25, 28
Belgium, 155
Ben-David, J., 302
Berg, E., 96, 105
Bernal, J.D., 138, 309, 315, 340
Berrill, K., 318
Bessemer, H., 72, 113, 182, 192, 195, 305, 324
bicycles, 16, 18, 23, 24, 27, 29, 30, 31, 112, 153, 157, 294, 305
biology, 136-7
Bishop, J.L., 296
bituminous coal, *see* coal
Blanchard, T., 19, 42, 292
Blanchard lathe, 297, 298

blast furnace, 182, 183, 184, 185
Blaug, M., 335
Bloch, M., 321, 325
Bober, M., 138, 315
Bolles, A.S., 296
Boorstein, E., 240, 254, 259, 310, 336, 337, 338
Booth, W.H., 318
bottlenecks, 125
Boulton, F., 309, 321, 326
Bowden, W., 328
Bowley, A., 300
Bowman, M.J., 320, 325
Brackenbury, H., 306
Brady, D., 322, 324, 335
Brassey, T., 308
Brazil, 247, 311
brewing industry, 251, 272, 273, 326
bridgebuilding, 124, 202
Bright, A., 70
Britain, see United Kingdom
Britannia bridge, 124
British consulting engineer, 160
Brown, David, 22
Brown, Joseph R., 22
Brown, M., 300, 316
Brown, R.G., 339
Brown, W., 311
Brown and Sharpe Manufacturing Company, 22, 23, 25, 293, 294
Burke, J., 329
Burlingame, L., 293
Burlingame, R., 296
business cycles, 260

Cadillac Automobile Company, 27, 31, 220
Cairncross, A., 102, 105
Calhoun, D., 318
Callahan, J., 337
Campbell, A., 194
Campbell, R., 105, 304
Canada, 122, 293
capital deepening, 85
capital goods industries, machine tools, 10-11, 17; importance of, 99-100, 135; responsiveness of, 112, 263; and economic growth, 141-150, 315-317; and the transfer of technology, 152, 156-168, 170, 320; in steam and iron, importance to the diffusion of technology, 199-202; see also Producer goods industries, Machine tool industries
capital intensive production, 100, 310, 317; see also labor saving technology
capital saving technology, 141, 142, 146, 147, 148, 149, 165, 166, 247, 314
capital stock, 9, 10, 102, 148, 261

capitalism, role of technology under, 126-138, 260
capitalism, social structure, 127
capitalist apologist, 245
Carothers, W.H., 72
Carr, J., 324
Carter, A., 239, 318, 336
Carter, C., 316
Cartwright, E., 112
Caspian Sea, 334
Census of U.S., of 1810, 296; of 1840, 46; of 1850, 47; of 1860, 48, 293, 296; of 1880, 294; of 1900, 294, 295; of 1905, 291, 294, 305; of 1910, 294; of 1914, 292
Centennial Exhibition of 1876, 22, 24; see also Philadelphia Exhibition
Ceylon, 316, 334
chain, flat link, 25
Chaloner, W., 324
charcoal, 182, 204
chemists, 269, 270, 282, 320, 340
Chi, H., 105, 304
China, 286, 341
Chipman, J., 329
Christy, F., 240, 336
Cincinnati Milling Machine Company, 115, 295
Cipolla, C., 326, 329
Clapham, J., 294, 299, 322, 325
Clark, D., 297
Clark, E., 310
Clark, J.B., 230, 335
Clark, J.M., 145
Clark, V.S., 36, 52, 56, 57, 296, 299, 323, 330
class struggle, 307
classical economic theory, 82, 88, 92, 93, 233, 234; see also Malthus, Ricardo
Clemenceau, G., 1, 5
clothing manufacture, 158
clustering of innovation, 67, 284; see also complementarities, interdependence
coal, economics of extraction, 243; Jevons on, 231, 330, 335; Pinchot on, 232; replaced by petroleum, 254; source of air pollution, 220; substitution for wood, 251, 253, 272, 273; use in railroads, 181; use in steel manufacture, 184, 185, 204, 324, 340
Cockerill, James, 155
Cockerill, John, 155
coke smelting, 182-185
Cole, W., 322, 325
Coleman, D., 308
collectivization of agriculture, 98
colonial economies, 83

Colt, S., 19, 20, 21, 154; *see also* firearms
Colt armory, 292, 295
Colvin, F., 294, 305
Comanche Indians, 326, 327
Communism, 304; *see also* U.S.S.R., China
comparative advantage, 230
competitive framework, *see* neoclassical models
complementarity and substitution in technology, 11, 75, 81, 241, 271; *see also* technological complementarities
computers, 163, 200, 276, 329
conservationism, 218, 231, 232, 242, 334, 335
consumer durables, 103
consumer goods industry, 142, 143, 150
contraceptive technology, 288
Conway Tubular Bridge, 307
corn-hog cycle, 304
Cort, H., 192, 198, 273
cotton, 252, 275, 276, 277, 303, 304, 307, 322, 325, 329
Coventry Sewing Machine Company, 294
Coxe, T., 296
Crimean War, firearm development in, 113, 119, 306; *see also* firearms
Crombie, A., 306
Crouzet, F., 317
Cuba, 97
cutlery, 158

Dales, J., 337
Dalrymple, D., 321
Damocles, sword of, 124
Daniels traverse planing machine, 41
Darby, A., 182, 183, 192, 195, 273, 324, 326
David, P., 236, 305, 307, 327, 329, 335
da Vinci, L., 200, 275
Davis, K. 335
Davis, L., 336, 340
Davy, H., 71
D.D.T., 225, 226, 334
Deane, P., 322, 325
Defebaugh, J., 38, 296, 297
de Guise, duc, 307
De Leeuw, A., 115
demand, structure of, 104, 240; income factors, 289, 337; links to inventive activity, 262, 263, 265, 267, 268, 271-274, 277, 278, 341; market size requirements for capital goods industries, 143; for scientific activity, 129, 130; U.S. and U.K. compared, 158, 159, 162, 164; *see also* distribution of income

demographic shifts, 94, 95, 175, 225, 233, 245, 246, 288; *see also* population growth
demonstration effect, 25, 90
Department of Commerce, 259
Department of Defense, 223
Depew, C., 293
Derry, T., 310
De Solla Price, D., 318
Dewhurst, J., 243, 337
Deyrup, F., 293
Dickinson, H., 173, 190, 307, 309, 310, 325,
Dickinson, R., 309
diesel engines, 73, 164, 254
diffusion of innovation, rate of, 11, 12, 261; the machine tool industry, 18, 19; in bicycles and automobiles, 25-28; Schumpeterian heritage, 67-68, 292; concepts of the innovative process, 68-75, 285, 322, 341; importance to economics, 75-79; in the social sciences, 79-81; popular, 104; in steam and iron, 173-185, 275, 322; factors affecting, 189-210, 321, 325, 326, 327; in agriculture, 236; human implications of, 280-281; international, 76, 119, 120, 151-172, 255, 316, 317, 318, 321
distribution of income, 287; *see also* demand, structure of
division of labor, 17, 92, 131, 313
Dixon, J., 155, 316
Dobb, M., 322, 326
Domar, E., 302
Dorfman, R., 311
Douglas, N., 155
drilling machines, 14
drop forging, 292
Duesenberry, J., 125
Duhring, E., 138
Du Pont laboratories, 72
Durfee, W., 292
Dutch, on Manhattan Island, 37
dynamite, *see* ordnance

Eagly, R., 138
Eames, A., 293
Easterlin, R., 325, 336, 338, 341
Eckaus, R., 316
Eckstein, A., 97, 105
ecological aspects of technology, *see* environmental aspects of technology
econometric models, 81, 82
economic development, *see* economic growth and development

economic growth and development, 61, 302; and capital goods industry, 141-150, 315, 320; class attitudes toward, 92; classical models of, 231; composition of resource use, 94; and diffusion of innovations, 317; and direction of technical change, 108, 260-279; long-term, 83; and natural resource constraints, 240-242, 256, 280; neglected dimensions of, 85-90
economic incentive, 89, 90, 92, 95, 130
economic role of science, 126-138
economic sociology, 88
economies of specialization, 144, 145, 149
Eden, Sir Frederick, 235
Edison, Thomas, 75, 197
Egypt, 128, 312
Einstein, S., 295, 306, 320
Eisenbud, M., 332
electric lamp and lighting, 68, 71, 75
electric power and utilities, 220, 256, 302, 333
electronics industry, 163, 264
Ellison, T., 309
Ellsworth, L., 156, 318
Enfield Arsenal, 120, 154, 298
Engels, F., 127, 129, 136, 137, 138, 311, 312, 315, 340
Engerman, S., 126, 204, 209, 301, 330, 331, 332
England, see United Kingdom
Enke, S., 332
Enos, J., 68, 69-70, 194, 195, ·196, 300, 301, 326, 327
entrepreneur, entrepreneurship, 63, 67, 74, 82, 92, 98, 100, 125, 162, 166, 167, 175, 178, 285, 292, 310
environmental aspects of technology, 213-228, 229, 240, 332
Erie Canal, 243
Evans, O., 176, 178
exogeneity of technical change, 262
external economies, 17, 19, 95, 144, 156, 214, 216, 317, 324, 331, 332, 337; see also negative externalities

Fabricant, S., 105, 304
factor endowments, 88, 95, 141, 142, 149
factor prices, 62, 95, 100, 108, 150
factor saving biases, 108, 109; see also labor saving technology, capital saving technology
factor substitution, 64, 65, 66, 85, 86, 109, 146, 308, 316
factory system, 133
Fairbairn, W., 118, 119, 203, 308, 309, 320
Farey, J., 325

Fawcetts, 154
feedback effects in economic development, 87, 88, 93, 100, 101, 104; see also labor skill acquisition, "learning by doing"
Feldman, A., 100, 106
Feller, I., 336
Fellner, W., 109, 304
Fenichel, A., 322, 323, 330
fertilizer industry, 169, 170, 218, 226, 238, 271, 309, 321, 330, 334, 337
Feuerbach, L., 128
Finch, W., 204, 325, 330
firearms, breechloading cannon, 200, 276, 329; gunstocks, 42, 292; and machine tool diffusion, 19-21, 23, 30, 291, 293, 295, 305; rivalry between offensive and defensive weapons, 116, 306, 307; skills in production, 119; and steel production, 113, 124, 308; technological convergence with typewriters, bicycles, and sewing machines, 16, 153, 154, 157, 159
Fisher, J., 240, 254, 259, 336, 337, 338
fishing industry, 238
Fishlow, A., 323, 324, 329, 330
Fitch, C., 292, 293, 298
Fitch, J., 178
Fitch, S., 20
Flanders, R., 295
flatboat, 186
Flinn, M., 324
Fogel, R., 204, 209, 296, 301, 330, 331, 332
Ford, H., 160, 287
Ford Foundation, 227
foreign aid, 151, 167, 185
foreign trade, see international trade
forestry, see timber industry
forming tool, 29, 112, 113, 293, 305
Forrester, J., 244, 337
fossil fuels, 28; see also coal, petroleum
Fourdrinier machine, 308
Fourneyron, B., 323
France, 51, 53, 121, 155, 171, 282, 283, 293, 299, 315, 323, 330, 335, 341
Frankel, M., 316
Freeman, C., 162, 163, 318, 320
free trade, 230
Frisch, R., 81
Fulton, R., 178

Galileo, G., 71
Gallman, R., 10, 290
Gaskell, P., 307
gear cutting machinery, 22, 25, 27, 293
General Motors, 164
Germany, 152, 155, 282, 283, 325
Gerschenkron, A., 317

Gibb, G., 291
Giedion, S., 38, 158, 296, 318
Gilbert, K., 305
Gilchrist, C., 324
Gilfillan, S., 300, 331
Gille, B., 317
Gilman, F., 296
Goddard, D., 297
Goodman, J., 308
Graham, C., 205, 331
Granick, D., 316, 320
Greece, 229
"Gresham's Law of planning", 125
Griliches, Z., 191, 235, 236, 261, 301, 302, 335, 336, 339
grinders, precision, 13, 22, 25, 26, 27, 116
gross national product, see national product
gunstocks, 42, 298
gyrocompass, 68, 71

Habakkuk, H., 43, 52, 57, 58, 105, 190, 290, 299, 300, 304, 305, 308, 309, 310, 317, 319, 320, 322, 326, 335
Haber, L., 240, 309
Hagerstrand, T., 320
Haites, E., 331
Hall, A., 307, 317
Hall, J., 20
handicraft production, 130, 131, 132, 158, 312, 313, 314
Hannibal, 308
Hardie, D., 309
Harley, C., 331
Harper's Ferry Armory, 20, 292
Harris, J., 322, 326
Harrod, R., 81, 241, 336
Hartness, J., 293
Hawkins, Colonel, 158
Hayami, Y., 236, 237, 238, 300, 335
Hayes, J., 291
Hayes, S., 335
Haynes, W., 309
Hegel, G., 315
Henderson, W., 318
Hendrickson, F., 295
Hertz, R., 327
Hewitt, A., 323
Hichens, Captain, 293
Hicks, J., 81, 108, 109, 142, 149, 304, 315, 316
Higgins, B., 316
high-speed steel, 24, 29, 113, 114
Hills, R., 330
Hinde, H., 295
Hindle, B., 318
Hirsch, W., 329

Hirschman, A., 108, 112, 304, 305, 311, 321
historical materialism, 128, 135, 315
Hoglund, A., 299
Hole, F., 310
Hollander, S., 300
Holley, A., 113
Hollingdale, S., 329
"Horndal Effect," 197
households, 94, 101
Howe, F., 20, 23
Hubbard, G., 293, 294, 295, 306
Hughes, J., 305
human capital, 102, 331
human fertility, see population growth
Hume, D., 89, 90, 91, 105, 303
Hunter, L., 44, 299, 323, 327
Hutchinson, W., 307
Hyde, C., 329
hydraulics, 293, 294

income per capita, 9, 86, 94, 102, 103, 281
India, 44, 221, 225, 226, 335, 340
indivisibilities, 143
Indonesia, 333
induced invention, 109, 300
industrialization, 156, 247
industrial revolution, 189, 250, 251, 305, 313, 314, 326, 330, 336, 340
industrial wastes, 218
Inkerman, 306
innovation models, 68-75
innovation in the social sciences, 79-84
input-output models, 79
institutional organization, 97, 223, 246, 247, 248, 331, 332, 334
interchangeable parts, 286
International Rice Research Institute, 227, 330
international trade, 88, 90, 336, 338
inventive activity, see technological alternatives, generation of
investment, compositional shift, 10
investment goods industry, see capital goods industries
investment, rates of, 10; see also capital stock, growth of
iron and steel industry, 275, 284, 302, 305, 314, 329; alloys, 73, 78, 116, 241, 294; coke smelting, 195, 272, 340; competition with substitutes, 33, 240, 244, 251, 253; in English Industrial Revolution, 326; ore, 232, 243; oxygen method of steel production, 72; structural properties of steel, 122; and railroads, 202, 296, 310, 330; in the United States 1800-1870, 173-188, 321, 323, 324

Isard, W., 324
isoquants, 62, 63, 193, 253; *see also* neo-classical economic theory
Jackson, J., 155
Jacquard loom, 329
Japan, 44, 152, 163, 170, 282, 283, 285, 316, 321, 341
Jarrett, H., 333
Jeans, J., 319
Jefferson, T., 317, 323
Jenkins, R., 309
Jenks, J., 36
Jevons, W.S., 87, 105, 204, 231, 330, 335
Jewkes, J., 70, 71, 194, 302
Jin-bee, O., 333
joining machinery, 41
Jones and Lamson Machine Company, 21, 293

Kalahari Bushmen, 282
Kaldor, N., 315
Karpovich, M., 328
Kay, J., 29
Kay's flying shuttle, 29
Keirstead, B., 122, 309
Kennedy, C., 317
Keynes, J., 81, 82, 225, 302, 311
"Keynesian Revolution," 81, 82
Kindleberger, C., 97, 105
King, G., 80
Kirk, D., 341
Klein, B., 319
Klein, J., 335
Kneese, A., 333
knowledge, 62, 65, 129, 130, 149, 155, 157, 226, 237, 264, 269, 270, 334
Kranzberg, M., 317
Kuhn, T., 340
Kuznets, S., 10, 80, 94, 103, 105, 106, 150, 235, 247, 290, 302, 304, 317, 327, 328, 335, 338, 339

"laboring poor," 92
labor intensive technology, 146, 149, 252; *see also* capital saving technology
labor saving technology, 141, 142, 147, 149, 247, 275, 291, 298, 311; *see also* capital intensive technology
labor skills, acquisition, 95, 96, 97, 98, 104, 131, 149, 317
labor supply, domestic workers, 56, 141; farm laborers, 56, 57; common laborers, 56; skilled mechanics, 56; skilled labor, 56, 95
Lake Baikal, 334
Lake Erie, 219

Lampman, R., 332
Lanchester, F., 162
land tenure, systems of, 97
Landes, D., 155, 159, 274, 309, 311, 318, 324, 326, 329, 340
Landsberg, H., 337
Lange, O., 81
lathes, 13, 19, 20, 21, 25, 27, 114, 293, 294, 297, 298, 314
lathes, stocking, 19, 292
law of diminishing returns, 229
Lazarsfeld, P., 335
"learning by doing," 87, 100, 123, 133, 144, 197, 198, 281, 329
Lebergott, S., 299, 310
Leblanc process, 121
Lee, C., 328
Leibenstein, H., 301, 335
Leland and Faulconer Company, 25, 27, 31
Leontief, Wassily, 79
Lewis, D., 328
Lewis, W., 90, 95, 106, 290, 304
Lincoln, J., 291
Lindblom, C., 305
Livingston, R., 323
Lloyd, G., 158, 318
Locks and Canals Company, 291
locomotives, 13, 23, 73, 160, 161, 164, 291, 294, 319, 323, 324, 328
Loehr, R., 296, 297
Logan Airport, 215
London Exhibition of 1851, 293
Loomis, R., 335
Lord, J., 321, 326
Louisiana Purchase, 323
Lowell Machine Shop, 13, 291
Lowell Mills, 13
lubrication, 28
lumber industry, *see* timber industry
lumber, wholesale price index, 1798-1869, 49; per capita consumption, 252
Lynn, F., 302, 327

MacDougall, D., 293
machine tool industry, 113, 114, 115, 143, 154, 316; American, 9-31, 284, 292, 305, 318
machine tools, 115, 161, 247, 294, 314, 329; *see also* by type
Machlup, F., 101, 106
Maclaurin, W., 70
Mak, J., 196, 327, 331
Malaya, 333
Malthus, T., 4, 92, 106, 229, 230, 231, 232-233, 234, 241, 244, 245, 246, 249, 288, 289, 327, 335
manager, *see* entrepreneur

Manners, R., 97, 106
Mansfield, E., 191, 301, 326
Mantoux, P., 70, 305
manufacturing, 88, 95; in Marx, 130, 131-132, 134, 312, 313, 314
March, J., 125, 311
market failures, 219, 223, 224; institutional remedies, 223, 224; *see also* "negative externalities," pollution, neo-classical economic theory
Markham, J., 309
"marginal revolution," 87, 230
Marshall, A., 105, 106, 209, 310
Martin, H., 106, 304
Marx, K., 3, 9, 82, 118, 126-138, 146, 147, 165, 286, 290, 300, 307, 308, 309, 311, 312, 313, 314, 315, 317, 319, 327, 328
Massell, B., 290
mass production technology, 120, 159, 286
mathematics, 311
Mathiessen, P., 333
Maudsley, H., 135
McCloskey, D., 331
McKeown, T., 339
Meadows, D., 335
Meadows Report, 233
Mechanics' Magazine, 116
medicine and medical technology, 268, 270, 274, 278
Meiji Restoration, 316, 321
Mendelejeff, D., 269, 283
Mesabi Range, 243, 257
Mesopotamia, 222, 310
metallurgy, 40, 182, 183, 187, 189, 199
metalworking machinery, 11, 156; *see also* machine tool industry
Metraux, G., 317
Middle East, 222; Mideast oil, 281
military establishment, American, 223
Mill, J., 106, 303, 334
Miller, S., 39
milling machines, 13, 14, 20, 23, 27, 292, 293, 295
Mirsky, J., 290
Mitchell, B., 325
Modigliani, F., 81
Molesworth, G., 296, 298
monoculture, 98
Moore, W., 100, 106
Morris, W., 161, 287
Morse, C., 238, 241, 334, 335, 336, 337
Moseman, A., 340
mule, self-acting, 118, 290, 307
Mulhall, M., 190, 322, 326
multinational firm, 157
multiproduct firm, 292

Mulvany, W., 155
Mumford, L., 33, 296
Musson, A., 322, 326, 328, 329
Myint, H., 88, 96, 97, 106

nailmaking machinery, 37
Napoleon, 192
Napoleonic Wars, 311
Nasmyth, J., 120, 307, 308
National Aeronautics and Space Administration, 223
national income accounting, 80, 81
national product, 103, 228, 240, 250, 302, 335, 338
National Science Foundation, 103, 106
natural gas, 220, 254
Needham, J., 286, 341
Nef, J., 195, 251, 309, 326, 340
"negative externalities", 213, 215
Nelson, R., 335, 339
neoclassical economic theory, 82, 87, 93, 110, 111, 125, 214, 216-217, 219, 237, 272
Netherlands, 303
Netschert, B., 254, 259
Nevins, A., 290
Newberry, W., 40
Newcomen, T., 191
Newcomen engine, 186, 192, 321
Newton, I., 269, 283, 315
Niles' Weekly Register, 296, 297
Nobel, A., 122
Nobel Prize, 62
nonprofit corporation, 171
Nordhaus, W., 80
Norris Locomotive Works, 180
North, D., 207, 208, 209, 210, 331, 332
North, S., 20
Norton, C., 25
Nuffield, Lord, *see* William Morris

O.E.C.D., 76, 77, 283, 302, 341
Ohlin, B., 81
oil-tube drill, 25, 29, 113
optimizing behavior, *see* neoclassical economic theory
Orcutt, H., 319
Ordnance, 116, 124, 276, 291, 306-307, 326; *see also* firearms
Oshima, H., 106, 304
output, composition of, 94, 101
Owen, C., 295
Owen, R., 121, 309
Oxford, C., 295, 305

Paget, F., 293
Paley Commission, 240, 243, 258, 335
Paris Exhibition of 1867, 293

Parker, W., 235, 236, 275, 335, 336, 340
Parkhurst, E., 293
Parry, A., 319, 324
Passer, H., 302
Pasteur, L., 137
patents, 266, 267
Paton, W., 316
Perkins, J., 37
petroleum industry, 68, 71, 196, 243, 247, 254, 263, 316, 326, 327
pharmaceuticals industry, 247
Philippines, 227, 333
Phillips, A., 302
phonograph, 197
Pickard, J., 122
Pinchot, G., 232, 335
Pirenne, H., 310
planing machinery, 40, 41, 314
plastics, 240, 244, 255, 256, 319
Pole, W., 308, 330
Polham, C., 200, 276
pollution, 215, 219, 220, 226, 245, 246, 287, 334
Pope, A., 24, 294
population growth, 225, 227, 228, 230, 232-233, 244, 246, 281, 287-289, 337, 338, 341
Postan, M., 309, 310, 335
Potter, N., 240, 336
poverty, 217, 332
power generation, techniques of, 189
Pratt and Whitney Company, 30
preindustrial societies, 32, 250
President's Materials Policy Commission, see Paley Commission
prestressed concrete, 33
primary commodities, production of, 95
producer goods, see capital goods industries
production functions, 62-66, 86, 141, 155, 237, 300, 303
productivity of resources, 62
profit margins, 205
Prosser, R., 308
Providence Tool Company, 22
public education, 93
public sector, 103
pulp and paper industry, 122, 255, 312
Pursell, C., 317, 323

Quesnay, F., 79

radio, 197, 264, 327
Rae, J., 294
rails, iron, 202
railroads, 197, 323; bridges in Britain, 310; and diffusion of steam engines, 177, 180, 182; freight cost in U.S., 146, 243, 290; fuel sources for, 253; incremental technological change in, 201, 329, 330; Marx on 314; and rates of inventive activity, 262, 263; in Russia, 324
railroad companies, 164
Ramsey, J., 178
RAND Corporation, 171
Ranis, G., 316
Rapping, L., 329
Rasmussen, W., 336
raw materials, see resource endowments
Ray, G., 317
Record, R., 339
recycling of scrap or waste materials, 256
Reed, E., 306
Remington and Sons, 295
Research and Development, 76, 77, 103, 111, 260, 284, 319, 321; see also scientific knowledge and technological change
"Residual", 83, 207
resource endowments, 32, 33, 45, 78, 87, 229-248, 261, 280; comparative position of 19th century England and America, 35, 37; comparative resource-intensity of British and American technologies, 33, 35, 37, 38, 43; resource-intensive goods, 250; resource-saving innovations, 247, 249-259, 281
resource exhaustion, 74, 121, 229-248, 249-259, 280, 282, 334, 335, 336, 341; see also synthetic rubber, synthetic fibers, technological alternatives
reverberatory furnace, 188
Ricardo, D., 90, 229, 231, 234, 241, 250, 309, 327, 335
Ricardo, M., 309
rice, 304, 330, 334
Richards, J., 252, 259, 296, 297, 299
Richardson, J., 319
riveting machine, 118-119
Robbins and Lawrence Company, 20, 21, 30, 293
Roberts, R., 118, 307
Robinson, E., 322, 326, 329
Robinson, E.A.G., 338
Robinson, J., 317
Rockefeller Foundation, 227
Roe, J., 291, 292, 293, 317
Rogers, E., 302
Roman Empire, 312
Root, E., 20, 292
Rosenberg, N., 100, 106, 296, 297, 298, 300, 301, 302, 303, 305, 316, 317, 318, 319, 320, 325, 329, 337
Rosovsky, H., 305, 307, 327, 335
Rossman, J., 310

Rostow, W., 85, 106, 290, 304, 311; *see also* "take-off"
Rothschild, K., 315
Rotwein, E., 303
Ruttan, V., 236, 237, 238, 320, 321, 335, 337

sailing ship, 205, 206, 208, 331, 332
Salter, W., 65, 109, 300, 304
Samuelson, P., 109, 304
satisficing, 125
Saul, S., 295, 318, 319
Sawers, D., 70
sawmills, 37, 38
scale economies, 143, 144, 316, 331
Schmookler, J., 260-279, 339
Schultz, T., 64, 106, 107, 235, 241, 300, 304, 335, 336
Schumpeter, J., 9, 63, 66-68, 74, 75, 82, 166, 260, 292, 300, 301
Schumpeterian theory of technological development, 66-68, 74, 166, 167, 169, 170, 210, 292, 325
Schurr, S., 254, 259, 337
scientific development, history of, in Marx, 136, 137, 312
scientific knowledge and technological change, 260-279, 339; comparison of Britain, France, Germany, U.S., U.S.S.R., Japan, 282-283; changing relation over time, 282-286
Scitovsky, T., 305
Scott, A., 240, 336
Scott, J., 306
Scoville, W., 317
screws, 20, 21, 25
Select Committee on Artizans and Machinery, 54
Sellers, W., 154
semiconductor industry, 72
Seth Thomas Clock Company, 294
sewing machines, 13, 16, 19, 21, 22, 23, 30, 154, 157, 159, 161, 292, 293, 294
shapers, 14, 23
Sharps Rifle Manufacturing Company, 30
Shoemaker, F., 302
Simon, H., 125, 311
Simpson, E., 306
Singer, C., 305, 307, 329, 340, 341
Singer, H., 95, 106, 107, 316
"slash and burn" cultivation in tropics, 282, 333
slavery, 304
slide rest, 135
Smiles, S., 118, 120, 307, 309, 329
Smith, Adam, 17, 89, 91, 92, 93, 107, 143, 303, 304

Smith, T., 321
Smolensky, E., 126, 332
solar energy, 280
solid waste disposal, 216
Solow, R., 1, 83, 107, 207, 290
Southeast Asia, 222, 330
specialization of machinery, 42
Spencer, C., 21
Spencer repeating rifle, 21
Spengler, J., 241
Springfield Armory, 20, 23, 43
"Sputnik syndrome", 334
Sraffa, P., 309
Stanhope, Lord, 328
status anxieties of the economist, 61
steamboat, 177, 178, 179, 180, 186, 196, 298, 299, 323, 327
steam engines, 13, 122, 204, 308, 321, 325, 328, 330; fuel sources, 253, 254; learning by doing, 198; Marx on 290, 314; and oceanic shipping, 73, 276; and technological complementarities 112, 199; and technological diffusion, 190, 191, 203, 309, 326, 327; in U.S. 1800-1870, 173-188
steamship, 205, 206, 331
steel industry, *see* iron and steel industry
Steer, H., 34
Stephenson, G., 328
Stevens, J., 178
Steward, J., 97, 106
Stewart and Hill sawmill, 296
Stigler, G., 16, 17, 147, 233, 292, 317, 335
Stillerman, R., 70
Stocking, G., 309
Stone, H., 21
Stover, C., 138, 306
Strassmann, W., 163, 290, 305, 319, 325
strikes as cause of technical change, 117, 118, 120
sub-contractors, use of by American and European firms, 163
supersonic transport, 215, 332
supply factors in technological change, *see* technological alternatives, generation of
supply of labor, 90
supply of skilled labor, 54; comparison of Europe, U.S., 54
Sutton, A., 321
Svennilson, I., 155, 318
Sweden, 122
Sweezy, P., 138, 313, 314
synthetic fibers, 72; *see also* textile industry
synthetic rubber, 74

"take-off", 10, 85; *see also* capital stock growth, per capita income

Tangyes, 154
Taplin, W., 324
Taylor, F., 29, 114, 295
technological alternatives, 64, 224, 287; generation of, 63, 68, 127, 137, 260-279, 280, 281; inducements, 109-125, 280-286; see also Research and Development, resource exhaustion
technological change, continuous nature, 166, 191, 193, 292
technological change, discontinuous, 166, 192, 292
technological change, transmission of, see diffusion of innovation
technological complementarities, 37, 112, 169, 170, 317; see also technological interdependence
technological convergence, 16, 17, 18, 19, 21, 28, 30, 31, 157, 294
technological disequilibrium, 29, 113, 116, 117, 307
technological dynamism, 127, 134, 150; institutionalization of, 157, 164, 170, 171, 172, 284-286
"technological fix," 151
technological imbalance, see technological disequilibrium
technological innovation, conceptual models, 61-84
technological interdependence, 29, 112, 115
technological responsiveness, 281, 282, 284, 286; see also technological alternatives, technological dynamism, Research and Development, resource exhaustion
telephone exchange, 163
television, 163, 264
Temin, P., 296, 305, 322, 323, 324, 330
Tennessee Valley Authority, 171, 321
Texas Rangers, 326
textile industry, 12, 13, 23, 72, 112, 173, 190, 247, 274, 322
textile machinery, 162, 180, 299, 316, 322, 329
Thomas, D., 184, 324
Thomas, E., 155
timber industry, 32, 78, 238, 239, 243-244, 248, 251, 252, 253, 272, 296, 297, 337
tin, 241, 273
Tinbergen, J., 81
Tobin, J., 80
Tootill, G., 329
Toynbeean theory of social change, 314
Trafalgar, 306
transfer of innovations, see diffusion of innovations
transportation revolution, 145, 290, 295, 314

Trevithick, R., 73, 176
Truman, H., 226
turbines, 330; steam, 13
Turner, F., 232
typewriters, 23, 153, 159, 295

underdeveloped economies, 85-107, 141, 142, 146, 150, 151, 165, 169, 174, 224, 247, 261, 316, 317, 335, 341
United Kingdom, woodworking in, 34-43 passim, 297; and the transfer of technology, 152-165, 174, 319; steam power in, 176, 178, 190, 321, 322, 323, 326, 331; railroad and iron technology, 180-186, 323, 324; bridgebuilding in, 202; new and old technologies in, 204-205, 325, 326; materials shortages in, 229, 231, 335, 336; scarcity of timber, 250-252, 255; premature inventions in, 276; industry without science, 283, 325, 341; technology subservient to cultural values, 286-287; bicycles, firearms, and sewing machines, 292, 293, 294, 326; Marx on mechanics in, 314, 315
United Nations, 97, 107
urbanization, 214, 215, 246
Ure, A., 118, 307
Usher, A., 9, 18, 80, 192, 292, 317, 328
U.S.S.R., 98, 152, 163, 167, 182, 283, 319, 321, 324; central economic planning, 163; metalworking industries, 316; pollution, 334

Vanek, J., 259, 336, 338
Verne, J., 328
vertical disintegration, 16, 28, 292
vertical integration, 162

Waddingtons, 155
wage differentials, Anglo-American, 50-58, 300
wage rates, U.K., France and U.S.A., 51; U.K., France, U.S.A. and Holland, 53; skilled and unskilled, 55, 56
Walden Pond, 215
Wallis, G., 291
Walton, G., 196, 327, 331, 332
war as cause of resource substitution, 121-122
Warner, T., 20
Waterloo, 306
waterpower, 175, 176, 177, 186, 203, 254, 323
Waters, A., 297
Watkins, M., 309
Watt, J., 112, 122, 173, 186, 187, 190, 191, 192, 199, 200, 305, 309, 321, 326, 328

Webb, W., 327
Weed Sewing Machine Company, 294
Wells, H., 328
Wheeler and Wilson Sewing Machine Company, 293
Whistler, G., 324
White, L., 306, 332
Whitehead, A., 79
Whitmore's Card Machine, 298
Whitney, E., 19, 20, 290, 301
Whitworth, J., 42, 116, 291, 297
Wik, R., 322
Wilcox and Gibbs sewing machine, 22, 23
Wilkinson, J., 112, 199, 200, 305
Wilkinson's boring mill, 199
Williams, B., 316
Williams, T., 310
Wilson, C., 341

Wilson, T., 155
windmills, 187, 204, 325, 330
Wolman, A., 333
Woodbury, R., 115, 290, 292, 293, 294, 295, 306
wood, see timber industry
wood saws 38, 39, 40
woodworking industry, 33-49
woodworking machinery, 35, 36, 38, 40, 41, 252, 296, 297; see also planing machinery, joining machinery
Woodworth planing machine, 41
Wrigley, E., 336, 340

Young, A., 17, 292
Youngson, A., 335

Zvorikine, A., 138, 315